T0396872

Modern Approaches to Differential Geometry and its Related Fields

Modern Approaches to Differential Geometry and its Related Fields

Proceedings of the 7th International Colloquium on Differential Geometry and its Related Fields

8–11 September 2023 Patras, Greece

Editors

Yusuke Sakane
Osaka University, Japan

Toshiaki Adachi
Nagoya Institute of Technology, Japan

Hideya Hashimoto
Meijo University, Japan

NEW JERSEY • LONDON • SINGAPORE • GENEVA • BEIJING • SHANGHAI • TAIPEI • CHENNAI

Published by

World Scientific Publishing Co. Pte. Ltd.
5 Toh Tuck Link, Singapore 596224
USA office: 27 Warren Street, Suite 401-402, Hackensack, NJ 07601
UK office: 57 Shelton Street, Covent Garden, London WC2H 9HE

British Library Cataloguing-in-Publication Data
A catalogue record for this book is available from the British Library.

Cover image:
(front) View of Patras from St. Nicholas street
(back) Rion-Antirion bridge
Photographed by Toshiaki ADACHI

MODERN APPROACHES TO DIFFERENTIAL GEOMETRY AND ITS RELATED FIELDS
Proceedings of the 7th International Colloquium on Differential Geometry and its Related Fields

ISBN 978-981-12-9670-3 (hardcover)
ISBN 978-981-12-9671-0 (ebook for institutions)
ISBN 978-981-12-9672-7 (ebook for individuals)

For any available supplementary material, please visit
https://www.worldscientific.com/worldscibooks/10.1142/13946#t=suppl

PREFACE

The 7th International Colloquium on Differential Geometry and its Related Fields (ICDG2023) was held at the Auditoriums of Natural Sciences, Auditorium 9 ($A\Theta E$ 9), University of Patras, Rion, Greece, during the period of 8th–11th, September, 2023.

ICDG2023 was not a regular conference. The series of colloquiums ICDG started in 2008 as an academic program under an agreement between St. Cyril and St. Methodius University of Veliko Tarnovo, Bulgaria, and Nagoya Institute of Technology, Japan to exchange mathematical ideas of East Europe and Japan. ICDG has been held every two years. In 2018, Nagoya Institute of Technology also made an agreement with Gheorghe Asachi Technical University of Iasi, Romania. To extend our program, we were supposed to have ICDG at Iasi, Romania. But because of the COVID-19 pandemic, we could not have ICDG2020 and ICDG2022. During these years, a political situation also occurred, and Western airplanes now cannot fly over Russia. The phrase "far east" has been brought back to our attention.

Since one of the editors has close relationships with Greek mathematicians, with his mediation, this time, the Colloquium was co-organized with University of Patras, which is located 7 km north-east of the city of Patras, in the suburb of Rion. We had newcomers from Germany, Greece and India as well as participants from Bulgaria and Japan. In the large and beautiful campus of the University of Patras, participants spent several nice days and had many fruitful discussions and exchanges of ideas. The organizers believe that participants could make up for the previous four years of absence. Standing in the nation of the birthplace of the Olympic games, we hope that we will have peaceful and calm days to study mathematics.

This volume contains original research papers, an announcement of recent work and survey reports which were contributed by participants of the conference. These cover modern approaches on minimal surfaces with ends, Einstein or Einstein-like metrics, complex structures of T^2-bundles over the Hirzebruch surface, Sasakian magnetic fields, F-Yang-Mills connections, two weight projective codes, biconservative hypersurfaces, canonical form theory in geometry, *-Ricci solitons and so on. The editors expect these articles will provide significant information for researchers and a good guide for graduate students on modern approaches to Differential Geometry and its related fields. We thank all participants for their contributions in the conference and to this proceedings. We also acknowledge the scientific reviewers who read the articles carefully and gave many important suggestions to the authors.

Finally, we would like to express our hearty thanks to Professors Andreas Arvanitoyeorgos and Marina Statha for their help in coordinating the conference and their warm hospitality. We would also like to acknowledge the partial financial support from ELKE of University of Patras.

The Editors
24 May, 2024

The 7th International Colloquium on Differential Geometry and its Related Fields

8–11 September, 2023 – Patras, Greece

ORGANIZING COMMITTEE

T. Adachi	– Nagoya Institute of Technology, Nagoya, Japan
A. Arvanitoyeorgos	– University of Patras, Patras, Greece
H. Hashimoto	– Meijo University, Nagoya, Japan
Y. Sakane	– Osaka University, Osaka, Japan
M. Statha	– University of Thessaly, Lamia, Greece

Biology/Mathematics Building, University of Patras, Rio, Greece, September 2023

In front of a monument of Prof. Keiji KOKUBU, Olympia, Greece, September 11, 2023

PRESENTATIONS

at Auditoriums of Natural Sciences, Auditorium 9 ($A\Theta E$ 9),
Biology/Mathematics Building, University of Patras

1. **Yusuke Sakane** (Osaka Univ., Japan),
   Invariant Einstein metrics on compact simple Lie groups
2. **George Kaimakamis** (Hellenic Army Academy, Greece),
   Results on $*$-Ricci solitons
3. **Kurando Baba** (Tokyo Univ. Sci., Japan),
   On instability of F-Yang-Mills connections
4. **Georgios Kydonakis** (Univ. Patras, Greece),
   Nonabelian Hodge theory over compact or noncompact curves
5. **Takahiro Hashinaga** (Saga Univ., Japan),
   On sectional curvatures of some Einstein solvmanifolds
6. **Yusei Aoki** & **Toshiaki Adachi** (Nagoya Inst. Tech., Japan),
   Estimates on characteristic magnetic focal values for Kähler manifolds
7. **Misa Ohashi** (Nagoya Inst. Tech., Japan) &
   Hideya Hashimoto (Meijo Univ., Japan),
   Cocycle constructions of $S^3/\mathbb{Z}_m$ in Hirzebruch surfaces
8. **Athanasios Chatzikaleas** (Univ. Muenster, Germany),
   Unique continuation for nonlinear waves in asymptotically anti-de Sitter spacetimes
9. **Naoya Ando** (Kumamoto Univ., Japan),
   The $SO(3,1)$-orbits in the light cone of the 2-fold exterior power of the Minkowski 4-space
10. **Panagiotis Sakkalis** (Agricultural Univ. Athens, Greece),
    Proper polynomial maps and the fundamental theorem of algebra
11. **Prasad Sachchidanand**
    (International Center for Theoretical Sciences, Bangalore, India),
    On manifold homeomorphic to n-sphere

12. **Hiroshi Matsuzoe** (Nagoya Inst. Tech., Japan),
Invariant dually flat structures for q-exponential families
13. **Norihiro Nakashima** (Nagoya Inst. Tech., Japan),
Higher-order freeness for hyperplane arrangements
14. **Paskal Nikoraev Piperkov** (Univ. Veliko Tarnovo, Bulgaria),
Two weight projective codes and some combinatorial objects
15. **Toshihiro Shoda** (Kansai Univ., Japan),
Geometric quantities of a triply periodic minimal surface of genus four
16. **Andreas Arvanitoyeorgos** (Univ. Patras, Greece),
Biconservative hypersurfaces in space form with harmonic curvature

CONTENTS

Preface v

Organizing Committee vii

Presentations xi

Nonorientable minimal surfaces with various types of ends 1
Kohei HAMADA and Shin KATO

The $SO(3,1)$-orbits in the light cone of the 2-fold exterior power of the Minkowski 4-space 17
Naoya ANDO

On sectional curvatures of some Einstein solvmanifolds 33
Takahiro HASHINAGA and Akira KUBO

Einstein-like metrics on flag manifolds 45
Andreas ARVANITOYEORGOS, Yusuke SAKANE and Marina STATHA

Representation of the complex structure of the T^2 fibre bundle over the Hirzebruch surface $\mathbb{CP}^2\#\overline{\mathbb{CP}^2}$ 73
Hideya HASHIMOTO and Misa OHASHI

Non-horocyclic unbounded trajectories on a complex hyperbolic space are not expressed by those on tubes of type (A) 85
Yusei AOKI and Toshiaki ADACHI

On instability of F-Yang-Mills connections 97
Kurando BABA

Two weight projective codes and some combinatorial objects 111
Paskal PIPERKOV and Mariya DZHUMALIEVA-STOEVA

Kähler graphs whose principal graphs are of tensor product type 123
Kohei SHIOTANI and Toshiaki ADACHI

Biconservative hypersurfaces in $\mathbb{E}^5_s$ 137
Ram Shankar GUPTA, Andreas ARVANITOYEORGOS, Savita RANI and Marina STATHA

Canonical form theory in geometry — Foundations and applications 159
Osamu IKAWA

An overview on *-Ricci solitons 183
George KAIMAKAMIS and Konstantina PANAGIOTIDOU

Invariant dually flat structures on q-exponential families 195
Hiroshi MATSUZOE

Afterword 211

Modern Approaches to Differential Geometry
and its Related Fields 1 – 16

NONORIENTABLE MINIMAL SURFACES WITH VARIOUS TYPES OF ENDS

Kohei HAMADA

Osaka Ibaraki High School,
12-1 Shinjocho, Ibaraki, Osaka, 567-8523, Japan
E-mail: k.hamada3221@gmail.com

Shin KATO

Department of Mathematics, Osaka Metropolitan University,
3-3-138 Sugimoto, Sumiyoshi-ku, Osaka 558-8585, Japan
E-mail: shinkato@omu.ac.jp

In this paper, we show the existence of complete $\mathbf{Z}_N$-invariant conformal minimal immersions from finitely punctured real projective planes into $\mathbf{R}^3$ with various types of ends. In particular, there exists such a surface with N embedded ends and 1 non-embedded (or embedded) end for any $d = \deg g \geq 2N+1$ if $\gcd(d, N) = 1$.

Keywords: Minimal surface; nonorientable.

1. Introduction

Let M be a nonorientable surface, $\widetilde{M}$ be a Riemann surface, and $\pi : \widetilde{M} \to M$ be a double covering map. Let $X : M \to \mathbf{R}^3$ be a complete conformal minimal immersion with finite total curvature. Then a natural lift $\widetilde{X} := X \circ \pi : \widetilde{M} \to \mathbf{R}^3$ is also a complete conformal minimal immersion with finite total curvature. By Osserman [8], $\widetilde{M}$ is conformally equivalent with a compact Riemann surface $\overline{\widetilde{M}}$ punctured at a finite number of points, and the Gauss map $G : \widetilde{M} \to \mathbf{S}^2$ of $\widetilde{X}$ extends holomorphically on $\overline{\widetilde{M}}$. We call each of these puncturing points, or the image of some neighbourhood of the point, an end of $\widetilde{X}$. In the same way, we define an end of X also. Since π is a double covering map, the number of the ends of $\widetilde{X}$ is even.

Denote the degree of G by d. Since the total curvature $TC(\widetilde{X})$ of $\widetilde{X}$ coincides with $-4d\pi$, the total curvature $TC(X)$ of X is $-2d\pi$. In general, it is known as a special case of Chern-Osserman's inequality that, if $\overline{\widetilde{M}} = \mathbf{S}^2$ and the number of the ends is $\widetilde{n}$, then $TC(\widetilde{X}) \leq -4(\widetilde{n}-1)\pi$

holds. This implies the inequality $TC(X) \leq -2(2n-1)\pi$ for conformal minimal immersions defined on $\mathbf{R}P^2 \setminus \{n \text{ points}\}$. In particular, for each inequality, the equality, that is equivalent with $d = 2n-1$, holds if and only if all the ends are embedded ends, that is to say, catenoidal or planer ends. We call such a surface an n-noid.

Since catenoids and Enneper's surfaces only have the total curvature -4π, $d=1$ cannot be attained by any nonorientable one. Meeks [5, Theorem 2] gave the first example of a complete conformal minimal immersion defined on a real projective plane $\mathbf{R}P^2 \setminus \{1 \text{ point}\}$, that is a Möbius strip, with total curvature $-2 \cdot 3\pi = -6\pi$. Meeks [5, Lemma 3] also proved that there are no examples defined on $\mathbf{R}P^2 \setminus \{2 \text{ points}\}$ with total curvature -6π.

After [5], Oliveira [7, Proposition 2.4] proved that d must be odd. Oliveira [7, Theorems 2.6, 2.8, 2.9] and Barros [1, Theorem 1.1 (b), (c)] proved that, for any $n \in \{1,2,3\}$ and $d \geq 2n+1$, there exists an example defined on $\mathbf{R}P^2 \setminus \{n \text{ points}\}$ with total curvature $-2d\pi$. See also [1, 3, 6] etc. for other examples defined on $\mathbf{R}P^2 \setminus \{n \text{ points}\}$ with $n \in \{1,2\}$ and $d \geq 2n+1$.

On the other hand, Kusner [4, Theorem B] gave an example of n-noid defined on punctured $\mathbf{R}P^2$ all of whose ends are planer ends for any odd integer $n \geq 3$, and the authors [2, Theorem 6.6] showed the existence of a 1-parameter family of n-noids defined on punctured $\mathbf{R}P^2$ all of whose ends are catenoidal ends for any even integer $n \geq 4$. Both of these examples satisfy $d = 2n-1$.

Now, it is natural to expect the existence of complete conformal minimal immersions defined on $\mathbf{R}P^2 \setminus \{n \text{ points}\}$ with total curvature $-2d\pi$ for any remaining pair (d,n) satisfying $d \geq 2n+1$. In this paper, we give three families of $\mathbf{Z}_N$-invariant examples with $N+1$ (or N) ends, which include many known examples as special cases. Among these facts, we get the following:

Theorem 1.1. *For any positive integer N, there exists a $\mathbf{Z}_N$-invariant complete conformal minimal immersion defined on $\mathbf{R}P^2$ punctured at $N+1$ points with total curvature $-2d\pi$ if $d \geq 2(N+1)-1 = 2N+1$ and $\gcd(d,N)=1$. In particular, in the case that $d \geq 2(N+1)+1 = 2N+3$, there exists such a surface as above, all of whose ends are embedded ends except for one end.*

2. Basic facts on nonorientable minimal surfaces

In this section, we summarize basic facts on nonorientable minimal surfaces studied mainly by Meeks [5] and Oliveira [7].

Let M, π, $\widetilde{M}$ and $\overline{\widetilde{M}}$ be as in §1. Let $\widetilde{X} : \widetilde{M} = \overline{\widetilde{M}} \setminus \{q_1, \ldots, q_{2n}\} \to \mathbf{R}^3$ be a complete conformal minimal immersion with finite total curvature. We use the Enneper-Weierstrass representation formula of the following type:

$$\widetilde{X}(z) = \operatorname{Re} \int^z {}^t(1-g^2, \sqrt{-1}(1+g^2), 2g)\eta,$$

where g is a meromorphic function on $\overline{\widetilde{M}}$, and η is a meromorphic 1-form on $\overline{\widetilde{M}}$ such that both η and $g^2\eta$ is holomorphic on $\widetilde{M}$. We call (g, η) the Weierstrass data of $\widetilde{X}$. This $\widetilde{X}$ is well-defined on $\widetilde{M}$ if and only if

$$\operatorname{Re} \int_C {}^t(1-g^2, \sqrt{-1}(1+g^2), 2g)\eta \;=\; 0 \tag{1}$$

holds for any loop C in $\widetilde{M}$. On the other hand, $\widetilde{X}$ has no branch points if and only if η and $g^2\eta$ have no common zeroes. In the case that $\overline{\widetilde{M}} = \mathbf{S}^2$, the condition (1) is equivalent with the following condition described by the residues of ends:

$$\overline{R_0(q_j)} + R_2(q_j) = 0, \quad \overline{R_1(q_j)} = R_1(q_j) \qquad (j = 1, \ldots, 2n), \tag{2}$$

where we set $R_i(q_j) := \operatorname{Res}_{z=q_j} g^i\eta$ $(i = 0, 1, 2\,;\ j = 1, \ldots, 2n)$.

Now, let $I : \widetilde{M} \to \widetilde{M}$ be the covering transformation corresponding to π. Then I is an antiholomorphic involution which satisfies $\pi \circ I = \pi$, $\partial I = 0$ and $I^2 = id_{\widetilde{M}}$. In particular, I extends on $\overline{\widetilde{M}}$ and $\{q_1, \ldots, q_{2n}\}$ coincides with $\{q_1, \ldots, q_n, I(q_1), \ldots, I(q_n)\}$ by permutating the ends if necessary. $\widetilde{X}$ is a lift of some complete conformal minimal immersion $X : M \to \mathbf{R}^3$ if and only if the Weierstrass data (g, η) of $\widetilde{X}$ satisfies the following condition (cf. Meeks [5, Proposition 1]):

$$g \circ I = -\frac{1}{\overline{g}}, \qquad I^*\eta = -\overline{g^2\eta}. \tag{3}$$

Note here that

$$\operatorname*{Res}_{z=I(q_j)} \overline{I^*(g^i\eta)} = \overline{\operatorname*{Res}_{z=q_j} g^i\eta} \qquad (i = 0, 1, 2\,;\ j = 1, \ldots, 2n)$$

holds in general. The conditions (3) and (2) imply the following equalities:

$$\overline{R_0(q_j)} = -R_2(I(q_j)), \quad \overline{R_1(q_j)} = R_1(I(q_j)) \qquad (j = 1, \ldots, 2n). \tag{4}$$

In this paper, we consider the case that $\overline{M} = \mathbf{R}P^2$ and $\overline{\widetilde{M}} = \mathbf{S}^2$. In this case, M and $\widetilde{M}$ is given by $M := \mathbf{R}P^2 \setminus \{[q_1], \ldots, [q_n]\}$ and $\widetilde{M} := \mathbf{S}^2 \setminus \{q_1, \ldots, q_n, I(q_1), \ldots, I(q_n)\}$. If we identify $\mathbf{S}^2$ with $\hat{\mathbf{C}} = \mathbf{C} \cup \{\infty\}$, then we can express the antiholomorphic involution as $I(z) = -1/\overline{z}$. By the first identity in the criterion (3), we see that, for any $z_0 \in \hat{\mathbf{C}}$, z_0 is a zero of g if and only if $I(z_0) = -1/\overline{z_0}$ is a pole of g. On the other hand, we also see that, if $\widetilde{X}$ has no branch points, then, for any $z_\infty \in \hat{\mathbf{C}}$ which is not an end of $\widetilde{X}$, z_∞ is a pole of g of order M_∞ if and only if z_∞ is a zero of η of order $2M_\infty$. Hence, if we assume $g(0) \neq \infty$, then the Weierstrass data of $\widetilde{X}$ is of the following form (cf. Oliveira [7, §2]):

$$g(z) = \frac{\alpha z^K \prod_{J=1}^{L} (z - r_J)^{M_J}}{\prod_{J=1}^{L} (\overline{r_J}\, z + 1)^{M_J}}, \qquad \eta = \frac{a \prod_{J=1}^{L} (\overline{r_J}\, z + 1)^{2M_J}}{z^k \prod_{j=1}^{\ell} \{(z - q_j)^{m_j} (\overline{q_j}\, z + 1)^{m_j}\}}\, dz,$$

where

$$k \geq 0,\ \ K \geq 0,\ \ m_j \geq 1\ (j = 1, \ldots, \ell),\ \ M_J \geq 1\ (J = 1, \ldots, L),$$
$$r_J \ \neq\ 0,\ r_{J'},\ -1/\overline{r_{J'}}\ (\text{for every } J' (\neq J))\ (J = 1, \ldots, L).$$

This assertion is valid even if q_j or $-1/\overline{q_j}$ coincides with $-1/\overline{r_J}$ for some j and J, that is, some pole of g is an end of $\widetilde{X}$.

Note here that

$$g^2\eta = \frac{a\alpha^2 z^{2K-k} \prod_{J=1}^{L} (z - r_J)^{2M_J}}{\prod_{j=1}^{\ell} \{(z - q_j)^{m_j} (\overline{q_j}\, z + 1)^{m_j}\}}\, dz.$$

If some pole of η, $g\eta$ or $g^2\eta$ is of order 1, then the condition (1) does not hold for any loop surrounding the pole, and hence $\widetilde{X}$ is not well-defined. Therefore we must assume that $k = 0$ or $k \geq 2$, and $m_j \geq 2$ $(j = 1, \ldots, \ell)$. $k = 0$ is the case that $[0]$ is not an end but a regular point of X. Set $m_0 := \sum_{j=1}^{\ell} m_j$ and $M_0 := \sum_{J=1}^{L} M_J$.

Some additional restrictions are necessary for the parameters in the Weierstrass data of the form as above to satisfy the condition (3). The first equality $g \circ I = -1/\overline{g}$ holds if and only if $(-1)^{K+M_0}\alpha = -1/\overline{\alpha}$, that is $|\alpha|^2 = (-1)^{1-K-M_0}$. This condition can be rewritten as follows:

$$|\alpha| = 1, \quad d = \deg g = K + M_0\ :\ \text{odd}. \tag{5}$$

On the other hand, the second equality $I^*\eta = -\overline{g^2\eta}$ holds if and only if $(-1)^{-k-m_0}a = -\overline{a\alpha^2}$ and $k+2m_0-2M_0-2 = 2K-k$. Under the condition (5), these conditions can be rewritten as follows:

$$a\alpha \in \sqrt{-1}\mathbf{R}, \quad k+m_0 \ : \ \text{even}, \quad K = k+m_0-M_0-1. \tag{6}$$

In particular, if both the set of ends and the set of zeroes of g is symmetric with respect to the origin, then we have the following:

Lemma 2.1. *Assume that the set of ends satisfies*

$$\{q_j, I(q_j) \mid j=1,\dots,\ell\} = \{-q_j, -I(q_j) \mid j=1,\dots,\ell\},$$

and that the orders of q_j and $-q_j$ as poles of η and/or $g^2\eta$ are the same with each other for $j=1,\dots,\ell$. Moreover assume that the set of zeroes of g satisfies

$$\{r_J \mid J=1,\dots,L\} = \{-r_J \mid J=1,\dots,L\},$$

and that the orders of r_J and $-r_J$ as zeroes of g are the same with each other for $J=1,\dots,L$. Then the following equalities hold for any j such that $|q_j|=1$:

$$\overline{R_0(q_j)} = (-1)^k R_2(q_j), \qquad \overline{R_1(q_j)} = (-1)^k R_1(q_j).$$

Proof. Under the assumption, it holds that

$$g^i\eta\big|_{z=-z'} = (-1)^{k-1+i}\eta\big|_{z=z'} \quad (i=0,1,2)$$

which implies

$$R_i(-q_j) = (-1)^{k-1+i}R_i(q_j) \quad (i=0,1,2;\ j=1,\dots,\ell).$$

By combining (4) and this equality, we have

$$\overline{R_0(q_j)} = (-1)^k R_2(-I(q_j)), \qquad \overline{R_1(q_j)} = (-1)^k R_1(-I(q_j)).$$

Hence, if $|q_j|=1$, then, since $-I(q_j)=q_j$, we get our assertion. □

We note here that we use the following form of η in §4:

$$\eta = \frac{a'\displaystyle\prod_{J=1}^{L}(\overline{r_J}\,z+1)^{2M_J}}{z^k\displaystyle\prod_{j=1}^{\ell}\left\{(z-q_j)^{m_j}\left(z+\frac{1}{\overline{q_j}}\right)^{m_j}\right\}}\,dz,$$

where we set $a' := a\Big/\displaystyle\prod_{j=1}^{\ell}\overline{q_j}^{\,m_j}$.

3. Lemmas for calculations of residues

In this section, we prepare lemmas for calculations of residues which we use in §4.

Lemma 3.1. *For any positive integers κ, μ and ν, the following equality holds*:

$$\operatorname*{Res}_{z=0} \frac{1}{z^\kappa (z^\nu - 1)^\mu} = \begin{cases} (-1)^\mu {}_\mu H_{(\kappa-1)/\nu} & \left(\dfrac{\kappa-1}{\nu} \in \mathbf{N} \cup \{0\}\right), \\ 0 & \left(\dfrac{\kappa-1}{\nu} \notin \mathbf{N} \cup \{0\}\right). \end{cases}$$

Proof. For any z such that $|z| < 1$, it holds that $1/(z^\nu - 1) = -\sum_{i=0}^{\infty} (z^\nu)^i$. Hence we see that

$$\frac{1}{(z^\nu - 1)^\mu} = (-1)^\mu \left\{ \sum_{i=0}^{\infty} (z^\nu)^i \right\}^\mu = (-1)^\mu \sum_{i=0}^{\infty} {}_\mu H_i (z^\nu)^i,$$

where ${}_\mu H_i = {}_{\mu+i-1} C_i = (\mu + i - 1)!/\{(\mu - 1)! i!\}$. Our assertion follows from this expansion. □

Lemma 3.2. *For any integer γ, and positive integers μ and ν, the following equality holds*:

$$\operatorname*{Res}_{z={\zeta_\nu}^{j-1}} \frac{z^\gamma}{(z^\nu - 1)^\mu} = {\zeta_\nu}^{(j-1)(\gamma+1)} \frac{\prod_{i=1}^{\mu-1} (\gamma + 1 - i\nu)}{\nu^\mu (\mu - 1)!},$$

where we set $\zeta_\nu := e^{2\pi\sqrt{-1}/\nu}$.

Proof. Set $H_{\mu,\gamma}(z) := z^\gamma/(z^\nu - 1)^\mu$. Then it holds that

$$H'_{\mu-1,\gamma+1}(z) = \{\gamma + 1 - (\mu - 1)\nu\} H_{\mu-1,\gamma}(z) - (\mu - 1)\nu H_{\mu,\gamma}(z),$$

from which it also follows that

$$\operatorname*{Res}_{z={\zeta_\nu}^{j-1}} H_{\mu,\gamma}(z) = \frac{\gamma + 1 - (\mu - 1)\nu}{(\mu - 1)\nu} \operatorname*{Res}_{z={\zeta_\nu}^{j-1}} H_{\mu-1,\gamma}(z)$$

for $\mu \geq 2$. By using this equality repeatedly, we have

$$\operatorname*{Res}_{z={\zeta_\nu}^{j-1}} H_{\mu,\gamma}(z) = \frac{\prod_{i=1}^{\mu-1} (\gamma + 1 - i\nu)}{\nu^{\mu-1} (\mu - 1)!} \operatorname*{Res}_{z={\zeta_\nu}^{j-1}} H_{1,\gamma}(z).$$

On the other hand, since

$$H_{1,\gamma}(z) = \frac{z^\gamma}{z^\nu - 1} = \frac{z^\gamma}{\nu}\sum_{j=1}^{\nu} \frac{{\zeta_\nu}^{j-1}}{z - {\zeta_\nu}^{j-1}},$$

we see that

$$\operatorname*{Res}_{z={\zeta_\nu}^{j-1}} H_{1,\gamma}(z) = \operatorname*{Res}_{z={\zeta_\nu}^{j-1}} \frac{{\zeta_\nu}^{j-1} z^\gamma}{\nu(z - {\zeta_\nu}^{j-1})} = \frac{{\zeta_\nu}^{(j-1)(\gamma+1)}}{\nu}.$$

Hence we get our assertion. □

Lemma 3.3. *For any integer γ, any positive integer ν, and any unit complex number such that $q^\nu \neq \pm 1$, the following equality holds:*

$$\begin{aligned}\operatorname*{Res}_{z=q{\zeta_\nu}^{j-1}} & \frac{z^\gamma}{(z^\nu - q^\nu)^2(z^\nu - \overline{q}^\nu)^2} \\ &= \frac{{\zeta_\nu}^{(j-1)(\gamma+1)} q^{\gamma+1-2\nu}}{\nu^2(q^\nu - \overline{q}^\nu)^3}\{(\gamma+1-3\nu)q^\nu - (\gamma+1-\nu)\overline{q}^\nu\}.\end{aligned}$$

Proof. Note here that

$$\begin{aligned}\frac{1}{(z^\nu - q^\nu)^2(z^\nu - \overline{q}^\nu)^2} = \frac{1}{(q^\nu - \overline{q}^\nu)^2}&\left\{\frac{1}{(z^\nu - q^\nu)^2} + \frac{1}{(z^\nu - \overline{q}^\nu)^2}\right\} \\ - \frac{2}{(q^\nu - \overline{q}^\nu)^3}&\left\{\frac{1}{z^\nu - q^\nu} - \frac{1}{z^\nu - \overline{q}^\nu}\right\}.\end{aligned} \tag{7}$$

By applying Lemma 3.2 and the equality

$$\operatorname*{Res}_{z=q{\zeta_\nu}^{j-1}} \frac{z^\gamma}{(z^\nu - q^\nu)^\mu} = q^{\gamma+1-\mu\nu} \operatorname*{Res}_{z={\zeta_\nu}^{j-1}} \frac{z^\gamma}{(z^\nu - 1)^\mu}$$

for $\mu = 1, 2$, we have

$$\begin{aligned}\operatorname*{Res}_{z=q{\zeta_\nu}^{j-1}} & \frac{z^\gamma}{(z^\nu - q^\nu)^2(z^\nu - \overline{q}^\nu)^2} \\ &= \operatorname*{Res}_{z=q{\zeta_\nu}^{j-1}} \frac{z^\gamma}{(q^\nu - \overline{q}^\nu)^3}\left\{\frac{q^\nu - \overline{q}^\nu}{(z^\nu - q^\nu)^2} - \frac{2}{(z^\nu - q^\nu)}\right\} \\ &= \frac{1}{(q^\nu - \overline{q}^\nu)^3}\left\{(q^\nu - \overline{q}^\nu)q^{\gamma+1-2\nu}{\zeta_\nu}^{(j-1)(\gamma+1)}\frac{\gamma+1-\nu}{\nu^2(2-1)!}\right. \\ &\qquad\qquad \left. -2q^{\gamma+1-\nu}{\zeta_\nu}^{(j-1)(\gamma+1)}\frac{1}{\nu}\right\} \\ &= \frac{{\zeta_\nu}^{(j-1)(\gamma+1)} q^{\gamma+1-2\nu}}{\nu^2(q^\nu - \overline{q}^\nu)^3}\{(\gamma+1-3\nu)q^\nu - (\gamma+1-\nu)\overline{q}^\nu\}\end{aligned}$$

for any $\gamma \in \mathbf{Z}$. □

4. Main results

Now, let N be a positive integer, and let us consider the case that the number of ends of X is N or $N+1$, and that $X(M)$ is invariant under the action of $\mathbf{Z}_N$, where we denote a cyclic subgroup of $SO(3)$ of order N by $\mathbf{Z}_N$. We may assume that the axis of the symmetry is the x_3-axis without loss of generality if we identify the surfaces congruent with each other.

In particular, we consider the following situation:

(I) The set of ends of X is $\{[0]\} \cup \{[{\zeta_{2N}}^{j-1}] \mid j = 1, \ldots, N\}$ (or is $\{[{\zeta_{2N}}^{j-1}] \mid j = 1, \ldots, N\}$), and the order of ${\zeta_{2N}}^{j-1}$ as a pole of η is m for any $j = 1, \ldots, 2N$, where we set $\zeta_{2N} := e^{\pi\sqrt{-1}/N}$.
(II) The set of limit normals of $\widetilde{X}$ is $\{0, \infty\} \cup \{{\zeta_{2N}}^{j-1} \mid j = 1, \ldots, 2N\}$ (or is $\{{\zeta_{2N}}^{j-1} \mid j = 1, \ldots, 2N\}$).
(III) L is a positive multiple of N, and the set of zeroes of g is $\{0\} \cup \{r_J = r{\zeta_L}^{j-1} \mid J = 1, \ldots, L\}$, and the order M_J of r_J as a zero of g is 1 for any $J = 1, \ldots, L$, where we set $\zeta_L := e^{2\pi\sqrt{-1}/L}$.

If an end with horizontal limit normal is a catenoidal end, then it cannot be mapped to itself by a half turn with respect to the x_3-axis, the axis of $\mathbf{Z}_N$ symmetry. Based on this fact essentially, the authors [2, Theorem 7.4] proved the nonexistence of $\mathbf{Z}_N$-invariant nonorientable $(N+1)$-noids defined on punctured $\mathbf{R}P^2$ all of whose ends are catenoidal ends for any even integer N.

On the other hand, if an end with horizontal limit normal is a planar end (or a non-embedded end whose flux vector is $\mathbf{0}$), then it may be mapped to itself by the half turn. Hence we may consider the situation with the following (I') and (II') instead of (I) and (II):

(I') N is even, and the set of ends of X is $\{[0]\} \cup \{[q{\zeta_N}^{j-1}], [\overline{q}{\zeta_N}^{j-1}] \mid j = 1, \ldots, N/2\}$, and the orders of $q{\zeta_N}^{j-1}$ and $\overline{q}{\zeta_N}^{j-1}$ as poles of η are m for any $j = 1, \ldots, N$, where q is a unit complex number such that $q^N \neq \pm 1$, and we set $\zeta_N := e^{2\pi\sqrt{-1}/N}$.
(II') The set of limit normals of $\widetilde{X}$ is $\{0, \infty\} \cup \{p{\zeta_N}^{j-1}, \overline{p}{\zeta_N}^{j-1} \mid j = 1, \ldots, N\}$, where p is also a unit complex number such that $p^N \neq \pm 1$.

Under the conditions (I), (II), (III) (resp. (I'), (II'), (III)), the Weierstrass data of $\widetilde{X}$ is given by

$$g(z) = \frac{\alpha z^K (z^L - \beta)}{-(-1)^L\{\overline{\beta} z^L - (-1)^L\}},$$

$$\eta = \frac{a'\{\overline{\beta}z^L - (-1)^L\}^2}{z^k(z^{2N}-1)^m}\,dz \left(\text{resp. } \frac{a'\{\overline{\beta}z^L - (-1)^L\}^2}{z^k(z^N-q^N)^m(z^N-\overline{q}^N)^m}\,dz\right),$$

where $K = k + mN - L - 1$ and we set $\beta := r^L$. Since $a' = (-1)^{(N-1)m/2}a$, the first condition in (6) is rewritten as follows:

$$a'\alpha \in \begin{cases} \mathbf{R} & \text{if } N \text{ is even and } m \text{ is odd,} \\ \sqrt{-1}\,\mathbf{R} & \text{if otherwise stated.} \end{cases}$$

By the second condition in (6), $k + m_0 = k + mN$ must be even. Hence any possible choice of parities of (k, m, N) is one of the following cases:

(even, even, even), (even, even, odd), (even, odd, even), (odd, odd, odd).

We note here that $d = \deg g = K + L = k + mN - 1$. In each case above, we assume $\gcd(d, N)(= \gcd(k-1, N)) = 1$ to exclude the case that the given surface coincides with a covering of another surface, even in the case that the corresponding surface is well-defined.

Under the assumptions (I), (II), (III) above with $L = 2N$, we can construct a family of nonorientable minimal surfaces with various pairs of orders of ends as follows. This family includes examples by Oliveira [7, Theorems 2.8, 2.9] and Barros [1, Theorem 1.1 (b), (c)] as special cases.

Theorem 4.1. *Assume* (I), (II), (III) *and* $\gcd(k-1, N) = 1$. *Then the data*

$$g(z) = \frac{\alpha z^{k+(m-2)N-1}(z^{2N}-\beta)}{-(\overline{\beta}z^{2N}-1)}, \quad \eta = \frac{a'(\overline{\beta}z^{2N}-1)^2}{z^k(z^{2N}-1)^m}\,dz \tag{8}$$

realizes a well-defined nonorientable minimal surface if one of the following condition holds:

(i) (k, m, N) *is one of the cases (even, even, even), (even, even, odd), (even, odd, even), and satisfies* $k = 0$ or $k \geq 2$, $m \geq 3$, $N \geq 1$, *and*

$$\begin{cases} a'\alpha \in (-1)^{(m-1)/2}\mathbf{R} \setminus \{0\}, \\ D_1(\beta) := (k-1)(k-1-2N)\beta^2 \\ \qquad - 2(k-1)\{k-1+(2m-4)N\}\beta \\ \qquad + \{k-1+(2m-2)N\}\{k-1+(2m-4)N\} = 0. \end{cases} \tag{9}$$

(ii) (k, m, N) *is the case (odd, odd, odd), and satisfies* $k \geq 3$, $m \geq 3$, $N \geq 3$, *and*

$$\begin{cases} a'\alpha \in \sqrt{-1}\mathbf{R} \setminus \{0\}, \\ D_2(\beta) := (m-2)(|\beta|^2+1) + m(\beta+\overline{\beta}) = 0. \end{cases}$$

In particular in the case (ii), *there exist a* 1*-parameter family of nonorientable minimal surfaces for each* (k, m, N).

Proof. By direct computation, we have

$$\eta = \frac{a'(\overline{\beta}^2 z^{-k+4N} - 2\overline{\beta} z^{-k+2N} + z^{-k})}{(z^{2N}-1)^m}\, dz,$$

$$g\eta = \frac{-a'\alpha\{\overline{\beta} z^{(m+2)N-1} - (|\beta|^2+1) z^{mN-1} + \beta z^{(m-2)N-1}\}}{(z^{2N}-1)^m}\, dz,$$

$$g^2\eta = \frac{a'\alpha^2(z^{k+2mN-2} - 2\beta z^{k+(2m-2)N-2} + \beta^2 z^{k+(2m-4)N-2})}{(z^{2N}-1)^m}\, dz.$$

Since we assume $m \geq 3$, it holds that $(m-2)N - 1 \geq (3-2)\cdot 1 - 1 = 0$, and hence 0 is not a pole of $g\eta$ and $g^2\eta$. Therefore we see that $R_1(0) = R_2(0) = 0$. Since we assume $\gcd(k-1, N) = 1$, and $N \geq 3$ if k is odd, we find that $(\kappa - 1)/(2N)$ is not an integer for $\kappa = k, k-2N, k-4N$. Hence, by applying Lemma 3.1 with $\nu = 2N$, we also see that $R_0(0) = 0$.
(i) Since k is even, by Lemma 2.1, we have $R_1({\zeta_{2N}}^{j-1}) \in \mathbf{R}$. On the other hand, by Lemma 2.1 again, we also have $\overline{R_0({\zeta_{2N}}^{j-1})} = R_2({\zeta_{2N}}^{j-1})$. Hence $\overline{R_0({\zeta_{2N}}^{j-1})} + R_2({\zeta_{2N}}^{j-1}) = 0$ holds if and only if $R_2({\zeta_{2N}}^{j-1}) = 0$. By applying Lemma 3.2 with $\nu = 2N$, we see that

$$R_2({\zeta_{2N}}^{j-1}) = \frac{a'\alpha^2 {\zeta_{2N}}^{(j-1)(k-1)}}{(2N)^m \cdot (m-1)!} D_1(\beta) \prod_{i=1}^{m-3} (k-1+2iN).$$

Now, by the assumption $D_1(\beta) = 0$, we have $R_2({\zeta_{2N}}^{j-1}) = 0$.

(ii) Since k is odd, by Lemma 2.1, we have $\overline{R_0({\zeta_{2N}}^{j-1})} + R_2({\zeta_{2N}}^{j-1}) = 0$. On the other hand, by applying Lemma 3.2 with $\nu = 2N$, we see that

$$R_1({\zeta_{2N}}^{j-1}) = -\frac{(-1)^{j-1} a'\alpha}{2^m N \cdot (m-1)!} D_2(\beta) \prod_{i=1}^{m-2} (m-2i).$$

Now, by the assumption $D_2(\beta) = 0$, we have $R_1({\zeta_{2N}}^{j-1}) = 0$. □

In particular in the case $N = 1$ or 2, we see that, for any

$$(k, m, N) = \begin{cases} (k, m, 1) & (k \geq 2, m \geq 4; k, m : \text{even}), \\ (k, m, 2) & (k \geq 2, m \geq 3; k : \text{even}), \end{cases}$$

the Weierstrass data (8) with the condition (9) realizes a nonorientable minimal surfaces with two or three ends. Two families of examples by Oliveira [7, Theorems 2.8, 2.9] and Barros [1, Theorem 1.1 (b), (c)] are

special cases of this family, which are given by $(k, m, N) = (k, 4, 1)$ and $(k, 3, 2)$ ($k \geq 2$; k : even).

In the case that $k = 2$, the end $[0]$ of each surface in Theorem 4.1(i) is an embedded end. Since $R_0(0) = R_1(0) = R_2(0) = 0$, the end $[0]$ is a planar end. However, since we have to assume $m \geq 3$ here, we cannot find a nonorientable $(N+1)$-noid in this family.

On the other hand, under the assumptions (I), (II), (III) above with $L = N$, we can construct another family of examples.

Theorem 4.2. *Assume* (I), (II), (III) *and* $\gcd(k-1, N) = 1$. *Then the data*

$$g(z) = \frac{\alpha z^{k+(m-1)N-1}(z^N - \beta)}{\overline{\beta} z^N + 1}, \qquad \eta = \frac{a'(\overline{\beta} z^N + 1)^2}{z^k (z^{2N} - 1)^m}\, dz \tag{10}$$

realizes a well-defined nonorientable minimal surface if (k, m, N) is the case (even, even, odd), and satisfies $k \geq 2$, $m \geq 2$, $N \geq 3$, and

$$a'\alpha \in \sqrt{-1}\mathbf{R} \setminus \{0\}, \quad D_3(\beta) := (k-1)\beta^2 + \{k-1+(2m-2)N\} = 0.$$

Proof. By direct computation, we have

$$\eta = \frac{a'(\overline{\beta}^2 z^{-k+2N} + 2\overline{\beta} z^{-k+N} + z^{-k})}{(z^{2N} - 1)^m}\, dz,$$

$$g\eta = \frac{a'\alpha\{\overline{\beta} z^{(m+1)N-1} - (|\beta|^2 - 1) z^{mN-1} - \beta z^{(m-1)N-1}\}}{(z^{2N} - 1)^m}\, dz,$$

$$g^2\eta = \frac{a'\alpha^2 (z^{k+2mN-2} - 2\beta z^{k+(2m-1)N-2} + \beta^2 z^{k+(2m-2)N-2})}{(z^{2N} - 1)^m}\, dz.$$

Since we assume $m \geq 2$, it holds that $(m-1)N - 1 \geq (2-1)\cdot 1 - 1 = 0$, and hence 0 is not a pole of $g\eta$ and $g^2\eta$. Therefore we see that $R_1(0) = R_2(0) = 0$. Since we assume $\gcd(k-1, N) = 1$ and $N \geq 3$, we find that $(\kappa - 1)/N$ is not an integer for $\kappa = k, k-N, k-2N$. Hence, by applying Lemma 3.1 with $\nu = 2N$, we also see that $R_0(0) = 0$.

On the other hand, by applying Lemma 3.2 with $\nu = 2N$ and the assumption that m is even, we see that

$$\begin{aligned}
&\overline{R_0({\zeta_{2N}}^{j-1})} + R_2({\zeta_{2N}}^{j-1}) \\
&= \frac{\alpha {\zeta_{2N}}^{(j-1)(k-1)}}{(2N)^m \cdot (m-1)!} \Bigg[(a'\alpha - \overline{a'\alpha}) D_3(\beta) \prod_{i=1}^{m-2} (k-1+2iN) \\
&\qquad - (-1)^{j-1}(a'\alpha + \overline{a'\alpha}) \cdot 2\beta \prod_{i=1}^{m-1} \{k-1+(2i-1)N\} \Bigg],
\end{aligned}$$

$$R_1({\zeta_{2N}}^{j-1}) = \frac{a'\alpha}{2^m N\cdot(m-1)!}\left\{-(|\beta|^2-1)\prod_{i=1}^{m-1}(m-2i)\right.$$
$$\left.+(-1)^{j-1}(\beta+\overline{\beta})\prod_{i=1}^{m-1}(m+1-2i)\right\}.$$

By the assumption $D_3(\beta)=0$ and $a'\alpha\in\sqrt{-1}\mathbf{R}$, we have $\overline{R_0({\zeta_{2N}}^{j-1})+R_2({\zeta_{2N}}^{j-1})}=0$. Since $m-2i=0$ holds for $i=m/2\in[1,m-1]\cap\mathbf{N}$, and since $\beta\in\sqrt{-1}\mathbf{R}$, we have $R_1({\zeta_{2N}}^{j-1})=0$. Hence we get our assertion. □

In particular, in the case that $k=m=2$, the data (10) with $\beta^2=-(2N+1)$ realizes a nonorientable $(N+1)$-noid all of whose ends are catenoidal ends except for one planar end (cf. [2, Lemma 5.2]).

Remark 4.1. If we consider the situation that $\gcd(k-1,N)=1$, (k,m,N) is the case (even, even, odd), and satisfies $k\geq N+3$, $m=0$ (namely $[{\zeta_{2N}}^{j-1}]$'s are not ends), $N\geq 1$ and $|\beta|=1$, then the data (10) is of the form

$$g(z)=\frac{\alpha z^{k-N-1}(z^N-\beta)}{\overline{\beta}z^N+1},\qquad \eta=\frac{a'(\overline{\beta}z^N+1)^2}{z^k}\,dz,$$

and realizes a $\mathbf{Z}_N$-invariant example with one end (which is independent of β). Indeed, since we assume that k is even, it holds that

$$-k+2N,\ -k+N,\ -k,\ k-2,\ k-N-2,\ k-2N-2\in\mathbf{Z}\setminus\{-1\},$$

and hence we see that $R_0(0)=R_2(0)=0$. On the other hand, since we assume $|\beta|=1$, we also see that $R_1(0)=-a'\alpha(|\beta|^2-1)=0$. In particular, if $N=1$ (resp. $N=k-3$), then the data realizes a family given by Meeks [5, Theorem 2] and Oliveira [7, Theorem 2.6] (resp. bent helicoids by Meeks-Weber [6]). Ishihara [3] gave a classification of surfaces in this class.

For any odd integer $N\geq 3$, by applying Theorem 4.2 for $(k,m,N)=(k,2,N)$ ($k\geq 2; k$: even), we see that there exists an example with $d=k+2N-1$ if $\gcd(d,N)=\gcd(k-1,N)=1$. Here

$$\{k+2N-1 \mid k\geq 2; k:\text{even}\}=\{d\mid d\geq 2N+1; d:\text{odd}\}.$$

On the other hand, since we must assume $m\geq 3$ in Theorem 4.1, we cannot construct any example with $2+2N-1=2N+1\leq d<3N+1=2+3N-1$ for any even number $N\geq 2$ as an application to Theorem 4.1.

Now, we construct another family of examples under the conditions (I'), (II'), (III), and the additional assumptions $q^N \neq \beta, \overline{\beta}$, that is, $g(q{\zeta_N}^{j-1}) \neq 0$, $g(\overline{q}{\zeta_N}^{j-1}) \neq 0$ $(j = 1, \ldots, N)$, $L = N$ and $m = 2$. This fact completes the proof for Theorem 1.1.

Theorem 4.3. *Assume* (I′), (II′), (III) *and* $\gcd(k-1, N) = 1$, *and set* $K := k + N - 1$. *Then the data*

$$g(z) = \frac{\alpha z^K (z^N - \beta)}{-(\overline{\beta} z^N - 1)}, \qquad \eta = \frac{a'(\overline{\beta} z^N - 1)^2}{z^k (z^N - q^N)^m (z^N - \overline{q}^N)^m}\, dz \tag{11}$$

realizes a well-defined nonorientable minimal surface if (k, m, N) *is the case* (*even*, *even*, *even*), *and satisfies* $k \geq 2$, $m = 2$, $N \geq 2$, *and*

$$\begin{cases} a'\alpha \in \sqrt{-1}\mathbf{R} \setminus \{0\}, \\ K^2 q^{4N} - 2(K^2 - 2N^2) q^{2N} + K^2 = 0, \\ (K - N)\beta^2 = K + N. \end{cases}$$

Proof. By direct computation, we have

$$\eta = \frac{a'(\overline{\beta}^2 z^{-k+2N} - 2\overline{\beta} z^{-k+N} + z^{-k})}{(z^N - q^N)^2 (z^N - \overline{q}^N)^2}\, dz,$$

$$g\eta = \frac{-a'\alpha\{\overline{\beta} z^{3N-1} - (|\beta|^2 + 1) z^{2N-1} + \beta z^{N-1}\}}{(z^N - q^N)^2 (z^N - \overline{q}^N)^2}\, dz,$$

$$g^2\eta = \frac{a'\alpha^2 (z^{K+3N-1} - 2\beta z^{K+2N-1} + \beta^2 z^{K+N-1})}{(z^N - q^N)^2 (z^N - \overline{q}^N)^2}\, dz.$$

Since 0 is not a pole of $g\eta$ and $g^2\eta$, $R_1(0) = R_2(0) = 0$. Since we assume that both k and N are even, we find that $(\kappa - 1)/N$ is not an integer for $\kappa = k, k - N, k - 2N$. Hence, by applying Lemma 3.1 with $\nu = N$ and the equality (7), we also see that $R_0(0) = 0$.

On the other hand, by Lemma 2.1, we have $R_1(q{\zeta_N}^{j-1}) \in \mathbf{R}$ and $R_1(\overline{q}{\zeta_N}^{j-1}) \in \mathbf{R}$.

By Lemma 2.1 again, we also have $\overline{R_0(q{\zeta_N}^{j-1})} = R_2(q{\zeta_N}^{j-1})$. Hence $\overline{R_0(q{\zeta_N}^{j-1})} + R_2(q{\zeta_N}^{j-1}) = 0$ holds if and only if $R_2(q{\zeta_N}^{j-1}) = 0$.

Set $F_1(q) := (K - 2N) q^{2N} - K$ and $F_2(q) := K q^{2N} - (K + 2N)$. Then, by applying Lemma 3.3, we see that

$$R_2(q{\zeta_N}^{j-1}) = \frac{a'\alpha^2 q^{K-2N} {\zeta_N}^{(j-1)K}}{N^2 (q^N - \overline{q}^N)^3} (\beta - q^N)\big(F_1(q)\beta - F_2(q) q^N\big).$$

Since we assume $q^N \neq \beta$, $R_2(q{\zeta_N}^{j-1}) = 0$ holds if and only if $F_1(q)\beta = F_2(q) q^N$. By replacing q by $\overline{q}$, we also see that $R_2(\overline{q}{\zeta_N}^{j-1}) = 0$ holds if and only if $F_1(\overline{q})\beta = F_2(\overline{q})\overline{q}^N$ which is equivalent with $F_1(q)\overline{\beta} = F_2(q)q^N$. Since $F_1(q) \neq 0$, both $F_1(q)\beta = F_2(q)q^N$ and $F_1(q)\overline{\beta} = F_2(q)q^N$ hold if and only if $\beta = \overline{\beta} = (F_2(q)q^N)/F_1(q)$. This equation has a solution (q, β) if and only if

$$\begin{aligned} 0 &= q^{3N}(\overline{q}^N \overline{F_2(q)}\, F_1(q) - q^N F_2(q)\overline{F_1(q)}) \\ &= (q^{2N} - 1)\left\{K^2 q^{4N} - 2(K^2 - 2N^2)q^{2N} + K^2\right\}. \end{aligned}$$

By solving the equation $K^2q^{4N} - 2(K^2 - 2N^2)q^{2N} + K^2 = 0$, we get

$$q^N = \pm\frac{\sqrt{K^2 - N^2} \pm N\sqrt{-1}}{K},$$

where the double signs do not correspond, and

$$\beta = \frac{\overline{F_1(q)}\, F_2(q)q^N}{|F_1(q)|^2} = \frac{4N^2(q^N + \overline{q}^N)}{8N^2(K - N)/K} = \pm\sqrt{\frac{K + N}{K - N}}.$$

□

Since $m = 2$, the ends $q\zeta_N{}^{j-1}$ and $\overline{q}\zeta_N{}^{j-1}$ are embedded ends, and hence each of their flux vectors is parallel to the corresponding limit normal. Moreover, since $|g(q\zeta_N{}^{j-1})| = |g(\overline{q}\zeta_N{}^{j-1})| = 1$, it holds that $R_1(q\zeta_N{}^{j-1}) = R_1(\overline{q}\zeta_N{}^{j-1}) = 0$.

In particular, in the case that $k = 2$, the data (11) with $\beta^2 = 2N + 1$ realizes a nonorientable flat-ended $(N + 1)$-noid which is not conformal with Kusner's one in [4, Theorem B]. Hence its compactification gives a new example of Willmore projective plane in $\mathbf{S}^3$.

Proof of Theorem 1.1. By applying Theorem 4.1(i) for $(k, m, N) = (2, m, 1)$ ($m \geq 4$; m : even), Theorem 4.3 for $(k, m, N) = (k, 2, N)$ ($k \geq 2, N \geq 2$; k, N : even), and Theorem 4.2 for $(k, m, N) = (k, 2, N)$ ($k \geq 2, N \geq 3; k$: even, N : odd), we get the assertion of Theorem 1.1. □

By the facts we have proved in this paper, we get Table 1. Each $(k_K, m, N)^*$ (resp. $(k_K, m, N)^{**}$, $(k_K, 2, N)^{***}$) is the data of a surface in Theorem 4.1 (resp. Theorem 4.2 and Remark 4.1, Theorem 4.3). Each of the other (k_K, m, N)'s also corresponds to some $\mathbf{Z}_N$-invariant surface whose η has 2 poles of order k and $2N$ poles of order m. But its Weierstrass data is of a form different from (8), (10) and (11). In the table, we show the reference after (k_K, m, N).

We note here that $(0, m, 1)^*$ ($m \geq 4; m$: even) does not coincide with $(k_K, 0, N)^{**}$ ($k \geq 4; k$: even, $K = k - N - 1$) except for the case that $m = k = 4$, since $g'(1) \neq 0$ holds for $m \geq 6$.

(#1) Since $\gcd(10-1,3) = 3 \neq 1$, $(10_6,0,3)^{**}$ is excluded. This data realizes a triple covering of $(4_2,0,1)^{**}(=(0,4,1)^{*})$.
(#2) Since $\gcd(4-1,3) = 3 \neq 1$, $(4_6,2,3)^{**}$ is excluded. This data realizes a triple covering of $(2_2,2,1)^{**}$ which is not well-defined.

References

[1] A. A. de Barros, Complete nonorientable minimal surfaces in $\mathbf{R}^3$ with finite total curvature, *An. Acad. Brasil. Cienc.* **59** (1987), 141–143.
[2] K. Hamada, S. Kato, Nonorientable minimal surfaces with catenoidal ends, *Ann. Mat. Pura Appl.* **200** (2021), 1573–1603.
[3] T. Ishihara, Complete Möbius strips minimally immersed in $\mathbf{R}^3$, *Proc. Amer. Math. Soc.* **3** (1989), 803–806.
[4] R. Kusner, Conformal geometry and complete minimal surfaces, *Bull. Amer. Math. Soc.* **17** (1987), 291–295.
[5] W.H. Meeks III, The classification of complete minimal surfaces in $\mathbf{R}^3$ with total curvature greater than -8π, *Duke Math. J.* **48** (1981), 523–535.
[6] W.H. Meeks III, M. Weber, Bending the helicoid, *Math. Ann.* **339** (2007), 783–798.
[7] E. G. G. de Oliveira, Some new examples of nonorientable minimal surfaces, *Proc. Amer. Math. Soc.* **98** (1986), 629–636.
[8] R. Osserman, *A survey of minimal surfaces*, Van Nostrand Reinhold Co., New York-London-Melbourne 1969.

Received December 19, 2023

Table 1. The list of (k_K, m, N) $(d \leq 13)$.

d	1	3	5	7	9	11	13	$\cdots$
$n \backslash TC$	-2π	-6π	-10π	-14π	-18π	-22π	-26π	$\cdots$
1	-	$(0,4,1)^*$ [5] $(4_2,0,1)^{**}$ [5]	$(0,6,1)^*$ $(6_4,0,1)^{**}$ [7] $(6_2,0,3)^{**}$ [6] $(6_*,0,*)$ [3]	$(0,8,1)^*$ $(8_6,0,1)^{**}$ [7] $(8_4,0,3)^{**}$ $(8_2,0,5)^{**}$ [6] $(8_*,0,*)$ [3]	$(0,10,1)^*$ $(10_8,0,1)^{**}$ [7] (#1) $(10_4,0,5)^{**}$ $(10_2,0,7)^{**}$ [6] $(10_*,0,*)$ [3]	$(0,12,1)^*$ $(12_{10},0,1)^{**}$ [7] $(12_8,0,3)^{**}$ $(12_6,0,5)^{**}$ $(12_4,0,7)^{**}$ $(12_2,0,9)^{**}$ [6] $(12_*,0,*)$ [3]	$(0,14,1)^*$ $(14_{12},0,1)^{**}$ [7] $(14_{10},0,3)^{**}$ $(14_8,0,5)^{**}$ $(14_6,0,7)^{**}$ $(14_4,0,9)^{**}$ $(14_2,0,11)^{**}$ [6] $(14_*,0,*)$ [3]	$\cdots$
2	-	- [5]	$\mathbf{(2_3,4,1)}^*$ [7] $(2_1,4,1)$ [1] $(0,3,2)^*$	$(4_5,4,1)^*$ [1] $\mathbf{(2_5,6,1)}^*$ $(0,4,2)^*$	$(6_7,4,1)^*$ [1] $(4_7,6,1)^*$ $\mathbf{(2_7,8,1)}^*$ $(0,5,2)^*$	$(8_9,4,1)^*$ [1] $(6_9,6,1)^*$ $(4_9,8,1)^*$ $\mathbf{(2_9,10,1)}^*$ $(0,6,2)^*$	$(10_{11},4,1)^*$ [1] $(8_{11},6,1)^*$ $(6_{11},8,1)^*$ $(4_{11},10,1)^*$ $\mathbf{(2_{11},12,1)}^*$ $(0,7,2)^*$	$\cdots$
3	-	-	$\mathbf{(2_3,2,2)}^{***}$ $(0,2,3)$ [4]	$\mathbf{(4_5,2,2)}^{***}$ $(2_3,3,2)^*$ [7]	$\mathbf{(6_7,2,2)}^{***}$ $(4_5,3,2)^*$ [1] $(2_5,4,2)^*$	$\mathbf{(8_9,2,2)}^{***}$ $(6_7,3,2)^*$ [1] $(4_7,4,2)^*$ $(2_7,5,2)^*$ $(0,4,3)^*$	$\mathbf{(10_{11},2,2)}^{***}$ $(8_9,3,2)^*$ [1] $(6_9,4,2)^*$ $(4_9,5,2)^*$ $(2_9,6,2)^*$	$\cdots$
4	-	-	-	$\mathbf{(2_4,2,3)}^{**}$ [2] $(2_1,2,3)$ [2]	(#2)	$\mathbf{(6_8,2,3)}^{**}$ $(3_5,3,3)^*$ $(0,3,4)^*$	$\mathbf{(8_{10},2,3)}^{**}$ $(5_7,3,3)^*$ $(2_{10},4,3)^{**}$ $(2_7,4,3)^*$	$\cdots$
5	-	-	-	-	$\mathbf{(2_5,2,4)}^{***}$ $(0,2,5)$ [4]	$\mathbf{(4_7,2,4)}^{***}$	$\mathbf{(6_9,2,4)}^{***}$ $(2_5,3,4)^*$	$\cdots$
6	-	-	-	-	-	$\mathbf{(2_6,2,5)}^{**}$ [2] $(2_1,2,5)$ [2]	$\mathbf{(4_8,2,5)}^{**}$	$\cdots$
7	-	-	-	-	-	-	$\mathbf{(2_7,2,6)}^{***}$ $(0,2,7)$ [4]	$\cdots$
$\vdots$	-	-	-	-	-	-	-	$\ddots$

Modern Approaches to Differential Geometry
and its Related Fields 17 – 31

THE $SO(3,1)$-ORBITS IN THE LIGHT CONE OF THE 2-FOLD EXTERIOR POWER OF THE MINKOWSKI 4-SPACE

Naoya ANDO

Faculty of Advanced Science and Technology, Kumamoto University,
2–39–1 Kurokami, Chuo-ku, Kumamoto, 860–8555, Japan
E-mail: andonaoya@kumamoto-u.ac.jp

Two special neutral hypersurfaces $\mathcal{L}_{\pm}$ in the light cone $L(\bigwedge^2 E_1^4)$ studied in [1], [2] are $SO(3,1)$-orbits. In this paper, we see that each $SO(3,1)$-orbit in $L(\bigwedge^2 E_1^4)$ is either a neutral hypersurface homothetic to one of $\mathcal{L}_{\pm}$ in $L(\bigwedge^2 E_1^4)$ or a hypersurface with a two-dimensional involutive distribution where the induced metric vanishes. The difference between these hypersurfaces can be understood in terms of the stabilizer and the r-slice of $L(\bigwedge^2 E_1^4)$ for $r > 0$.

Keywords: $SO(3,1)$-orbit; light cone; 2-fold exterior power; Minkowski 4-space; r-slice.

1. Introduction

The purpose of this paper is to study the $SO(3,1)$-orbits in the light cone $L(\bigwedge^2 E_1^4)$ of $\bigwedge^2 E_1^4$.

The $SO(3,1)$-action on the Minkowski 4-space E_1^4 yields a natural $SO(3,1)$-action on the 2-fold exterior power $\bigwedge^2 E_1^4$ of E_1^4. In particular, $SO(3,1)$ acts on the light cone $L(\bigwedge^2 E_1^4)$ of $\bigwedge^2 E_1^4$. Two special $SO(3,1)$-orbits in $L(\bigwedge^2 E_1^4)$ appear in studies of space-like or time-like surfaces in E_1^4 with zero mean curvature vector ([1]). Let M be a Riemann surface and $F : M \to E_1^4$ a space-like and conformal immersion of M into E_1^4 with zero mean curvature vector. Then the lifts $\Omega_{F,\pm}$ of F are maps from M into neutral hypersurfaces $\mathcal{L}_{\pm}$ of $L(\bigwedge^2 E_1^4)$ respectively, and they are holomorphic with respect to parallel almost complex structures of $\mathcal{L}_{\pm}$ ([1], [2]). If we suppose that M is a Lorentz surface and that $F : M \to E_1^4$ is a time-like and conformal immersion of M into E_1^4 with zero mean curvature vector, then we have $\Omega_{F,\pm} : M \to \mathcal{L}_{\pm}$ and they are holomorphic with respect to parallel almost paracomplex structures of $\mathcal{L}_{\pm}$ ([1], [2]). The hypersurfaces $\mathcal{L}_{\pm}$ are $SO(3,1)$-orbits in $L(\bigwedge^2 E_1^4)$.

In this paper, we will see that each $SO(3,1)$-orbit $\mathcal{L}$ in $L(\bigwedge^2 E_1^4)$ is either a neutral hypersurface which is homothetic to one of $\mathcal{L}_{\pm}$ in $L(\bigwedge^2 E^4_1)$ (Proposition 4.1, Theorem 5.2) or a hypersurface with a two-dimensional involutive distribution where the induced metric vanishes (Theorem 5.1). In order to understand the difference between these hypersurfaces, we will study the stabilizer of $SO(3,1)$ at a point of $\mathcal{L}$ and the invariant subspaces of the tangent space by the stabilizer (Proposition 6.1, Theorem 6.1). In addition, we will define the r-slice $L_r(\bigwedge^2 E^4_1)$ and study the intersection of the r-slice and each $SO(3,1)$-orbit in Section 7. The r-slice is a subset of $L(\bigwedge^2 E^4_1)$ defined for each $r > 0$ and each equivalence class in the set $\mathcal{B}(E^4_1)$ of ordered pseudo-orthonormal bases of E^4_1 giving the orientation with respect to an equivalence relation given in Section 2. For each neutral $SO(3,1)$-orbit $\mathcal{L}$ and each equivalence class in $\mathcal{B}(E^4_1)$, we will see that there exists a positive number $r_0 > 0$ such that the intersection $\mathcal{L} \cap L_r(\bigwedge^2 E^4_1)$ is not empty if and only if $r \geq r_0$, and that if $\mathcal{L} \cap L_r(\bigwedge^2 E^4_1) \neq \emptyset$, then $\mathcal{L} \cap L_r(\bigwedge^2 E^4_1)$ is diffeomorphic to $\boldsymbol{R}P^3$ or S^2 according to $r > r_0$ or $r = r_0$ (Theorem 7.1). We will see that if an $SO(3,1)$-orbit $\mathcal{L}$ has a two-dimensional involutive distribution where the induced metric vanishes, then $\mathcal{L} \cap L_r(\bigwedge^2 E^4_1)$ is not empty, and diffeomorphic to $\boldsymbol{R}P^3$ for any $r > 0$ (Theorem 7.2).

Remark 1.1. Holomorphicity of the lifts of space-like or time-like surfaces in E^4_1 with zero mean curvature vector is an analogue of holomorphicity of the Gauss maps of minimal surfaces in the Euclidean 4-space E^4 ([4, pp. 16–22]) and holomorphicity of the Gauss maps of space-like or time-like surfaces with zero mean curvature vector in E^4_2 ([3]).

Remark 1.2. In Proposition 1 of [1], it was asserted that $\mathcal{L}_{\pm}$ are flat. However, this is not true. In [2], the curvature tensors of $\mathcal{L}_{\pm}$ are explicitly represented and in particular, they do not vanish.

2. The light cone of $\bigwedge^2 E^4_1$

Let h be the metric of the Minkowski 4-space E^4_1:

$$h(x, y) := x^1y^1 + x^2y^2 + x^3y^3 - x^4y^4$$

for $x = {}^t(x^1\ x^2\ x^3\ x^4)$, $y = {}^t(y^1\ y^2\ y^3\ y^4) \in E^4_1$. Let $\hat{h}$ be the metric of the 2-fold exterior power $\bigwedge^2 E^4_1$ of E^4_1 induced by h:

$$\hat{h}(x \wedge y, u \wedge v) := h(x, u)h(y, v) - h(x, v)h(y, u)$$

for $x, y, u, v \in E^4_1$. Let (e_1, e_2, e_3, e_4) be an element of $\mathcal{B}(E^4_1)$, and suppose that e_1, e_2, e_3 are space-like and that e_4 is time-like. Then

$$\omega_{ij} := e_i \wedge e_j \quad (1 \leq i < j \leq 4)$$

form a pseudo-orthonormal basis of $\bigwedge^2 E^4_1$ with respect to $\hat{h}$ such that ω_{12}, ω_{13}, ω_{23} (respectively, ω_{14}, ω_{24}, ω_{34}) are space-like (respectively, time-like). In particular, we see that $\bigwedge^2 E^4_1$ is of dimension 6 and that $\hat{h}$ has signature $(3,3)$. For $\Omega = \sum_{1\leq i<j\leq 4} c_{ij}\omega_{ij}$, we set

$$A(\Omega) := \sum_{1\leq i<j\leq 3} c_{ij}^2, \quad B(\Omega) := \sum_{1\leq i\leq 3} c_{i4}^2.$$

The light cone $L(\bigwedge^2 E^4_1)$ of $\bigwedge^2 E^4_1$ is given by

$$L(\textstyle\bigwedge^2 E^4_1) := \left\{ \Omega \in \textstyle\bigwedge^2 E^4_1 \;\middle|\; A(\Omega) = B(\Omega) \neq 0 \right\}.$$

We see that $L(\bigwedge^2 E^4_1)$ does not depend on the choice of $(e_1, e_2, e_3, e_4) \in \mathcal{B}(E^4_1)$. We set

$$L_1(\textstyle\bigwedge^2 E^4_1) := \left\{ \Omega \in L(\textstyle\bigwedge^2 E^4_1) \;\middle|\; A(\Omega) = B(\Omega) = 1 \right\}.$$

Then $L_1(\bigwedge^2 E^4_1)$ depends on the choice of $(e_1, e_2, e_3, e_4) \in \mathcal{B}(E^4_1)$. For elements $e = (e_1, e_2, e_3, e_4)$, $e' = (e'_1, e'_2, e'_3, e'_4)$ of $\mathcal{B}(E^4_1)$, we write $e \sim e'$ if

$$(e'_1\ e'_2\ e'_3) = \varepsilon(e_1\ e_2\ e_3)U, \quad e'_4 = \varepsilon e_4$$

for an element U of $SO(3)$ and $\varepsilon \in \{1, -1\}$. Then $\sim$ is an equivalence relation in $\mathcal{B}(E^4_1)$. Let $\Omega = \sum_{i<j} c_{ij}\omega_{ij}$ be an element of $L_1(\bigwedge^2 E^4_1)$ for $e = (e_1, e_2, e_3, e_4) \in \mathcal{B}(E^4_1)$. We set

$$\begin{aligned} a_1 &:= c_{23}, & a_2 &:= -\,c_{13}, & a_3 &:= c_{12}, \\ b_1 &:= c_{14}, & b_2 &:= \,c_{24}, & b_3 &:= c_{34} \end{aligned}$$

and $\boldsymbol{a} := {}^t(a_1\ a_2\ a_3)$, $\boldsymbol{b} := {}^t(b_1\ b_2\ b_3)$. Then the lengths of $\boldsymbol{a}$, $\boldsymbol{b}$ are equal to one and Ω is represented as

$$\Omega = (\omega_{23}\ \omega_{31}\ \omega_{12})\boldsymbol{a} + (\omega_{14}\ \omega_{24}\ \omega_{34})\boldsymbol{b}. \tag{1}$$

If $e \sim e'$, then we see by (1) that Ω is represented as

$$\Omega = (\omega'_{23}\ \omega'_{31}\ \omega'_{12})\,{}^tU\boldsymbol{a} + (\omega'_{14}\ \omega'_{24}\ \omega'_{34})\,{}^tU\boldsymbol{b}, \tag{2}$$

where $\omega'_{ij} := e'_i \wedge e'_j$. Therefore $L_1(\bigwedge^2 E^4_1)$ is determined by each equivalence class with respect to $\sim$ and therefore we denote it by $L_1(\bigwedge^2 E^4_1, [e])$ for the equivalence class $[e]$ containing $e \in \mathcal{B}(E^4_1)$.

Proposition 2.1. *For each element* Ω *of* $L_1(\bigwedge^2 E_1^4, [e])$, *there exists an element* e' *of* $[e]$ *satisfying*

$$\Omega = (\cos\phi)\omega'_{12} + (\sin\phi)\omega'_{23} + \omega'_{34}$$

for a number $\phi \in [0, \pi]$.

Proof. We suppose that Ω is represented as in (1). We represent U as $U = (\boldsymbol{u}_1\ \boldsymbol{u}_2\ \boldsymbol{u}_3)$. If we set $\boldsymbol{u}_3 = \boldsymbol{b}$ and if we suppose $\langle \boldsymbol{u}_2, \boldsymbol{a}\rangle = 0$ for the standard inner product $\langle\ ,\ \rangle$ of $\boldsymbol{R}^3$, then we see from (2) that Ω is represented as

$$\Omega = \langle \boldsymbol{u}_1, \boldsymbol{a}\rangle\omega'_{23} + \langle \boldsymbol{u}_3, \boldsymbol{a}\rangle\omega'_{12} + \omega'_{34}.$$

Since $\langle \boldsymbol{u}_2, \boldsymbol{a}\rangle = 0$, we have $\langle \boldsymbol{u}_1, \boldsymbol{a}\rangle^2 + \langle \boldsymbol{u}_3, \boldsymbol{a}\rangle^2 = 1$. Therefore there exists a number $\phi \in \boldsymbol{R}$ satisfying $\cos\phi = \langle \boldsymbol{u}_3, \boldsymbol{a}\rangle$ and $\sin\phi = \langle \boldsymbol{u}_1, \boldsymbol{a}\rangle$. We can choose $\boldsymbol{u}_1$ satisfying $\langle \boldsymbol{u}_1, \boldsymbol{a}\rangle \geq 0$. Hence we obtain Proposition 2.1. □

From Proposition 2.1, we see that for each element Ω of $L(\bigwedge^2 E_1^4)$, there exist an element $e = (e_1, e_2, e_3, e_4)$ of $\mathcal{B}(E_1^4)$, $R\,(>0)$ and $\phi \in [0, \pi]$ satisfying

$$\Omega = R\big((\cos\phi)\omega_{12} + (\sin\phi)\omega_{23} + \omega_{34}\big). \tag{3}$$

3. The $SO(3,1)$-action on $L(\bigwedge^2 E_1^4)$

For $P \in SO(3,1)$, we set $\tilde{T}_P(\omega_{ij}) := Pe_i \wedge Pe_j$. For an element $\Omega = \sum_{1\leq i<j\leq 4} c_{ij}\omega_{ij}$ of $\bigwedge^2 E_1^4$, we set

$$\tilde{T}_P(\Omega) := \sum_{1\leq i<j\leq 4} c_{ij}\tilde{T}_P(\omega_{ij}).$$

Then we obtain a linear transformation $\tilde{T}_P$ of $\bigwedge^2 E_1^4$, which does not depend on the choice of $(e_1, e_2, e_3, e_4) \in \mathcal{B}(E_1^4)$. Since $\tilde{T}_P \circ \tilde{T}_{P'} = \tilde{T}_{PP'}$, we obtain an $SO(3,1)$-action on $\bigwedge^2 E_1^4$. We see that $\tilde{T}_P$ is an isometry of $\bigwedge^2 E_1^4$ with respect to the metric $\hat{h}$. In particular, if $\Omega \in L(\bigwedge^2 E_1^4)$, then $\tilde{T}_P(\Omega) \in L(\bigwedge^2 E_1^4)$, and therefore we obtain an $SO(3,1)$-action on $L(\bigwedge^2 E_1^4)$. Suppose that Ω is represented as in (3). The $SO(3,1)$-orbit through Ω is given by

$$\mathcal{L}(\Omega) = \{\tilde{T}_P(\Omega) \mid P \in SO(3,1)\}. \tag{4}$$

Although $SO(3,1)$ has just two connected components, $\mathcal{L}(\Omega)$ is connected. In the following, we suppose $R=\sqrt{2}$. As in [1], we set

$$P_{1,1}:=\begin{bmatrix}\cos\theta & -\sin\theta & 0 & 0\\ \sin\theta & \cos\theta & 0 & 0\\ 0 & 0 & 1 & 0\\ 0 & 0 & 0 & 1\end{bmatrix},\quad P_{1,2}:=\begin{bmatrix}1 & 0 & 0 & 0\\ 0 & 1 & 0 & 0\\ 0 & 0 & \cosh t & \sinh t\\ 0 & 0 & \sinh t & \cosh t\end{bmatrix},$$

$$P_{2,1}:=\begin{bmatrix}\cos\theta & 0 & -\sin\theta & 0\\ 0 & 1 & 0 & 0\\ \sin\theta & 0 & \cos\theta & 0\\ 0 & 0 & 0 & 1\end{bmatrix},\quad P_{2,2}:=\begin{bmatrix}1 & 0 & 0 & 0\\ 0 & \cosh t & 0 & \sinh t\\ 0 & 0 & 1 & 0\\ 0 & \sinh t & 0 & \cosh t\end{bmatrix},$$

$$P_{3,1}:=\begin{bmatrix}1 & 0 & 0 & 0\\ 0 & \cos\theta & -\sin\theta & 0\\ 0 & \sin\theta & \cos\theta & 0\\ 0 & 0 & 0 & 1\end{bmatrix},\quad P_{3,2}:=\begin{bmatrix}\cosh t & 0 & 0 & \sinh t\\ 0 & 1 & 0 & 0\\ 0 & 0 & 1 & 0\\ \sinh t & 0 & 0 & \cosh t\end{bmatrix}$$

for $\theta, t\in\boldsymbol{R}$. Then the connected component $SO_0(3,1)$ of the unit element of $SO(3,1)$ is generated by $P_{k,l}$ ($k=1,2,3,\ l=1,2,\ \theta,t\in\boldsymbol{R}$). We set

$$E_{\pm,1}:=\frac{1}{\sqrt{2}}(\omega_{12}\pm\omega_{34}),\quad E_{\pm,2}:=\frac{1}{\sqrt{2}}(\omega_{13}\pm\omega_{42}),\quad E_{\pm,3}:=\frac{1}{\sqrt{2}}(\omega_{14}\pm\omega_{23}).$$

Then we have

$$\begin{aligned}&[\tilde{T}_{P_{1,1}}(E_{\pm,1})\ \tilde{T}_{P_{1,1}}(E_{\pm,2})\ \tilde{T}_{P_{1,1}}(E_{\pm,3})]\\ &\quad=[E_{\pm,1}\ E_{\pm,2}\ E_{\pm,3}]\begin{bmatrix}1 & 0 & 0\\ 0 & \cos\theta & \mp\sin\theta\\ 0 & \pm\sin\theta & \cos\theta\end{bmatrix},\\ &[\tilde{T}_{P_{1,2}}(E_{\pm,1})\ \tilde{T}_{P_{1,2}}(E_{\pm,2})\ \tilde{T}_{P_{1,2}}(E_{\mp,3})]\\ &\quad=[E_{\pm,1}\ E_{\pm,2}\ E_{\mp,3}]\begin{bmatrix}1 & 0 & 0\\ 0 & \cosh t & \sinh t\\ 0 & \sinh t & \cosh t\end{bmatrix},\end{aligned}\tag{5}$$

$$\begin{aligned}&[\tilde{T}_{P_{2,1}}(E_{\pm,1})\ \tilde{T}_{P_{2,1}}(E_{\pm,2})\ \tilde{T}_{P_{2,1}}(E_{\pm,3})]\\ &\quad=[E_{\pm,1}\ E_{\pm,2}\ E_{\pm,3}]\begin{bmatrix}\cos\theta & 0 & \pm\sin\theta\\ 0 & 1 & 0\\ \mp\sin\theta & 0 & \cos\theta\end{bmatrix},\\ &[\tilde{T}_{P_{2,2}}(E_{\pm,1})\ \tilde{T}_{P_{2,2}}(E_{\pm,2})\ \tilde{T}_{P_{2,2}}(E_{\mp,3})]\\ &\quad=[E_{\pm,1}\ E_{\pm,2}\ E_{\mp,3}]\begin{bmatrix}\cosh t & 0 & \sinh t\\ 0 & 1 & 0\\ \sinh t & 0 & \cosh t\end{bmatrix}\end{aligned}\tag{6}$$

and

$$\begin{aligned}
&[\tilde{T}_{P_{3,1}}(E_{\pm,1})\ \tilde{T}_{P_{3,1}}(E_{\pm,2})\ \tilde{T}_{P_{3,1}}(E_{\pm,3})] \\
&\qquad = [E_{\pm,1}\ E_{\pm,2}\ E_{\pm,3}] \begin{bmatrix} \cos\theta & -\sin\theta & 0 \\ \sin\theta & \cos\theta & 0 \\ 0 & 0 & 1 \end{bmatrix}, \\
&[\tilde{T}_{P_{3,2}}(E_{\pm,1})\ \tilde{T}_{P_{3,2}}(E_{\mp,2})\ \tilde{T}_{P_{3,2}}(E_{\pm,3})] \\
&\qquad = [E_{\pm,1}\ E_{\mp,2}\ E_{\pm,3}] \begin{bmatrix} \cosh t & \mp\sinh t & 0 \\ \mp\sinh t & \cosh t & 0 \\ 0 & 0 & 1 \end{bmatrix}.
\end{aligned} \tag{7}$$

Since

$$\begin{aligned}
\omega_{12} &= \frac{1}{\sqrt{2}}(E_{+,1} + E_{-,1}), & \omega_{34} &= \frac{1}{\sqrt{2}}(E_{+,1} - E_{-,1}), \\
\omega_{13} &= \frac{1}{\sqrt{2}}(E_{+,2} + E_{-,2}), & \omega_{42} &= \frac{1}{\sqrt{2}}(E_{+,2} - E_{-,2}), \\
\omega_{14} &= \frac{1}{\sqrt{2}}(E_{+,3} + E_{-,3}), & \omega_{23} &= \frac{1}{\sqrt{2}}(E_{+,3} - E_{-,3}),
\end{aligned} \tag{8}$$

applying (5) and (8) to (3) with $R = \sqrt{2}$, we obtain

$$\begin{aligned}
\tilde{T}_{P_{1,1}}(\Omega) &= (\cos\phi + 1)E_{+,1} + (\cos\phi - 1)E_{-,1} \\
&\quad - \sin\phi\sin\theta(E_{+,2} + E_{-,2}) \\
&\quad + \sin\phi\cos\theta(E_{+,3} - E_{-,3}), \\
\tilde{T}_{P_{1,2}}(\Omega) &= (\cos\phi + 1)E_{+,1} + (\cos\phi - 1)E_{-,1} \\
&\quad - \sin\phi\sinh t(E_{+,2} - E_{-,2}) \\
&\quad + \sin\phi\cosh t(E_{+,3} - E_{-,3}).
\end{aligned} \tag{9}$$

Similarly, applying (6), (7) and (8) to (3) with $R = \sqrt{2}$, we obtain

$$\begin{aligned}
\tilde{T}_{P_{2,1}}(\Omega) &= (\cos(\phi - \theta) + \cos\theta)E_{+,1} \\
&\quad + (\cos(\phi - \theta) - \cos\theta)E_{-,1} \\
&\quad + (\sin(\phi - \theta) - \sin\theta)E_{+,3} \\
&\quad - (\sin(\phi - \theta) + \sin\theta)E_{-,3}, \\
\tilde{T}_{P_{2,2}}(\Omega) &= ((\cos\phi + 1)\cosh t - \sin\phi\sinh t)E_{+,1} \\
&\quad + ((\cos\phi - 1)\cosh t + \sin\phi\sinh t)E_{-,1} \\
&\quad + ((\cos\phi - 1)\sinh t + \sin\phi\cosh t)E_{+,3} \\
&\quad + ((\cos\phi + 1)\sinh t - \sin\phi\cosh t)E_{-,3}
\end{aligned} \tag{10}$$

and

$$
\begin{aligned}
\tilde{T}_{P_{3,1}}(\Omega) &= (\cos\phi+1)\cos\theta E_{+,1} + (\cos\phi-1)\cos\theta E_{-,1} \\
&\quad + (\cos\phi+1)\sin\theta E_{+,2} + (\cos\phi-1)\sin\theta E_{-,2} \\
&\quad + \sin\phi(E_{+,3}-E_{-,3}), \\
\tilde{T}_{P_{3,2}}(\Omega) &= (\cos\phi+1)\cosh t E_{+,1} + (\cos\phi-1)\cosh t E_{-,1} \\
&\quad + (\cos\phi-1)\sinh t E_{+,2} - (\cos\phi+1)\sinh t E_{-,2} \\
&\quad + \sin\phi(E_{+,3}-E_{-,3}).
\end{aligned} \tag{11}
$$

4. Tangent vectors of $SO(3,1)$-orbits

By (9), (10) and (11), we obtain

$$
\begin{aligned}
\left.\frac{\partial}{\partial\theta}\right|_{\theta=0} \tilde{T}_{P_{1,1}}(\Omega) &= -\sin\phi(E_{+,2}+E_{-,2}), \\
\left.\frac{\partial}{\partial t}\right|_{t=0} \tilde{T}_{P_{1,2}}(\Omega) &= -\sin\phi(E_{+,2}-E_{-,2}), \\
\left.\frac{\partial}{\partial\theta}\right|_{\theta=0} \tilde{T}_{P_{2,1}}(\Omega) &= \sin\phi(E_{+,1}+E_{-,1}) \\
&\quad - (\cos\phi+1)E_{+,3} + (\cos\phi-1)E_{-,3}, \\
\left.\frac{\partial}{\partial t}\right|_{t=0} \tilde{T}_{P_{2,2}}(\Omega) &= -\sin\phi(E_{+,1}-E_{-,1}) \\
&\quad + (\cos\phi-1)E_{+,3} + (\cos\phi+1)E_{-,3}, \\
\left.\frac{\partial}{\partial\theta}\right|_{\theta=0} \tilde{T}_{P_{3,1}}(\Omega) &= (\cos\phi+1)E_{+,2} + (\cos\phi-1)E_{-,2}, \\
\left.\frac{\partial}{\partial t}\right|_{t=0} \tilde{T}_{P_{3,2}}(\Omega) &= (\cos\phi-1)E_{+,2} - (\cos\phi+1)E_{-,2}.
\end{aligned} \tag{12}
$$

From (12), we obtain

Proposition 4.1. *The tangent space $\mathcal{T}_\Omega(\mathcal{L}(\Omega))$ of $\mathcal{L}(\Omega)$ at Ω is spanned by*

$$
\begin{aligned}
X_\pm &:= (\sin\phi)E_{\pm,1} \mp (\cos\phi)E_{\pm,3} - E_{\mp,3}, \\
Y_\pm &:= E_{\pm,2}.
\end{aligned}
$$

In addition,

(a) *if $\phi = 0$ or π, then $\mathcal{T}_\Omega(\mathcal{L}(\Omega))$ is spanned by $E_{\pm,2}$, $E_{\pm,3}$, i.e., by ω_{13}, ω_{42}, ω_{14}, ω_{23};*

(b) *if $\phi \neq \pi/2$, then the metric $\hat{h}^\top$ of $\mathcal{L}(\Omega)$ induced by $\hat{h}$ is neutral;*

(c) *if* $\phi = \pi/2$, *then* $\hat{h}^\top$ *is degenerate on* $\mathcal{L}(\Omega)$ *so that every vector of a two-dimensional subspace* W_0 *of* $\mathcal{T}_\Omega(\mathcal{L}(\Omega))$ *spanned by* $X_\pm$ *is normal to all the vectors of* $\mathcal{T}_\Omega(\mathcal{L}(\Omega))$ *with respect to* $\hat{h}^\top$.

Corollary 4.1. *If* $\phi \neq \pi/2$, *then*

$$\begin{aligned} X_1 &:= \frac{1}{2\cos\phi}(\cos\phi\sin\phi\,\omega_{12} - (1+\cos^2\phi)\omega_{23} - \sin\phi\,\omega_{34}), \\ X_2 &:= \frac{1}{2\cos\phi}(\cos\phi\sin\phi\,\omega_{12} - 2\cos\phi\,\omega_{14} + \sin^2\phi\,\omega_{23} + \sin\phi\,\omega_{34}), \\ Y_1 &:= \omega_{13}, \\ Y_2 &:= \omega_{42} \end{aligned}$$

form a pseudo-orthonormal basis of $\mathcal{T}_\Omega(\mathcal{L}(\Omega))$ *satisfying that* X_1, Y_1 *are space-like and that* X_2, Y_2 *are time-like.*

5. A special surface in an $SO(3,1)$-orbit

Suppose $\phi \neq \pi/2$. We set

$$\mathcal{S} := \{\tilde{T}_{P_{2,1}} \circ \tilde{T}_{P_{2,2}}(\Omega) \mid \theta, t \in \boldsymbol{R}\}. \tag{13}$$

Noticing that $\tilde{T}_{P_{2,1}}$ and $\tilde{T}_{P_{2,2}}$ are commutative, we see that $\mathcal{S}$ is a surface in $\mathcal{L}(\Omega)$, i.e., a two-dimensional submanifold of $\mathcal{L}(\Omega)$. Let $\tilde{X}_\pm$, $\tilde{Y}_\pm$ be vector fields along $\mathcal{S}$ given by

$$\tilde{X}_\pm = \tilde{T}_{P_{2,1}} \circ \tilde{T}_{P_{2,2}}(X_\pm), \quad \tilde{Y}_\pm = \tilde{T}_{P_{2,1}} \circ \tilde{T}_{P_{2,2}}(Y_\pm).$$

Then on $\mathcal{S}$, $\tilde{X}_\pm$ (respectively, $\tilde{Y}_\pm$) are tangent (respectively, normal) to $\mathcal{S}$.

Let $\hat{\nabla}$, $\hat{\nabla}^\top$ be the Levi-Civita connections of $\hat{h}$, $\hat{h}^\top$ respectively. Since

$$\tilde{T}_{P_{2,1}}(E_{\pm,2}) = \tilde{T}_{P_{2,2}}(E_{\pm,2}) = E_{\pm,2}$$

from (6), $\tilde{Y}_\pm$ are parallel along $\mathcal{S}$, that is, $\hat{\nabla}^\top_X \tilde{Y}_\pm = 0$ for any tangent vector X to $\mathcal{S}$. By (6), we obtain

$$\begin{aligned} \tilde{T}_{P_{2,1}}(X_\pm) &= \sin(\phi-\theta)E_{\pm,1} \mp \cos(\phi-\theta)E_{\pm,3} \\ &\quad \pm \sin\theta E_{\mp,1} - \cos\theta E_{\mp,3}, \\ \tilde{T}_{P_{2,2}}(X_\pm) &= (\sin\phi\cosh t - \sinh t)E_{\pm,1} \mp \cos\phi\cosh t E_{\pm,3} \\ &\quad \mp \cos\phi\sinh t E_{\mp,1} + (\sin\phi\sinh t - \cosh t)E_{\mp,3}. \end{aligned} \tag{14}$$

If we set

$$N_\pm := -\cos\phi E_{\pm,1} \mp \sin\phi E_{\pm,3} \pm E_{\mp,1},$$

then from (12) and (14), we obtain

$$\hat{\nabla}_{X_+ + X_-}\tilde{X}_\pm = \left.\frac{\partial}{\partial\theta}\right|_{\theta=0} \tilde{T}_{P_{2,1}}(X_\pm) = N_\pm,$$
$$\hat{\nabla}_{-X_+ + X_-}\tilde{X}_\pm = \left.\frac{\partial}{\partial t}\right|_{t=0} \tilde{T}_{P_{2,2}}(X_\pm) = \pm N_\mp.$$

We see from Proposition 4.1 that $N_\pm$ are normal to $\mathcal{L}(\Omega)$ at Ω with respect to $\hat{h}$. Since

$$\frac{\partial}{\partial\theta}\tilde{T}_{P_{2,1}} \circ \tilde{T}_{P_{2,2}}(X_\pm) = \tilde{T}_{P_{2,1}} \circ \tilde{T}_{P_{2,2}}(N_\pm),$$
$$\frac{\partial}{\partial t}\tilde{T}_{P_{2,1}} \circ \tilde{T}_{P_{2,2}}(X_\pm) = \tilde{T}_{P_{2,1}} \circ \tilde{T}_{P_{2,2}}(\pm N_\mp),$$

we obtain $\hat{\nabla}^\top_X \tilde{X}_\pm = 0$ for any tangent vector X to $\mathcal{S}$ and this means that $\tilde{X}_\pm$ are parallel along $\mathcal{S}$ in $\mathcal{L}(\Omega)$. Hence we obtain

Proposition 5.1. *If $\phi \neq \pi/2$, then $\mathcal{S}$ is a time-like surface and $\tilde{X}_\pm$, $\tilde{Y}_\pm$ are parallel along $\mathcal{S}$ in $\mathcal{L}(\Omega)$.*

Remark 5.1. Suppose $\phi \neq \pi/2$. Then by Proposition 5.1, $\mathcal{S}$ is flat, and in addition, $\mathcal{S}$ is totally geodesic and has flat normal connection in $\mathcal{L}(\Omega)$.

In the case of $\phi = \pi/2$, we obtain $N_+ = X_-$ and $N_- = -X_+$. Therefore we obtain

Theorem 5.1. *If $\phi = \pi/2$, then there exists a two-dimensional involutive distribution $\mathcal{D}$ on $\mathcal{L}(\Omega)$ where $\hat{h}^\top$ vanishes, and the covariant derivatives of $\tilde{X}_\pm$ (respectively, $\tilde{Y}_\pm$) along each integral surface $\mathcal{S}$ of $\mathcal{D}$ are zero or light-like, and tangent to $\mathcal{S}$ (respectively, vanish).*

Noticing $\mathcal{L}_\pm = \{\tilde{T}_P(E_{\pm,1}) \mid P \in SO(3,1)\}$, we will prove

Theorem 5.2. *If $\phi \neq \pi/2$, then $\mathcal{L}(\Omega)$ is homothetic to either $\mathcal{L}_+$ or $\mathcal{L}_-$ in $L(\bigwedge^2 E^4_1)$.*

In order to prove Theorem 5.2, we have only to show the following:

Proposition 5.2. *Suppose $\phi \neq \pi/2$. Let $\mathcal{S}$ be a surface in $\mathcal{L}(\Omega)$ given by (13). Then $\mathcal{S}$ has a unique element in the form of $R(\omega_{12} + \varepsilon\omega_{34})$ for $R \in (0, \sqrt{2}]$ and $\varepsilon \in \{1, -1\}$.*

Proof. If $\phi = 0$ or π, then we immediately obtain Proposition 5.2. In the following, suppose $\phi \neq 0,\ \pi/2,\ \pi$. The following holds:

$$\begin{aligned}\frac{1}{\sqrt{2}}\tilde{T}_{P_{2,1}} \circ \tilde{T}_{P_{2,2}}(\Omega) = &(\cos(\phi-\theta)\cosh t - \sin\theta \sinh t)\omega_{12} \\ &+ (-\sin(\phi-\theta)\sinh t + \cos\theta\cosh t)\omega_{34} \\ &+ (\cos(\phi-\theta)\sinh t - \sin\theta\cosh t)\omega_{14} \\ &+ (\sin(\phi-\theta)\cosh t - \cos\theta\sinh t)\omega_{23}.\end{aligned} \tag{15}$$

Both the coefficients of ω_{14}, ω_{23} in the right side of (15) vanish if and only if ϕ, θ, t satisfy

$$\begin{aligned}\tan 2\theta &= \tan\phi, \\ \sin\theta\cosh t &= \cos(\phi-\theta)\sinh t.\end{aligned} \tag{16}$$

In addition, the first relation in (16) is equivalent to $\theta = (\phi/2) + (k\pi/2)$ for an integer $k \in \boldsymbol{Z}$. Suppose $\theta = (\phi/2) + \ell\pi$ for $\ell \in \boldsymbol{Z}$. Then the second relation in (16) is rewritten into $\tan\phi/2 = \tanh t$.

(i) If $\phi \in (0, \pi/2)$, then there exists a unique number $t \in \boldsymbol{R}$ satisfying $\tan\phi/2 = \tanh t$;
(ii) if $\phi \in (\pi/2, \pi)$, then $\tan\phi/2 = \tanh t$ does not hold for any $t \in \boldsymbol{R}$.

If ϕ, t satisfy $\tan\phi/2 = \tanh t$, then from (15), we obtain

$$\tilde{T}_{P_{2,1}} \circ \tilde{T}_{P_{2,2}}(\Omega) = \sqrt{2}(-1)^{\ell}\frac{\cos(\phi/2)}{\cosh t}(\omega_{12} + \omega_{34}). \tag{17}$$

Suppose $\theta = \phi/2 + (\ell + (1/2))\pi$ for $\ell \in \boldsymbol{Z}$. Then the second relation in (16) is rewritten into $\cot\phi/2 = \tanh t$.

(i) If $\phi \in (0, \pi/2)$, then $\cot\phi/2 = \tanh t$ does not hold for any $t \in \boldsymbol{R}$;
(ii) if $\phi \in (\pi/2, \pi)$, then there exists a unique number $t \in \boldsymbol{R}$ satisfying $\cot\phi/2 = \tanh t$.

If ϕ, t satisfy $\cot\phi/2 = \tanh t$, then from (15), we obtain

$$\tilde{T}_{P_{2,1}} \circ \tilde{T}_{P_{2,2}}(\Omega) = \sqrt{2}(-1)^{\ell}\frac{\sin(\phi/2)}{\cosh t}(\omega_{12} - \omega_{34}). \tag{18}$$

From (17) and (18), we immediately obtain Proposition 5.2. □

From Proposition 5.2, we see that there exist $R \in (0, \sqrt{2}]$ and $\varepsilon \in \{1, -1\}$ such that $\Omega_0 := R(\omega_{12} + \varepsilon\omega_{34})$ is an element of $\mathcal{L}(\Omega)$. Therefore $\mathcal{L}(\Omega)$ is an $SO(3,1)$-orbit through Ω_0. Hence we obtain Theorem 5.2.

6. The stabilizers

Let Ω be as in (3) with $R = \sqrt{2}$. Suppose $\phi \neq \pi/2$. Then as was already shown in Proposition 5.2, $\mathcal{L}(\Omega)$ has an element Ω_0 in the form of $R(\omega_{12} + \varepsilon\omega_{34})$. The stabilizer of $SO(3,1)$ at Ω_0 is generated by $P_{1,1}$, $\pm P_{1,2}$ (θ, $t \in \boldsymbol{R}$). Let W_+ (respectively, W_-) be a two-dimensional subspace of $\mathcal{T}_{\Omega_0}(\mathcal{L}(\Omega))$ generated by $E_{+,2}+E_{-,3}$, $E_{-,2}+E_{+,3}$ (respectively, $E_{+,2}-E_{-,3}$, $E_{-,2} - E_{+,3}$). We will prove

Proposition 6.1. *The proper invariant subspaces of $\mathcal{T}_{\Omega_0}(\mathcal{L}(\Omega_0))$ by the stabilizer are given by $W_\pm$.*

Proof. Let Ω_0 be an element of $L(\bigwedge^2 E_1^4)$ in the form of $R(\omega_{12} + \varepsilon\omega_{34})$ for $R > 0$ and $\varepsilon \in \{1, -1\}$. Then $E_{\pm,2}$, $E_{\pm,3}$ form a basis of the tangent space $\mathcal{T}_{\Omega_0}(\mathcal{L}(\Omega_0))$ of $\mathcal{L}(\Omega_0)$ at Ω_0. Since the stabilizer of $SO(3,1)$ at Ω_0 is generated by $P_{1,1}$, $\pm P_{1,2}$ (θ, $t \in \boldsymbol{R}$), the derivatives $d\tilde{T}_{P_{1,1}}$, $d\tilde{T}_{P_{1,2}}$ of $\tilde{T}_{P_{1,1}}$, $\tilde{T}_{P_{1,2}}$ respectively give linear transformations of $\mathcal{T}_{\Omega_0}(\mathcal{L}(\Omega_0))$. Let W be an invariant subspace of $\mathcal{T}_{\Omega_0}(\mathcal{L}(\Omega_0))$ by the stabilizer. Let

$$E = aE_{+,2} + bE_{-,2} + cE_{+,3} + dE_{-,3}$$

be an element of W. Then we have

$$d\tilde{T}_{P_{1,1}} \circ d\tilde{T}_{P_{1,2}}(E) = (E_{+,2}\ E_{-,2}\ E_{+,3}\ E_{-,3})\boldsymbol{a},$$

where

$$\begin{aligned} \boldsymbol{a} :=& \cos\theta \cosh t\, \boldsymbol{a}_1 + \cos\theta \sinh t\, \boldsymbol{a}_2 \\ &+ \sin\theta \cosh t\, \boldsymbol{a}_3 + \sin\theta \sinh t\, \boldsymbol{a}_4 \end{aligned}$$

and

$$\boldsymbol{a}_1 := \begin{bmatrix} a \\ b \\ c \\ d \end{bmatrix}, \quad \boldsymbol{a}_2 := \begin{bmatrix} d \\ c \\ b \\ a \end{bmatrix}, \quad \boldsymbol{a}_3 := \begin{bmatrix} -c \\ d \\ a \\ -b \end{bmatrix}, \quad \boldsymbol{a}_4 := \begin{bmatrix} -b \\ a \\ d \\ -c \end{bmatrix}.$$

We set $A := [\boldsymbol{a}_1\ \boldsymbol{a}_2\ \boldsymbol{a}_3\ \boldsymbol{a}_4]$. Then we have

$$\det A = -\big((a-d)^2 + (b-c)^2\big)\big((a+d)^2 + (b+c)^2\big).$$

Therefore we see that if $\det A \neq 0$, then W coincides with $\mathcal{T}_{\Omega_0}(\mathcal{L}(\Omega_0))$. Suppose $\det A = 0$. Then we have either $(a,b) = (d,c)$ or $(a,b) = (-d,-c)$. Therefore we see that if W has a nonzero element, then $W = W_+$ or W_-, where $W_\pm$ are as in the beginning of this section. Hence we obtain Proposition 6.1. □

Remark 6.1. As in [1], [2], there exist an almost complex structure $\mathcal{I}$ and an almost paracomplex structure $\mathcal{J}$ on $\mathcal{L}(\Omega_0)$ parallel with respect to the Levi-Civita connection $\hat{\nabla}^{\top}$. We see that $W_{\pm}$ are invariant by $\mathcal{I}$, $\mathcal{J}$.

In the following, suppose $\phi = \pi/2$. Then we have

$$\Omega = E_{+,1} - E_{-,1} + E_{+,3} - E_{-,3}. \tag{19}$$

For $k = 1, 2, 3$, we denote $P_{k,1}$, $P_{k,2}$ by $P_{k,1,\theta}$, $P_{k,2,t}$ respectively. For $t \in \boldsymbol{R}$, we set

$$\theta(t) := \sin^{-1}(\tanh t), \quad s(t) := -\log(\cosh t) \tag{20}$$

and

$$\begin{aligned} U_t &:= P_{3,2,t} P_{1,1,-\theta(t)} P_{2,2,s(t)}, \\ V_t &:= P_{1,2,t} P_{3,1,\theta(t)} P_{2,2,s(t)}. \end{aligned} \tag{21}$$

We will prove

Theorem 6.1. *The stabilizer of* $SO(3,1)$ *at* Ω *is generated by* U_t, V_t *in* (21) ($t \in \boldsymbol{R}$) *and each element of the stabilizer fixes each vector of the two-dimensional subspace* W_0 *of* $\mathcal{T}_{\Omega}(\mathcal{L}(\Omega))$ *spanned by* $X_{\pm}$. *In addition,*

(a) *a one-dimensional subspace of* $\mathcal{T}_{\Omega}(\mathcal{L}(\Omega))$ *is invariant by the stabilizer if and only if it is contained in* W_0;
(b) W_0 *is a unique invariant two-dimensional subspace of* $\mathcal{T}_{\Omega}(\mathcal{L}(\Omega))$ *by the stabilizer*;
(c) *a three-dimensional subspace of* $\mathcal{T}_{\Omega}(\mathcal{L}(\Omega))$ *is invariant by the stabilizer if and only if it contains* W_0.

Proof. From (10) with $\phi = \pi/2$ and (19), we have $\tilde{T}_{P_{2,2,s}}(\Omega) = e^{-s}\Omega$ for any $s \in \boldsymbol{R}$. Therefore by (9) with $\phi = \pi/2$, we have

$$\begin{aligned} &\tilde{T}_{P_{1,1,\theta}} \circ \tilde{T}_{P_{2,2,s}}(\Omega) \\ &= e^{-s}\big(E_{+,1} - E_{-,1} - \sin\theta(E_{+,2} + E_{-,2}) + \cos\theta(E_{+,3} - E_{-,3})\big) \end{aligned} \tag{22}$$

for any $\theta \in \boldsymbol{R}$. By (7) and (22), we have

$$\begin{aligned} &\tilde{T}_{P_{3,2,t}} \circ \tilde{T}_{P_{1,1,\theta}} \circ \tilde{T}_{P_{2,2,s}}(\Omega) \\ &= e^{-s}\big((\cosh t + \sin\theta \sinh t)(E_{+,1} - E_{-,1}) \\ &\qquad - (\sinh t + \sin\theta \cosh t)(E_{+,2} + E_{-,2}) \\ &\qquad + \cos\theta(E_{+,3} - E_{-,3})\big) \end{aligned}$$

for any $t \in \boldsymbol{R}$. Let $\theta(t)$, $s(t)$ be as in (20). Then noticing

$$e^{s(t)} = \cosh t + \sin(-\theta(t)) \sinh t = \cos\theta(t) = \frac{1}{\cosh t}$$

and

$$\sinh t + \sin(-\theta(t)) \cosh t = 0,$$

we obtain

$$\tilde{T}_{U_t}(\Omega) = \tilde{T}_{P_{3,2,t}} \circ \tilde{T}_{P_{1,1,-\theta(t)}} \circ \tilde{T}_{P_{2,2,s(t)}}(\Omega) = \Omega.$$

Similarly, we obtain

$$\tilde{T}_{V_t}(\Omega) = \tilde{T}_{P_{1,2,t}} \circ \tilde{T}_{P_{3,1,\theta(t)}} \circ \tilde{T}_{P_{2,2,s(t)}}(\Omega) = \Omega.$$

Therefore U_t, V_t ($t \in \boldsymbol{R}$) are elements of the stabilizer of $SO(3,1)$ at Ω. We set

$$U'_x := \begin{bmatrix} 1 & x & 0 & x \\ -x & 1-\frac{x^2}{2} & 0 & -\frac{x^2}{2} \\ 0 & 0 & 1 & 0 \\ x & \frac{x^2}{2} & 0 & 1+\frac{x^2}{2} \end{bmatrix},$$

$$V'_x := \begin{bmatrix} 1 & 0 & 0 & 0 \\ 0 & 1-\frac{x^2}{2} & -x & -\frac{x^2}{2} \\ 0 & x & 1 & x \\ 0 & \frac{x^2}{2} & x & 1+\frac{x^2}{2} \end{bmatrix}.$$

Then we have $U'_{\tanh t} = U_t$, $V'_{\tanh t} = V_t$. We see that $\{U'_x \mid x \in \boldsymbol{R}\}$, $\{V'_x \mid x \in \boldsymbol{R}\}$ are one-parameter subgroups of $SO(3,1)$ which are different from each other. Since the dimension of the stabilizer of $SO(3,1)$ at Ω is equal to two, it is generated by U_t, V_t in (21) ($t \in \boldsymbol{R}$). We set

$$\Omega' := E_{+,1} + E_{-,1} - E_{+,3} - E_{-,3}.$$

Then U_t, V_t fix Ω'. Since Ω, Ω' form a basis of W_0, each element of the stabilizer fixes each vector of W_0. Noticing

$$\tilde{T}_{U_t}(E_{\pm,2}) = E_{\pm,2} - \frac{\tanh t}{2}(\Omega \mp \Omega')$$

and

$$\tilde{T}_{V_t}(E_{\pm,2}) = E_{\pm,2} \mp \frac{\tanh t}{2}(\Omega \pm \Omega'),$$

we obtain (a), (b), (c) of Theorem 6.1. □

Remark 6.2. By direct computations, we see that U'_x and V'_y are commutative for arbitrary x, $y \in \boldsymbol{R}$.

7. The r-slice of $L(\bigwedge^2 E_1^4)$

For an equivalence class $[e]$ in $\mathcal{B}(E_1^4)$ and each $r > 0$, the *r-slice* of $L(\bigwedge^2 E_1^4)$ given by

$$L_r(\bigwedge^2 E_1^4, [e]) := \left\{ \Omega \in L(\bigwedge^2 E_1^4) \;\middle|\; A(\Omega) = B(\Omega) = r^2 \right\}$$

is well-defined. The 1-slice of $L(\bigwedge^2 E_1^4)$ coincides with $L_1(\bigwedge^2 E_1^4, [e])$ given in Section 2. The r-slice of $L(\bigwedge^2 E_1^4)$ depends on the choice of an equivalence class $[e]$ in $\mathcal{B}(E_1^4)$. In the following, we fix an equivalence class $[e]$ and we simply denote the r-slice of $L(\bigwedge^2 E_1^4)$ by $L_r(\bigwedge^2 E_1^4)$. Let Ω be as in (3) with $R = \sqrt{2}$ and $\mathcal{L}(\Omega)$ as in (4). We set

$$\mathcal{L}_r(\Omega) := \mathcal{L}(\Omega) \cap L_r(\bigwedge^2 E_1^4).$$

For $\phi \in [0, \pi]$, set

$$r(\phi) := \inf\{r > 0 \mid \mathcal{L}_r(\Omega) \neq \emptyset\}.$$

For $\phi \in [0, \pi] \setminus \{\pi/2\}$, set

$$t(\phi) := \begin{cases} \tanh^{-1}(\tan(\phi/2)) & (\phi \in [0, \pi/2)), \\ \tanh^{-1}(\cot(\phi/2)) & (\phi \in (\pi/2, \pi]). \end{cases}$$

We have $t(\pi - \phi) = t(\phi)$ for $\phi \in [0, \pi/2)$. We will prove

Theorem 7.1. *For $\phi \in [0, \pi] \setminus \{\pi/2\}$, $r(\phi)$ is positive and given by*

$$r(\phi) = \begin{cases} \sqrt{2}\,\dfrac{\cos(\phi/2)}{\cosh t(\phi)} & (\phi \in [0, \pi/2)), \\ \sqrt{2}\,\dfrac{\sin(\phi/2)}{\cosh t(\phi)} & (\phi \in (\pi/2, \pi]). \end{cases} \tag{23}$$

In particular, a function r of ϕ satisfies $r(\pi - \phi) = r(\phi)$ and $\lim_{\phi \to \pi/2} r(\phi) = 0$, and it is monotonically decreasing on $[0, \pi/2)$. In addition,

(a) *$\mathcal{L}_r(\Omega)$ is not empty, and diffeomorphic to $\boldsymbol{R}P^3$ for any $r > r(\phi)$;*
(b) *$\mathcal{L}_{r(\phi)}(\Omega)$ is not empty, and diffeomorphic to S^2.*

Proof. We denote by r_0 the number given by the right side of (23). Then noticing (17) and (18), we see that $\mathcal{L}_{r_0}(\Omega)$ is not empty, diffeomorphic to S^2 and given by

$$\mathcal{L}_{r_0}(\Omega) = \{r_0((\omega_{23}\ \omega_{31}\ \omega_{12}) + \varepsilon(\omega_{14}\ \omega_{24}\ \omega_{34}))\boldsymbol{a} \mid \langle \boldsymbol{a}, \boldsymbol{a} \rangle = 1\}$$

for $\varepsilon \in \{1, -1\}$. If $r > r_0$, then we obtain $\mathcal{L}_r(\Omega) \neq \emptyset$ using $\tilde{T}_{P_{1,2}}$, $\tilde{T}_{P_{2,2}}$, $\tilde{T}_{P_{3,2}}$, and noticing that $\mathcal{L}_r(\Omega)$ is represented as

$$\mathcal{L}_r(\Omega) = \{r((\omega_{23}\ \omega_{31}\ \omega_{12})U\boldsymbol{a} + (\omega_{14}\ \omega_{24}\ \omega_{34})U\boldsymbol{b}) \mid U \in SO(3)\}$$

with $\boldsymbol{b} \neq \pm \boldsymbol{a}$, we see that $\mathcal{L}_r(\Omega)$ is diffeomorphic to $\boldsymbol{R}P^3$. On a neighborhood of an element of $\mathcal{L}_{r_0}(\Omega)$ in $\mathcal{L}(\Omega)$, there exist no elements of $\mathcal{L}_r(\Omega)$ with $r < r_0$. Therefore we obtain $\mathcal{L}_r(\Omega) = \emptyset$ for $r \in (0, r_0)$. Hence we obtain Theorem 7.1. □

For $R > 0$ and $\varepsilon \in \{1, -1\}$, an element $R(\omega_{12} + \varepsilon\omega_{34})$ is contained in a neutral $SO(3,1)$-orbit. This means that an $SO(3,1)$-orbit with a two-dimensional involutive distribution where $\hat{h}^\top$ vanishes does not contain an element in the form of $R(\omega_{12} + \varepsilon\omega_{34})$ for $R > 0$ and $\varepsilon \in \{1, -1\}$. By (10) with $\phi = \pi/2$, we obtain $r(\pi/2) = 0$ and we see that Ω with $\phi = \pi/2$ satisfies $\mathcal{L}_r(\Omega) \neq \emptyset$ for any $r > 0$. Hence we obtain

Theorem 7.2. *Suppose $\phi = \pi/2$. Then $r(\phi) = 0$, and $\mathcal{L}_r(\Omega)$ is not empty, and diffeomorphic to $\boldsymbol{R}P^3$ for any $r > 0$.*

Remark 7.1. Let Ω be as in (3) with $R = \sqrt{2}$. Then $\phi \in [0, \pi]$ depends on the choice of an equivalence class $[e]$ in $\mathcal{B}(E^4_1)$. However, noticing Theorem 7.1 and Theorem 7.2, we see that whether ϕ is equal to $\pi/2$ or not does not depend on the choice of $[e]$.

Acknowledgments

The author is grateful to the reviewer for valuable comments. The author is also grateful to Professors Osamu Ikawa and Takahiro Hashinaga for valuable discussions and comments. This work was supported by JSPS KAKENHI Grant Number JP21K03228.

References

[1] N. Ando, Isotropicity of surfaces in Lorentzian 4-manifolds with zero mean curvature vector, *Abh. Math. Semin. Univ. Hambg.* **92** (2022), 105–123.

[2] N. Ando, Correction to: Isotropicity of surfaces in Lorentzian 4-manifolds with zero mean curvature vector, *Abh. Math. Semin. Univ. Hambg.* **93** (2023), 163–166.

[3] N. Ando, The lifts of surfaces in neutral 4-manifolds into the 2-Grassmann bundles, *Differ. Geom. Appl.* **91** (2023), Paper no. 102073, 25pp.

[4] D. A. Hoffman and R. Osserman, The geometry of the generalized Gauss map, *Memoirs of A.M.S.* **236**, 1980.

Received November 27, 2023
Revised January 26, 2024

Modern Approaches to Differential Geometry
and its Related Fields 33 – 44

ON SECTIONAL CURVATURES OF SOME EINSTEIN SOLVMANIFOLDS

Takahiro HASHINAGA
Faculty of Education, Saga University,
Saga, 840-8502, Japan
E-mail: hashinag@cc.saga-u.ac.jp

Akira KUBO
Department of Food Sciences and Biotechnology, Hiroshima Institute of Technology,
Hiroshima, 731-5193, Japan
E-mail: a.kubo.3r@cc.it-hiroshima.ac.jp

In this paper, we study Einstein solvmanifolds constructed from parabolic subalgebras of real special linear Lie algebras. We calculate sectional curvatures of these Einstein solvmanifolds, and show that they do not have nonpositive sectional curvature unless they are not isometric to symmetric spaces.

Keywords: Einstein solvmanifolds; solvable parts of parabolic subalgebras; sectional curvatures.

1. Introduction

A Riemannian manifold which admits a transitive solvable Lie group of isometries is called a *solvmanifold*, which provides interesting examples in differential geometry and harmonic analysis. Especially, Einstein solvmanifolds have been deeply investigated (see [2, 4, 9, 12] for instance), and recently the so-called Alekseevskii conjecture has been solved affirmatively by Böhm and Lafuente ([1]). Note that it follows that any connected noncompact homogeneous Einstein manifold is isometric to a solvmanifold.

Among known examples of Einstein solvmanifolds, ones constructed by Tamaru ([13]) would be interesting. We review them here briefly. Let $\mathfrak{g}$ be a semisimple Lie algebra, and $\mathfrak{q}_\Phi$ be a parabolic subalgebra of $\mathfrak{g}$, where Φ is a subset of a set of simple roots Λ of the restricted root system of $\mathfrak{g}$. By considering the Langlands decomposition $\mathfrak{q}_\Phi = \mathfrak{m}_\Phi \oplus \mathfrak{a}_\Phi \oplus \mathfrak{n}_\Phi$, we obtain a solvable subalgebra $\mathfrak{s}_\Phi := \mathfrak{a}_\Phi \oplus \mathfrak{n}_\Phi$, which we call the *solvable part* of the parabolic subalgebra $\mathfrak{q}_\Phi$ of $\mathfrak{g}$. Let S_Φ denote the corresponding simply-connected Lie group. Tamaru proved that, for any semisimple Lie algebra $\mathfrak{g}$ and any parabolic subalgebra $\mathfrak{q}_\Phi$, S_Φ has always an Einstein left-invariant metric g. Note that if $\Phi = \emptyset$, then $(S_\emptyset, g)$ is isometric to a symmetric space of noncompact type. And conversely, this class of Einstein solvmanifolds contains all Riemannian symmetric spaces of noncompact type.

We are interested in geometry of the solvmanifolds (S_Φ, g), and in this paper, we focus on the sectional curvatures of them. Note that symmetric ones have nonpositive sectional curvature. Furthermore, Mori studied a subclass of (S_Φ, g), and gave nonsymmetric examples with nonpositive sectional curvature ([11]).

We then deal with the case where $\mathfrak{g} = \mathfrak{sl}(n, \mathbb{R})$. The main result of this paper is as follows.

Theorem 1.1. *Let us consider the case where* $\mathfrak{g} = \mathfrak{sl}(n, \mathbb{R})$. *Then, the following holds.*

(1) *If* $\Phi = \emptyset$, *then the Einstein solvmanifold* $(S_\emptyset, g)$ *is isometric to a Reimannian symmetric space* $\mathrm{SL}_n(\mathbb{R})/\mathrm{SO}(n)$, *and therefore it has nonpositive sectional curvature.*
(2) *If* $\mathfrak{n}_\Phi$ *is abelian, then any Einstein solvmanifold* (S_Φ, g) *is isometric to a real hyperbolic space, and therefore it has negative sectional curvature.*
(3) *Otherwise, any Einstein solvmanifold* (S_Φ, g) *does not have nonpositive sectional curvature.*

We note that this theorem solves the problem whether the Einstein solvmanifold (S_Φ, g) is intrinsically a symmetric space or not.

Corollary 1.1. *Let us consider the case where* $\mathfrak{g} = \mathfrak{sl}(n, \mathbb{R})$. *Then, the Einstein solvmanifold* (S_Φ, g) *is not a symmetric space unless it is isometric to* $\mathrm{SL}_n(\mathbb{R})/\mathrm{SO}(n)$ *or a real hyperbolic space.*

2. The solvable parts of parabolic subalgebras of semisimple Lie algebras

In this section, we recall some notions and their basic properties of solvable parts $\mathfrak{s}_\Phi$ of parabolic subalgebras of semisimple Lie algebras $\mathfrak{g}$. Refer to [5, 6] for instance. See also [7, 13]. After that, we will mention $\mathfrak{s}_\Phi$ in the case of $\mathfrak{g} = \mathfrak{sl}(n, \mathbb{R})$.

2.1. *General settings*

Let $\mathfrak{g}$ be a real semisimple Lie algebra. Denote by θ a Cartan involution on $\mathfrak{g}$, and by $\mathfrak{g} = \mathfrak{k} \oplus \mathfrak{p}$ the corresponding Cartan decomposition as usual.

Let us take a maximal abelian subspace $\mathfrak{a}$ of $\mathfrak{p}$, and denote by $\mathfrak{a}^*$ the dual space of $\mathfrak{a}$. For $\lambda \in \mathfrak{a}^*$, we put

$$\mathfrak{g}_\lambda := \{X \in \mathfrak{g} \mid \operatorname{ad}_H X = \lambda(H)X \text{ for any } H \in \mathfrak{a}\},$$

and call $\lambda \in \mathfrak{a}^*$ the *(restricted) root* if $\lambda \neq 0$ and $\mathfrak{g}_\lambda \neq 0$. Denote by Σ the set of all roots. Let us take Λ as a set of all simple roots, and denote by Σ^+ the set of all positive roots with respect to Λ. Then, by putting

$$\mathfrak{n} := \bigoplus_{\lambda \in \Sigma^+} \mathfrak{g}_\lambda,$$

which is a nilpotent subalgebra of $\mathfrak{g}$, we obtain the Iwasawa decomposition $\mathfrak{g} = \mathfrak{k} \oplus \mathfrak{a} \oplus \mathfrak{n}$. And also, we define $\mathfrak{s} := \mathfrak{a} \oplus \mathfrak{n}$, which is a solvable subalgebra of $\mathfrak{g}$, and call it the *solvable part of the Iwasawa decomposition* of $\mathfrak{g}$.

Let us denote by B the Killing form on $\mathfrak{g}$, and put $B_{\theta,c}(X, Y) := -cB(X, \theta Y)$ for $X, Y \in \mathfrak{g}$, where c is a (suitable) positive constant. Then, we define an inner product $\langle\ ,\ \rangle$ on $\mathfrak{s}$ by

$$\langle\ ,\ \rangle = 2B_{\theta,c}|_{\mathfrak{a}\times\mathfrak{a}} + B_{\theta,c}|_{\mathfrak{n}\times\mathfrak{n}}.$$

And moreover, we define $H_\alpha \in \mathfrak{a}$ as the dual vector of $\alpha \in \mathfrak{a}^*$ with respect to $\langle\ ,\ \rangle$, that is,

$$\langle H_\alpha, H\rangle = \alpha(H) \qquad (\text{for } H \in \mathfrak{a}).$$

If $\alpha \in \Sigma$, we call H_α the *root vector* of $\alpha \in \Sigma$ with respect to $\langle\ ,\ \rangle$.

For a subset $\Phi \subsetneq \Lambda$, we put $\langle\Phi\rangle^+ := \operatorname{span}_{\mathbb{Z}}(\Phi) \cap \Sigma^+$, and define

$$\begin{aligned}
\mathfrak{a}_\Phi &:= \mathfrak{a} \ominus \operatorname{span}\{H_\alpha \mid \alpha \in \Phi\},\\
\mathfrak{n}_\Phi &:= \mathfrak{n} \ominus \textstyle\bigoplus_{\lambda \in \langle\Phi\rangle^+} \mathfrak{g}_\lambda = \bigoplus_{\mu \in \Sigma^+ \setminus \langle\Phi\rangle^+} \mathfrak{g}_\mu,\\
\mathfrak{s}_\Phi &:= \mathfrak{a}_\Phi \oplus \mathfrak{n}_\Phi.
\end{aligned}$$

Note that if $\Phi = \emptyset$, one has $\mathfrak{a}_\emptyset = \mathfrak{a}$, $\mathfrak{n}_\emptyset = \mathfrak{n}$, and hence, $\mathfrak{s}_\emptyset = \mathfrak{s}$. One can also show that all of them are subalgebras of $\mathfrak{s}$, and in particular $\mathfrak{n}_\Phi$ and $\mathfrak{s}_\Phi$ are ideals of $\mathfrak{s}$. We call $\mathfrak{s}_\Phi$ the *solvable part of the parabolic subalgebra* of $\mathfrak{g}$. Refer to [6] for parabolic subalgebras.

At the end of this subsection, we mention known results for geometry of $\mathfrak{s}_\Phi$. Since $\mathfrak{s}_\Phi$ is a subalgebra of the metric Lie algebra $(\mathfrak{s}, \langle\ ,\ \rangle)$, we consider the induced metric for $\mathfrak{s}_\Phi$, and hereafter denote it by the same symbol $\langle\ ,\ \rangle$. Let S_Φ and g be the corresponding simply-connected Lie group and left-invariant metric, respectively. In [13], Tamaru studied geometry of the solvmanifolds (S_Φ, g), in which the following result would be noteworthy.

Theorem 2.1 ([13, Theorem 5.3]). *All of the solvmanifolds (S_Φ, g) are Einstein manifolds.*

Remark 2.1. In his paper [13], Tamaru has also studied extrinsic geometry of $\mathfrak{s}_\Phi$, and showed that S_Φ is a minimal submanifold in (S, g). And furthermore, since (S, g) is isometric to some symmetric space, S_Φ can be considered minimal submanifolds in certain symmetric spaces (see Proposition 4.4 and Theorem 6.2 in [13]).

2.2. *The case of* $\mathfrak{g} = \mathfrak{sl}(n, \mathbb{R})$

In this subsection, we explain the solvable parts $\mathfrak{s}_\Phi$ of parabolic subalgebras of $\mathfrak{g} = \mathfrak{sl}(n, \mathbb{R})$.

First, we recall the solvable part of an Iwasawa decomposition of $\mathfrak{g} = \mathfrak{sl}(n, \mathbb{R})$. We set $\theta(X) := -{}^tX$ $(X \in \mathfrak{g})$ as a Cartan involution, and consider the Cartan decomposition $\mathfrak{g} = \mathfrak{k} \oplus \mathfrak{p}$, where

$$\mathfrak{k} = \{X \in \mathfrak{g} \mid X \text{ is skew-symmetric}\} = \mathfrak{o}(n),$$
$$\mathfrak{p} = \{X \in \mathfrak{g} \mid X \text{ is symmetric}\}.$$

Moreover, we take

$$\mathfrak{a} := \{X \in \mathfrak{p} \mid X \text{ is diagonal}\}$$

as a maximal abelian subspace of $\mathfrak{p}$. Note that the root system Σ is of type A_{n-1} in this case. By defining $\alpha_i \in \mathfrak{a}^*$ $(i = 1, \ldots, n-1)$ as

$$\alpha_i(\begin{pmatrix} a_{1,1} & & \\ & \ddots & \\ & & a_{n,n} \end{pmatrix}) = a_{i,i} - a_{i+1,i+1},$$

one can obtain a set of simple roots $\{\alpha_1, \ldots, \alpha_{n-1}\}$. We also see that the corresponding root space is given by

$$\mathfrak{g}_{\alpha_k + \cdots + \alpha_l} = \operatorname{span}\{E_{k,l+1}\}.$$

Therefore, we obtain

$$\mathfrak{n} = \{X \in \mathfrak{g} \mid X \text{ is strictly upper-triangular}\},$$

and conclude that the solvable part $\mathfrak{s}$ of the Iwasawa decomposition in this case is given by

$$\mathfrak{s} := \mathfrak{a} \oplus \mathfrak{n} = \{X \in \mathfrak{g} \mid X \text{ is upper-triangular}\}.$$

One knows that the Killing form B on $\mathfrak{g}$ is given by $B(X,Y) = 2n\,\mathrm{tr}(XY)$ for $X, Y \in \mathfrak{g}$. Therefore, by putting $c := (2n)^{-1}$, we can obtain that

$$\langle D + U, D' + U' \rangle = 2\,\mathrm{tr}(DD') + \mathrm{tr}(U^t U'), \tag{1}$$

where $D, D' \in \mathfrak{a}$ are diagonal matrices, and $U, U' \in \mathfrak{n}$ are strictly upper-triangular ones.

Example 2.1. Let us consider $\mathfrak{g} = \mathfrak{sl}(3, \mathbb{R})$. Recall that the corresponding root system is of type A_2 with $\Lambda = \{\alpha_1, \alpha_2\}$. In this case, one can see that the root vectors are given by

$$H_{\alpha_1} = \frac{1}{2}\begin{pmatrix} 1 & & \\ & -1 & \\ & & 0 \end{pmatrix}, \quad H_{\alpha_2} = \frac{1}{2}\begin{pmatrix} 0 & & \\ & 1 & \\ & & -1 \end{pmatrix}.$$

Note that $\langle H_{\alpha_1}, H_{\alpha_1} \rangle = \langle H_{\alpha_2}, H_{\alpha_2} \rangle = 1$, and $\langle H_{\alpha_1}, H_{\alpha_2} \rangle = -1/2$. On the other hand, $\{E_{12}, E_{23}, E_{13}\} \subset \mathfrak{n}$ is orthonormal. Therefore, for instance, $\{H_{\alpha_1}, H_{\alpha_1} + 2H_{\alpha_2}, E_{12}, E_{23}, E_{13}\}$ is an orthonormal basis of $\mathfrak{s}$ with respect to $\langle , \rangle$.

Remark 2.2. Let S be the simply-connected Lie group corresponding to $\mathfrak{s}$, and denote by g the left-invariant metric induced by $\langle , \rangle$. In this case, it is known that S acts on a Riemannian symmetric space $\mathrm{SL}(n, \mathbb{R})/\mathrm{SO}(n)$ simply-transitively, and (S, g) is isometric to it. For more details, refer to [13] and also, [3].

In the case of $\mathfrak{g} = \mathfrak{sl}(n, \mathbb{R})$, any solvable parts of parabolic subalgebras of $\mathfrak{g} = \mathfrak{sl}(n, \mathbb{R})$ can be constructed by block decompositions of matrices. In order to explain it, we will give some examples of $\mathfrak{s}_\Phi \subset \mathfrak{sl}(n, \mathbb{R})$.

Example 2.2. Let us consider $\mathfrak{g} = \mathfrak{sl}(5, \mathbb{R})$. Recall that the corresponding root system is of type A_4 with $\Lambda = \{\alpha_1, \alpha_2, \alpha_3, \alpha_4\}$, and the root vectors for the simple roots are given by $H_{\alpha_i} = (1/2)(E_{i,i} - E_{i+1,i+1})$.

(1) If we take $\Phi := \{\alpha_1\}$, then one can see that $\langle \Phi \rangle^+ = \Phi$, and also

$$\mathfrak{a}_\Phi = \left\{ \left(\begin{array}{cc|c|c|c} a & & & & \\ & a & & & \\ \hline & & b & & \\ \hline & & & c & \\ \hline & & & & d \end{array} \right) \in \mathfrak{g} \right\}, \quad \mathfrak{n}_\Phi = \left\{ \left(\begin{array}{cc|c|c|c} & & * & * & * \\ & & * & * & * \\ \hline & & & * & * \\ \hline & & & & * \\ \hline & & & & \end{array} \right) \right\}.$$

(2) If we take $\Phi := \{\alpha_1, \alpha_2\}$, then one can see that $\langle \Phi \rangle^+ = \{\alpha_1, \alpha_2, \alpha_1 + \alpha_2\}$, and also

$$\mathfrak{a}_\Phi = \left\{ \left(\begin{array}{ccc|c|c} a & & & & \\ & a & & & \\ & & a & & \\ \hline & & & b & \\ \hline & & & & c \end{array} \right) \in \mathfrak{g} \right\}, \qquad \mathfrak{n}_\Phi = \left\{ \left(\begin{array}{ccc|c|c} & & & * & * \\ & & & * & * \\ & & & * & * \\ \hline & & & & * \\ \hline & & & & \end{array} \right) \right\}.$$

(3) If we take $\Phi = \{\alpha_1, \alpha_3\}$, then one can see that $\langle \Phi \rangle^+ = \Phi$, and also

$$\mathfrak{a}_\Phi = \left\{ \left(\begin{array}{cc|cc|c} a & & & & \\ & a & & & \\ \hline & & b & & \\ & & & b & \\ \hline & & & & c \end{array} \right) \in \mathfrak{g} \right\}, \qquad \mathfrak{n}_\Phi = \left\{ \left(\begin{array}{cc|cc|c} & & * & * & * \\ & & * & * & * \\ \hline & & & & * \\ & & & & * \\ \hline & & & & \end{array} \right) \right\}.$$

(4) If we take $\Phi = \{\alpha_1, \alpha_4\}$, then one can see that $\langle \Phi \rangle^+ = \Phi$, and also

$$\mathfrak{a}_\Phi = \left\{ \left(\begin{array}{cc|c|cc} a & & & & \\ & a & & & \\ \hline & & b & & \\ \hline & & & c & \\ & & & & c \end{array} \right) \in \mathfrak{g} \right\}, \qquad \mathfrak{n}_\Phi = \left\{ \left(\begin{array}{cc|c|cc} & & * & * & * \\ & & * & * & * \\ \hline & & & * & * \\ \hline & & & & \\ & & & & \end{array} \right) \right\}.$$

(5) If we take $\Phi = \{\alpha_1, \alpha_2, \alpha_3\}$, then one can see that

$$\langle \Phi \rangle^+ = \{\alpha_1, \alpha_2, \alpha_3, \alpha_1 + \alpha_2, \alpha_2 + \alpha_3, \alpha_1 + \alpha_2 + \alpha_3\},$$

and also

$$\mathfrak{a}_\Phi = \left\{ \left(\begin{array}{cccc|c} a & & & & \\ & a & & & \\ & & a & & \\ & & & a & \\ \hline & & & & b \end{array} \right) \in \mathfrak{g} \right\}, \qquad \mathfrak{n}_\Phi = \left\{ \left(\begin{array}{cccc|c} & & & & * \\ & & & & * \\ & & & & * \\ & & & & * \\ \hline & & & & \end{array} \right) \right\}.$$

As we can see from the examples above, the solvable parts $\mathfrak{s}_\Phi$ of parabolic subgroups of $\mathfrak{g} = \mathfrak{sl}(n, \mathbb{R})$ correspond to the so-called *block decompositions* of matrices. Indeed, a block decomposition of type $(d_1, \ldots, d_k)$ corresponds to the subset $\Phi = \Lambda \setminus \{\alpha_{n_1}, \ldots, \alpha_{n_{k-1}}\}$, where we put $n_i := d_1 + \cdots + d_i$, and also to the solvable part $\mathfrak{s}_\Phi$. In the case of $\mathfrak{g} = \mathfrak{sl}(n, \mathbb{R})$, we denote by $\mathfrak{s}_{(d_1, \ldots, d_k)}$ the solvable part corresponding to a decomposition $(d_1, \ldots, d_k)$ hereafter.

Example 2.3. Let us consider $\mathfrak{g} = \mathfrak{sl}(5, \mathbb{R})$.

(1) The block decomposition of type $(1,1,1,1,1)$ corresponds to $\Phi = \Lambda \setminus \{\alpha_1, \alpha_2, \alpha_3, \alpha_4\} = \emptyset$, and hence, $\mathfrak{s}_{(1,1,1,1,1)} = \mathfrak{s}$.
(2) The block decomposition of type $(2,1,2)$ corresponds to $\Phi = \Lambda \setminus \{\alpha_2, \alpha_3\} = \{\alpha_1, \alpha_4\}$, and hence,

$$\mathfrak{s}_{(2,1,2)} = \mathfrak{s}_{\{\alpha_1,\alpha_4\}} = \left\{ \left(\begin{array}{cc|c|cc} a & & * & * & * \\ & a & * & * & * \\ \hline & & b & * & * \\ \hline & & & c & \\ & & & & c \end{array} \right) \in \mathfrak{g} \right\}.$$

Remark 2.3. It is known that $\mathfrak{s}_{(d_1,\ldots,d_k)}$ is isomorphic to $\mathfrak{s}_{(d_k,\ldots,d_1)}$, which is its transpose with respect to the anti-diagonal, and furthermore, the corresponding solvmanifolds are isometric to each other. Therefore, we can assume that $d_1 \geq d_k$.

At the end of this subsection, we mention an orthonormal basis of $(\mathfrak{s}_{(d_1,\ldots,d_k)}, \langle\ ,\ \rangle)$. Recall that $\langle\ ,\ \rangle$ is the inner product induced by Equation (1). For each $i \in \{1, 2, \ldots, k-1\}$, we put

$$A_i := \frac{1}{\sqrt{2n_i n_{i+1} d_{i+1}}} \left(n_{i+1} \sum_{j=1}^{n_i} E_{j,j} - n_i \sum_{j=1}^{n_{i+1}} E_{j,j} \right) \in \mathfrak{a},$$

where $n_i := d_1 + \cdots + d_i$. For convenience, we also set $n_0 := 0$. Then, the following holds.

Lemma 2.1. *One has that*

$$\{A_1, \ldots, A_{k-1}\} \\ \cup \left\{ E_{p,q} \;\middle|\; n_{i-1} < p \leq n_i < q \text{ for some } i = 1, \ldots, k-1 \right\}$$

is an orthonormal basis of $(\mathfrak{s}_{(d_1,\ldots,d_k)}, \langle\ ,\ \rangle)$.

Proof. By the definition of A_i, it is easily seen that the above is a basis of $\mathfrak{s}_{(d_1,\ldots,d_k)}$. Let us put

$$B_i := n_{i+1} \sum_{j=1}^{n_i} E_{j,j} - n_i \sum_{j=1}^{n_{i+1}} E_{j,j} \in \mathfrak{a}.$$

Then, one can also see that

$$\langle B_i, B_i \rangle = 2 \left(n_i n_{i+1}^2 - 2{n_i}^2 n_{i+1} + {n_i}^2 n_{i+1} \right) = 2 n_i n_{i+1} d_{i+1},$$

and if $i < j$,

$$\langle B_i, B_j \rangle = 2\left(n_i n_{i+1}(n_{j+1} - n_j) - n_i n_{i+1}(n_{j+1} - n_j)\right) = 0,$$

which means that $\{A_1, \ldots, A_{k-1}\}$ is orthonormal. Therefore, we obtain our assertion. □

Example 2.4. Let us consider the case where $(d_1, d_2, d_3) = (2, 1, 2)$. Then, one has $(n_1, n_2, n_3) = (2, 3, 5)$ and

$$A_1 = \frac{1}{\sqrt{12}} \left(\begin{array}{cc|c|cc} 1 & & & & \\ & 1 & & & \\ \hline & & -2 & & \\ \hline & & & 0 & \\ & & & & 0 \end{array}\right), \qquad A_2 = \frac{1}{\sqrt{60}} \left(\begin{array}{cc|c|cc} 2 & & & & \\ & 2 & & & \\ \hline & & 2 & & \\ \hline & & & -3 & \\ & & & & -3 \end{array}\right).$$

And moreover,

$$\{A_1, A_2, E_{13}, \ldots, E_{15}, E_{23}, \ldots, E_{25}, E_{34}, E_{35}\}$$

is the orthonormal basis of $(\mathfrak{s}_{(2,1,2)}, \langle\ ,\ \rangle)$ mentioned above.

3. Proof of the main theorem

In this section, we give the proof of Theorem 1.1 for $\mathfrak{g} = \mathfrak{sl}(n, \mathbb{R})$.

Let us consider the block decomposition of type $(d_1, \ldots, d_k)$, where $k \geq 2$ and $d_1 + \cdots + d_k = n$, and denote by $\mathfrak{s}_{(d_1,\ldots,d_k)}$ the corresponding solvable part of the parabolic subalgebra of $\mathfrak{g}$. Let $\langle , \rangle$ be the inner product on $\mathfrak{s}_{(d_1,\ldots,d_k)}$ induced by Equation (1). And also, let $S_{(d_1,\ldots,d_k)}$ denote the corresponding Lie group, and g denote the corresponding left-invariant metric. Now, we study the sectional curvatures of Einstein solvmanifolds $(S_{(d_1,\ldots,d_k)}, g)$.

Proposition 3.1. *The following holds.*

(1) *If $k = n$, then $(S_{(d_1,\ldots,d_n)}, g)$ is isometric to a symmetric space $\mathrm{SL}_n(\mathbb{R})/\mathrm{SO}(n)$, and hence it has nonpositive sectional curvature.*
(2) *If $k = 2$, then $(S_{(d_1,d_2)}, g)$ is isometric to a real hyperbolic space $\mathbb{R}\mathrm{H}^{d_1 d_2 + 1}$, and hence it has negative sectional curvature.*

Proof. First, we consider the case where $k = n$, namely, $(d_1, \ldots, d_k) = (1, \ldots, 1)$. Then, as mentioned above, $\mathfrak{s}_{(d_1,\ldots,d_k)} = \mathfrak{s}$, and $(S_{(d_1,\ldots,d_k)}, g)$ is isometric to a symmetric space $\mathrm{SL}(n, \mathbb{R})/\mathrm{SO}(n)$ (refer also to Remark 2.2). Next, we consider the case where $k = 2$, namely, the block decomposition is of type (d_1, d_2). Then, one can show that the orthonormal basis of $(\mathfrak{s}_{(d_1,d_2)}, \langle , \rangle)$

$$\{A_1\} \cup \{E_{p,q} \mid p \leq d_1 < q\}$$

satisfies that

$$[A_1, E_{p,q}] = \sqrt{\frac{n}{2d_1 d_2}} E_{p,q},$$

whereas the other relations vanish. This implies that $\mathfrak{s}_{(d_1,d_2)}$ is isomorphic to the so-called Lie algebra of a real hyperbolic space, and $(S_{(d_1,d_2)}, g)$ is isometric to the real hyperbolic space $\mathbb{R}\mathrm{H}^{d_1 d_2+1}$ with constant sectional curvature $-n/(2d_1 d_2)$. □

Proposition 3.2. *If* $3 \leq k \leq n-1$, *then* $(S_{(d_1,\ldots,d_k)}, g)$ *does not have nonpositive sectional curvature.*

Proof. Note that curvature properties of $(S_{(d_1,\ldots,d_k)}, g)$ can be calculated by the corresponding metric Lie algebra $(\mathfrak{s}_{(d_1,\ldots,d_k)}, \langle , \rangle)$, under the identification of the set of all left-invariant vector fields of $S_{(d_1,\ldots,d_k)}$ and the Lie algebra $\mathfrak{s}_{(d_1,\ldots,d_k)}$. According to [10, Lemma 1.1], the sectional curvature K_σ is given by

$$\begin{aligned}
K_\sigma &= \frac{1}{2}\langle \operatorname{ad}_X Y, -\operatorname{ad}_X Y + \operatorname{ad}_Y^* X + (-\operatorname{ad}_X^* Y)\rangle \\
&\quad - \frac{1}{4}\langle \operatorname{ad}_X Y - \operatorname{ad}_Y^* X + (-\operatorname{ad}_X^* Y), \operatorname{ad}_X Y + \operatorname{ad}_Y^* X - (-\operatorname{ad}_X^* Y)\rangle \\
&\quad - \langle \operatorname{ad}_X^* X, \operatorname{ad}_Y^* X\rangle \\
&= -\big\|[X,Y]\big\|^2 + \frac{1}{4}\big\|[X,Y] - \operatorname{ad}_X^* Y + \operatorname{ad}_Y^* X\big\|^2 \\
&\quad + \langle \operatorname{ad}_X^* Y, \operatorname{ad}_Y^* X\rangle - \langle \operatorname{ad}_X^* X, \operatorname{ad}_Y^* Y\rangle,
\end{aligned}$$

where σ is a two-dimensional subspace of $\mathfrak{s}_{(d_1,\ldots,d_k)}$ and $\{X, Y\}$ is its orthonormal basis. In order to show the assertion, we have only to take an orthonormal basis $\{X, Y\}$ such that $K_\sigma > 0$.

We prove the claim by considering two cases. Note that we can assume that $d_1 \geq d_k$ (see Remark 2.3). Then, since $k \neq n$, we can consider the following two cases:

(i) there exists $i \in \{2, 3, \ldots, k-1\}$ such that $d_i \geq 2$,
(ii) $d_1 \geq 2$.

First, we consider the case where $d_i \geq 2$ for some $i \in \{2, 3, \ldots, k-1\}$. Let us put $X := E_{n_{i-1}, n_{i-1}+1}$, $Y := E_{n_i, n_i+1} \in \mathfrak{s}_{(d_1, \ldots, d_k)}$. Then, by direct calculations, we see that $[X, Y] = \mathrm{ad}^*_X Y = \mathrm{ad}^*_Y X = 0$, and

$$\mathrm{ad}^*_X X = \sqrt{\frac{n_{i-2}}{2n_{i-1}d_{i-1}}}\, A_{i-2} - \sqrt{\frac{n_i}{2n_{i-1}d_i}}\, A_{i-1},$$

$$\mathrm{ad}^*_Y Y = \sqrt{\frac{n_{i-1}}{2n_i d_i}}\, A_{i-1} - \sqrt{\frac{n_{i+1}}{2n_i d_{i+1}}}\, A_i,$$

where we set $n_0 := 0$ and $A_0 := 0$ for the case of $i = 2$. Therefore, we obtain

$$K_\sigma(X, Y) = -\langle \mathrm{ad}^*_X X, \mathrm{ad}^*_Y Y \rangle = \sqrt{\frac{n_i}{2n_{i-1}d_i}} \sqrt{\frac{n_{i-1}}{2n_i d_i}} = \frac{1}{2d_i} > 0,$$

which proves our claim for Case (i). Next, we consider Case (ii) and suppose $n_1 = d_1 \geq 2$. Note that we may assume that $d_2 = d_3 = 1$. Then, since $E_{1,n_2}, E_{n_2,n_3}, E_{1,n_3} \in \mathfrak{s}_{(d_1, \ldots, d_k)}$, we take

$$X := (\cos\theta) E_{1,n_2} + (\sin\theta) E_{n_2,n_3}, \; Y := \frac{1}{\sqrt{2}} A_2 + \frac{1}{\sqrt{2}} E_{1,n_3},$$

where $\theta \in (0, \pi/2)$ is some constant. One has

$$[A_1, E_{1,n_2}] = \sqrt{\frac{1}{2n_1n_2d_2}} \cdot (1 - (-n_1)) E_{1,n_2} = \sqrt{\frac{n_2}{2n_1}}\, E_{1,n_2},$$

$$[A_1, E_{1,n_3}] = \sqrt{\frac{1}{2n_1n_2d_2}} \cdot (1 - 0) E_{1,n_3} = \sqrt{\frac{1}{2n_1n_2}}\, E_{1,n_3},$$

$$[A_1, E_{n_2,n_3}] = \sqrt{\frac{1}{2n_1n_2d_2}} \cdot ((-n_1) - 0) E_{n_2,n_3} = -\sqrt{\frac{n_1}{2n_2}}\, E_{n_2,n_3},$$

$$[A_2, E_{1,n_2}] = \sqrt{\frac{1}{2n_2n_3d_3}} \cdot (1 - 1) E_{1,n_2} = 0,$$

$$[A_2, E_{1,n_3}] = \sqrt{\frac{1}{2n_2n_3d_3}} \cdot (1 - (-n_2)) E_{1,n_3} = \sqrt{\frac{n_3}{2n_2}}\, E_{1,n_3},$$

$$[A_2, E_{n_2,n_3}] = \sqrt{\frac{1}{2n_2n_3d_3}} \cdot (1 - (-n_2)) E_{n_2,n_3} = \sqrt{\frac{n_3}{2n_2}}\, E_{n_2,n_3},$$

and $[A_i, E_{1,n_2}] = [A_i, E_{1,n_3}] = [A_i, E_{n_2,n_3}] = 0$ for $i = 3, \ldots, k-1$. These equations yield that

$$[X, Y] = -\sqrt{\frac{n_3}{4n_2}} (\sin\theta) E_{n_2,n_3},$$

and moreover,

$$\begin{aligned}
\operatorname{ad}_X^* Y &= \sum \langle Y, [X, Z_i] \rangle Z_i \\
&= \langle Y, [X, E_{1,n_2}] \rangle E_{1,n_2} + \langle Y, [X, E_{n_2,n_3}] \rangle E_{n_2,n_3} \\
&= \frac{1}{\sqrt{2}} \cdot \sin\theta \cdot \langle E_{1,n_3}, [E_{n_2,n_3}, E_{1,n_2}] \rangle E_{1,n_2} \\
&\quad + \frac{1}{\sqrt{2}} \cdot \cos\theta \cdot \langle E_{1,n_3}, [E_{1,n_2}, E_{n_2,n_3}] \rangle E_{n_2,n_3} \\
&= -\frac{1}{\sqrt{2}} (\sin\theta) E_{1,n_2} + \frac{1}{\sqrt{2}} (\cos\theta) E_{n_2,n_3},
\end{aligned}$$

where $\{Z_i\}$ is the orthonormal basis of $\mathfrak{s}_{(d_1,\ldots,d_k)}$ mentioned in Lemma 2.1. By similar calculations, we see that

$$\begin{aligned}
\operatorname{ad}_Y^* X &= \sqrt{\frac{n_3}{4n_2}} (\sin\theta) E_{n_2,n_3} = -[X, Y], \\
\operatorname{ad}_X^* X &= \left(-\sqrt{\frac{n_2}{2n_1}} \cos^2\theta + \sqrt{\frac{n_1}{2n_2}} \sin^2\theta \right) A_1 - \sqrt{\frac{n_3}{2n_2}} (\sin^2\theta) A_2, \\
\operatorname{ad}_Y^* Y &= -\frac{1}{2} \left(\sqrt{\frac{1}{2n_1 n_2}} A_1 + \sqrt{\frac{n_3}{2n_2}} A_2 - \sqrt{\frac{n_3}{2n_2}} E_{1,n_3} \right),
\end{aligned}$$

and hence,

$$\begin{aligned}
K_\sigma(X, Y) &= -\frac{n_3}{4n_2} \sin^2\theta + \frac{1}{8} + \sqrt{\frac{n_3}{8n_2}} \sin\theta\cos\theta \\
&\quad + \frac{1}{2} \left(-\frac{1}{2n_1} \cos^2\theta + \frac{1}{2n_2} \sin^2\theta - \frac{n_3}{2n_2} \sin^2\theta \right) \\
&= -\left(\sqrt{\frac{n_3}{4n_2}} \sin\theta - \frac{1}{2\sqrt{2}} \cos\theta \right)^2 + \frac{1}{8} \cos^2\theta \\
&\quad + \frac{1}{8} - \frac{1}{4n_1} \cos^2\theta - \frac{1}{4} \sin^2\theta \\
&= \frac{\cos^2\theta}{8} \left(-\left(\sqrt{\frac{2n_3}{n_2}} \tan\theta - 1 \right)^2 + 2 - \frac{2}{n_1} - \tan^2\theta \right).
\end{aligned}$$

Therefore, by defining $\theta \in (0, \pi/2)$ by $\tan\theta = (n_2/(2n_3))^{1/2}$, we can obtain that

$$K_\sigma(X, Y) = \frac{\cos^2\theta}{8} \left(2 - \frac{2}{n_1} - \frac{n_2}{2n_3} \right) = \frac{\cos^2\theta \cdot (3n_1^2 + 3n_1 - 8)}{16 n_1 n_3} > 0,$$

which proves our claim for Case (ii). This completes the proof. □

Acknowledgments

The authors would like to thank the referee for the careful review and the valuable comments for our paper. The second author was supported by JSPS KAKENHI Grant Number 22K13919.

References

[1] C. Böhm and R. A. Lafuente, Non-compact Einstein manifolds with symmetry. *J. Amer. Math. Soc.* **36** (2023), no. 3, 591–651.

[2] C. S. Gordon and M. R. Jablonski, Einstein solvmanifolds have maximal symmetry. *J. Differential Geom.* **111** (2019), no. 1, 1–38.

[3] T. Hashinaga, A. Kubo, Y. Taketomi, and H. Tamaru, A Lie theoretic interpretation of realizations of some contact metric manifolds. *New Horizons in Differential Geometry and its Related Fields*, World Sci. Publ. Co. Pte. Ltd., Hackensack, NJ, 2022, 71–90.

[4] J. Heber, Noncompact homogeneous Einstein spaces. *Invent. Math.* **133** (1998), no 2, 279–352.

[5] S. Helgason, Differential geometry, Lie groups, and symmetric spaces. *Grad. Stud. Math.*, **34**, American Mathematical Society, Providence, RI, 2001, xxvi+641 pp.

[6] A. W. Knapp, Lie groups beyond an introduction. 2nd edn. *Progr. Math.*, **140**, Birkhäuser Boston, Inc., Boston, MA, 2002, xviii+812 pp.

[7] A. Kubo and T. Tamaru, A sufficient condition for congruency of orbits of Lie groups and some applications. *Geom. Dedicata* **167** (2013), 233–238.

[8] J. Lauret, Einstein solvmanifolds and nilsolitons. *Contemp. Math.*, **491**, American Mathematical Society, Providence, RI, 2009, 1–35.

[9] J. Lauret, Einstein solvmanifolds are standard. *Ann. Math.* (2) **172** (2010), no. 3 1859–1877.

[10] J. Milnor, Curvatures of left invariant metrics on Lie groups. *Advances in Math.* **21** (1976), no. 3, 293–329.

[11] K. Mori, Einstein metrics on Boggino-Damek-Ricci type solvable Lie groups. *Osaka J. Math.* **39** (2002), no. 2, 345–362.

[12] Y. Nikolayevsky, Einstein solvmanifolds and the pre-Einstein derivation. *Trans. Amer. Math. Soc.* **363** (2011), no. 8, 3935–3958.

[13] H. Tamaru, Parabolic subgroups of semisimple Lie groups and Einstein solvmanifolds. *Math. Ann.* **351** (2011), no. 1, 51–66.

Received December 28, 2023
Revised February 20, 2024

Modern Approaches to Differential Geometry
and its Related Fields 45 – 71

EINSTEIN-LIKE METRICS ON FLAG MANIFOLDS

Andreas ARVANITOYEORGOS

*Department of Mathematics, University of Patras,
GR-26500 Rion, Greece, and
Hellenic Open University,
Aristotelous 18, GR-26335 Patras, Greece
E-mail: arvanito@math.upatras.gr*

Yusuke SAKANE

*Department of Pure and Applied Mathematics,
Graduate School of Information Science and Technology,
Osaka University, Suita, Osaka 565-0871, Japan
E-mail: sakane@math.sci.osaka-u.ac.jp*

Marina STATHA

*Department of Mathematics, University of Thessaly,
GR-35100 Lamia, Greece
E-mail: marinastatha@uth.gr*

We consider invariant Einstein-like metrics of type $\mathscr{B}$, that is, Ricci tensor r is a Codazzi tensor, on generalized flag manifolds. We show that, for generalized flag manifolds G/K with second Betti number $b_2(G/K) = 2$ and with ν isotropy summands ($\nu \leq 5$) or G_2-type, then invariant Einstein-like metrics of type $\mathscr{B}$ are Einstein.

Keywords: Homogeneous space; Einstein metric; Einstein-like metric; generalized flag manifold.

1. Introduction

A Riemannian manifold (M, g) is called *Einstein* if it has constant Ricci curvature, i.e. $\mathrm{Ric}_g = \lambda \cdot g$ for some $\lambda \in \mathbb{R}$. As a generalization of Einstein metrics, A. Gray introduced the notion of Einstein-like $\mathscr{A}$-metrics and $\mathscr{B}$-metrics on Riemannian manifolds in [12].

For a Riemannian manifold (M, g), the metric g is called an $\mathscr{A}$-metric if the Ricci tensor r satisfies

$$\nabla_X(r)(X, X) = 0$$

for all C^∞ vector fields X on M and the metric g is called a $\mathscr{B}$-metric if the Ricci tensor r satisfies

$$\nabla_X(r)(Y,Z) = \nabla_Y(r)(X,Z)$$

for all C^∞ vector fields X, Y, Z on M, that is, r is a Codazzi tensor [8]. If a Riemannian metric is either an $\mathscr{A}$-metric or a $\mathscr{B}$-metric, it is called an *Einstein-like metric.*

In their paper [17], C. Peng and C. Qian discussed a characterization of Einstein-like invariant metrics on homogeneous spaces and classified homogeneous $\mathscr{A}$-metrics and $\mathscr{B}$-metrics on spheres and projective spaces. In [18], C. Qian and A. Wu discussed Einstein-like invariant metrics on symmetric spaces M. Whenever there exists a closed proper subgroup G' of $G = \mathrm{Isom}_0(M)$ acting transitively on M, they found all the G'-invariant $\mathscr{A}$-metrics and $\mathscr{B}$-metrics on M.

More recently, F. Li, H. Chen and Z. Chen [14] obtained a characterization of Einstein-like invariant metrics on homogeneous space G/K and classified such invariant metrics on generalized Wallach spaces of exceptional type.

In the present paper, we investigate $\mathscr{B}$-metrics on generalized flag manifolds. For invariant metrics on generalized flag manifolds, Ricci tensor and Einstein metrics have been studied in [2, 4–7, 10, 13].

Our main results are the following:

Theorem 1.1. *For generalized flag manifolds G/K, we have*

(1) *If the second Betti number $b_2(G/K) = 1$, then $\mathscr{B}$-metrics on G/K are Einstein;*
(2) *If the second Betti number $b_2(G/K) = 2$ with ν isotropy summands ($\nu \leq 5$) or G_2-type, then $\mathscr{B}$-metrics on G/K are Einstein.*

2. The Ricci tensor for compact homogeneous spaces

We recall an expression for the Ricci tensor for a G-invariant Riemannian metric on a compact homogeneous space whose isotropy representation is decomposed into a sum of non equivalent irreducible summands.

Let G be a compact semi-simple Lie group, K a connected closed subgroup of G, and let $\mathfrak{g}$ and $\mathfrak{k}$ be the corresponding Lie algebras. The Killing form B of $\mathfrak{g}$ is negative definite, so we can define an $\mathrm{Ad}(G)$-invariant inner product $-B$ on $\mathfrak{g}$. Let $\mathfrak{g} = \mathfrak{k} \oplus \mathfrak{m}$ be a reductive decomposition of $\mathfrak{g}$ with respect to $-B$ so that $[\,\mathfrak{k}, \mathfrak{m}\,] \subset \mathfrak{m}$ and $\mathfrak{m} \cong T_o(G/K)$. We decompose $\mathfrak{m}$ into irreducible $\mathrm{Ad}(K)$-modules as follows:

$$\mathfrak{m} = \mathfrak{m}_1 \oplus \cdots \oplus \mathfrak{m}_\nu. \tag{1}$$

We assume that $\mathfrak{m}$ is decomposed into mutually non equivalent irreducible $\mathrm{Ad}(K)$-modules. Then any G-invariant metric on G/K can be expressed as

$$g = x_1(-B)|_{\mathfrak{m}_1} + \cdots + x_\nu(-B)|_{\mathfrak{m}_\nu}, \tag{2}$$

for positive real numbers $(x_1, \ldots, x_\nu) \in \mathbb{R}^\nu_+$. And Ricci tensor r of a G-invariant Riemannian metric $\langle\ ,\ \rangle$ on G/K is of the same form as (2), that is

$$r = z_1(-B)|_{\mathfrak{m}_1} + \cdots + z_\nu(-B)|_{\mathfrak{m}_\nu},$$

for some real numbers $z_1, \ldots, z_\nu$.

Let $\{e_\alpha\}$ be a $(-B)$-orthonormal basis adapted to the decomposition of $\mathfrak{m}$, i.e. $e_\alpha \in \mathfrak{m}_i$ for some i, and $\alpha < \beta$ if $i < j$. We put $C^\gamma_{\alpha\beta} = -B\left([e_\alpha, e_\beta], e_\gamma\right)$ so that $[e_\alpha, e_\beta]_{\mathfrak{m}} = \sum_\gamma C^\gamma_{\alpha\beta} e_\gamma$, and set $A_{ijk} = \sum (C^\gamma_{\alpha\beta})^2$, where the sum is taken over all indices α, β, γ with $e_\alpha \in \mathfrak{m}_i$, $e_\beta \in \mathfrak{m}_j$, $e_\gamma \in \mathfrak{m}_k$ (cf. [19]). Then the positive numbers A_{ijk} are independent of the $(-B)$-orthonormal bases chosen for $\mathfrak{m}_i, \mathfrak{m}_j, \mathfrak{m}_k$, and symmetric in all three indices

$$A_{ijk} = A_{jik} = A_{jki}.$$

Let $d_k = \dim \mathfrak{m}_k$. Then we have the following:

Lemma 2.1 ([16]). *The components $r_1, \ldots, r_\nu$ of the Ricci tensor r of the metric $\langle\ ,\ \rangle$ of the form (2) on G/K are given by*

$$r_k = \frac{1}{2x_k} + \frac{1}{4d_k} \sum_{i,j} \frac{x_k}{x_i x_j} A_{ijk} - \frac{1}{2d_k} \sum_{i,j} \frac{x_j}{x_k x_i} A_{kij} \quad (k = 1, \ldots, \nu), \tag{3}$$

where the sum is taken over $i, j = 1, \ldots, \nu$.

3. Characterization of Einstein-like metrics

For homogeneous space G/K, C. Peng and C. Qian [17] obtained a characterization of Einstein-like invariant metrics:

Theorem 3.1 ([17, Corollary 2.1]). *Let g be a G-invariant metric on G/K.*

1) *g is an $\mathscr{A}$-metric if and only if $r(U(X, X), X) = 0$ for all $X \in \mathfrak{m}$.*
2) *g is a $\mathscr{B}$-metric if and only if*

$$\frac{1}{2} r([Z, X]_{\mathfrak{m}}, Y) - \frac{1}{2} r([Y, X]_{\mathfrak{m}}, Z) + r(X, [Z, Y]_{\mathfrak{m}}) + r\big(U(Z, X), Y\big) - r\big(U(Y, X), Z\big) = 0,$$

for all X, Y, $Z \in \mathfrak{m}$.

In [14], F. Li, H. Chen and Z. Chen obtained characterizations of invariant $\mathscr{B}$-metrics using Theorem 3.1 (2).

Theorem 3.2 ([14, Theorem 3.4]). *Let g be a G-invariant metric on G/K of the form* (2). *The metric g is a $\mathscr{B}$-metric if and only if*

$$C_{\alpha\beta}^{\gamma}\big((x_j + x_i - x_k)r_j + (x_k + x_i - x_j)r_k - 2x_i r_i\big) = 0 \tag{4}$$

holds for any $i, j, k \in \{1, 2, \ldots, \nu\}$, $\alpha = 1, 2, \ldots, d_i$, $\beta = 1, 2, \ldots, d_j$ and $\gamma = 1, 2, \ldots, d_k$.

Corollary 3.1 ([14, Corollary 3.5]). *If $A_{iij} \neq 0$, then the Ricci components r_i, r_j of a $\mathscr{B}$-metric g of the form* (2) *satisfy $r_i = r_j$.*

Corollary 3.2. *If $A_{ijk} \neq 0$ and $r_j = r_k$, then the Ricci components of a $\mathscr{B}$-metric g of the form* (2) *satisfy $r_i = r_j = r_k$.*

Proof. From $A_{ijk} \neq 0$, we see that $C_{\alpha\beta}^{\gamma} \neq 0$ for some α, β, γ. Hence, by Theorem 3.2, we see that

$$(x_j + x_i - x_k)r_j + (x_k + x_i - x_j)r_k - 2x_i r_i = 0$$

and thus $0 = (x_j + x_i - x_k)r_j + (x_k + x_i - x_j)r_j - 2x_i r_i = 2x_i(r_j - r_i)$. Since $x_i > 0$, we have $r_i = r_j = r_k$. □

Corollary 3.3 ([14, Corollary 3.6]). *If $A_{ijk} \neq 0$, for i, j, k mutually distinct, and*

$${x_i}^2 - 2x_i x_j - 2x_i x_k + {x_j}^2 - 2x_j x_k + {x_k}^2 \neq 0,$$

then the Ricci components of a $\mathscr{B}$-metric g of the form (2) *satisfy $r_i = r_j = r_k$.*

4. Generalized flag manifolds

Let G be a compact semi-simple Lie group, $\mathfrak{g}$ the Lie algebra of G and $\mathfrak{h}$ a maximal abelian subalgebra of $\mathfrak{g}$. We denote by $\mathfrak{g}^{\mathbb{C}}$ and $\mathfrak{h}^{\mathbb{C}}$ the complexification of $\mathfrak{g}$ and $\mathfrak{h}$, respectively. We identify an element of the root system Δ of $\mathfrak{g}^{\mathbb{C}}$ relative to the Cartan subalgebra $\mathfrak{h}^{\mathbb{C}}$ with an element of $\mathfrak{h}_0 = \sqrt{-1}\mathfrak{h}$ by the duality defined by the Killing form of $\mathfrak{g}^{\mathbb{C}}$. Let $\Pi = \{\alpha_1, \ldots, \alpha_\ell\}$ be a fundamental system of Δ and $\{\Lambda_1, \ldots, \Lambda_\ell\}$ the fundamental weights of $\mathfrak{g}^{\mathbb{C}}$ corresponding to Π, that is

$$\frac{2(\Lambda_i, \alpha_j)}{(\alpha_j, \alpha_j)} = \delta_{ij} \qquad (1 \leq i, j \leq \ell).$$

Let Π_0 be a subset of Π and $\Pi \setminus \Pi_0 = \{\alpha_{i_1}, \ldots, \alpha_{i_r}\}$ $(1 \leq i_1 < \cdots < i_r \leq \ell)$. We put $[\Pi_0] = \Delta \cap \{\Pi_0\}_{\mathbb{Z}}$, where $\{\Pi_0\}_{\mathbb{Z}}$ denotes the subspace of $\mathfrak{h}_0$ generated by Π_0.

Consider the root space decomposition of $\mathfrak{g}^{\mathbb{C}}$ relative to $\mathfrak{h}^{\mathbb{C}}$:

$$\mathfrak{g}^{\mathbb{C}} = \mathfrak{h}^{\mathbb{C}} + \sum_{\alpha \in \Delta} \mathfrak{g}^{\mathbb{C}}_{\alpha}.$$

For a subset Π_0 of Π, we define a parabolic subalgebra $\mathfrak{u}$ of $\mathfrak{g}^{\mathbb{C}}$ by

$$\mathfrak{u} = \mathfrak{h}^{\mathbb{C}} + \sum_{\alpha \in [\Pi_0] \cup \Delta^+} \mathfrak{g}^{\mathbb{C}}_{\alpha},$$

where Δ^+ is the set of all positive roots relative to Π. We put $\Delta^+_{\mathfrak{m}} = \Delta^+ \setminus [\Pi_0]$. Then the nilradical $\mathfrak{n}$ of $\mathfrak{u}$ is given by

$$\mathfrak{n} = \sum_{\alpha \in \Delta^+_{\mathfrak{m}}} \mathfrak{g}^{\mathbb{C}}_{\alpha}.$$

Let $G^{\mathbb{C}}$ be a simply connected complex semi-simple Lie group whose Lie algebra is $\mathfrak{g}^{\mathbb{C}}$ and U the parabolic subgroup of $G^{\mathbb{C}}$ generated by $\mathfrak{u}$. Then the complex homogeneous manifold $G^{\mathbb{C}}/U$ is compact simply connected and G acts transitively on $G^{\mathbb{C}}/U$. Note also that $K = G \cap U$ is a connected closed subgroup of G, $G^{\mathbb{C}}/U = G/K$ as C^{∞}-manifolds, and $G^{\mathbb{C}}/U$ admits a G-invariant Kähler metric.

Let $\mathfrak{k}$ be the Lie algebra of K and $\mathfrak{k}^{\mathbb{C}}$ the complexification of $\mathfrak{k}$. Then we have a direct decomposition

$$\mathfrak{u} = \mathfrak{k}^{\mathbb{C}} \oplus \mathfrak{n}, \qquad \mathfrak{k}^{\mathbb{C}} = \mathfrak{h}^{\mathbb{C}} + \sum_{\alpha \in [\Pi_0]} \mathfrak{g}^{\mathbb{C}}_{\alpha}.$$

We put $\mathfrak{t} = \Big\{H \in \mathfrak{h}_0 \mid (H,\ \Pi_0) = (0)\Big\}$. Then $\{\Lambda_{i_1}, \ldots, \Lambda_{i_r}\}$ is a basis of $\mathfrak{t}$. Put $\mathfrak{s} = \sqrt{-1}\mathfrak{t}$. Then the Lie algebra $\mathfrak{k}$ is given by $\mathfrak{k} = \mathfrak{z}(\mathfrak{s})$ (the Lie algebra of centralizer of a torus S in G). Note that the second Betti number $b_2(G/K)$ of G/K is

$$b_2(G/K) = \dim \mathfrak{t} = (\text{the cardinality of } \Pi \setminus \Pi_0) = r$$

which is given by ([9]).

We consider the restriction map

$$\kappa : \mathfrak{h}_0^* \to \mathfrak{t}^* \quad \alpha \mapsto \alpha|_{\mathfrak{t}}$$

and set $\Delta_{\mathfrak{t}} = \kappa(\Delta)$. The elements of $\Delta_{\mathfrak{t}}$ are called $\mathfrak{t}$-roots ([1]).

There exists a one-to-one correspondence between $\mathfrak{t}$-roots ξ and irreducible submodules $\mathfrak{m}_\xi$ of the $\mathrm{Ad}_G(K)$-module $\mathfrak{m}^{\mathbb{C}}$ given by

$$\Delta_{\mathfrak{t}} \ni \xi \mapsto \mathfrak{m}_\xi = \sum_{\kappa(\alpha)=\xi} \mathfrak{g}_\alpha^{\mathbb{C}}.$$

Thus we have a decomposition of the $\mathrm{Ad}_G(K)$-module $\mathfrak{m}^{\mathbb{C}}$:

$$\mathfrak{m}^{\mathbb{C}} = \sum_{\xi \in \Delta_{\mathfrak{t}}} \mathfrak{m}_\xi.$$

Denote by $\Delta_{\mathfrak{t}}^+$ the set of all positive $\mathfrak{t}$-roots, that is, the restriction of the system Δ^+. Then $\mathfrak{n} = \sum_{\xi \in \Delta_{\mathfrak{t}}^+} \mathfrak{m}_\xi$.

Denote by τ the complex conjugation of $\mathfrak{g}^{\mathbb{C}}$ with respect to $\mathfrak{g}$ (note that τ interchanges $\mathfrak{g}_\alpha^{\mathbb{C}}$ and $\mathfrak{g}_{-\alpha}^{\mathbb{C}}$) and by $\mathfrak{v}^\tau$ the set of fixed points of τ in a (complex) vector subspace $\mathfrak{v}$ of $\mathfrak{g}^{\mathbb{C}}$. Thus we have a decomposition of $\mathrm{Ad}_G(K)$-module $\mathfrak{m}$ into irreducible submodules:

$$\mathfrak{m} = \sum_{\xi \in \Delta_{\mathfrak{t}}^+} (\mathfrak{m}_\xi + \mathfrak{m}_{-\xi})^\tau.$$

For non-negative integers $j_1, \ldots, j_r$ with $(j_1, \ldots, j_r) \neq (0, \ldots, 0)$, we put

$$\Delta(j_1, \ldots, j_r) = \left\{ \sum_{j=1}^{\ell} m_j \alpha_j \in \Delta^+ \;\middle|\; m_{i_1} = j_1, \ldots, m_{i_r} = j_r \right\}.$$

Then there exists a natural one-to-one correspondence between $\Delta_{\mathfrak{t}}^+$ and the set $\{\Delta(j_1, \ldots, j_r) \neq \emptyset\}$.

For a generalized flag manifold G/K, we have a decomposition of $\mathfrak{m}$ into mutually non-equivalent irreducible $\mathrm{Ad}_G(H)$-modules:

$$\mathfrak{m} = \sum_{\xi \in \Delta_{\mathfrak{t}}^+} (\mathfrak{m}_\xi + \mathfrak{m}_{-\xi})^\tau = \sum_{j_1, \ldots, j_r} \mathfrak{m}(j_1, \ldots, j_r).$$

Thus a G-invariant metric g on G/K can be written as

$$g = \sum_{\xi \in \Delta_{\mathfrak{t}}^+} x_\xi B|_{(\mathfrak{m}_\xi + \mathfrak{m}_{-\xi})^\tau} = \sum_{j_1, \ldots, j_r} x_{j_1 \cdots j_r} B|_{\mathfrak{m}(j_1, \ldots, j_r)} \tag{5}$$

for positive real numbers x_ξ, $x_{j_1 \cdots j_r}$.

5. $\mathscr{B}$-metrics on generalized flag manifolds

From now on we assume that the Lie group G is simple. We denote by ν the number of elements of $\Delta_{\mathfrak{t}}^+$ for a generalized flag manifold G/K, that is, the number of irreducible components of $\mathrm{Ad}_G(K)$-module $\mathfrak{m}$.

5.1. *The case* $r = 1$

Invariant metrics and their Ricci tensor on a generalized flag manifold G/K with the second Betti number $b_2(G/K) = r = 1$ have been studied in [2, 4, 10, 13] by A. Arvanitoyeorgos, I. Chrysikos, M. Kimura and Y. Sakane, and they classified all Einstein metrics. Note that we have $\nu \leq 6$ for the case $r = 1$.

Theorem 5.1. *Invariant $\mathscr{B}$-metrics g on generalized flag manifolds G/K with $r = 1$ are Einstein.*

Proof. If $\nu = 1$, then $\Delta_{\mathfrak{t}}^{+} = \{\,\xi\,\}$ and G/K is an irreducible Hermitian symmetric space with the symmetric pair $(\mathfrak{g}, \mathfrak{k})$. Thus invariant metrics are Einstein.

If $\nu = 2$, then we have $\mathfrak{m} = \mathfrak{m}(1) \oplus \mathfrak{m}(2) = \mathfrak{m}_1 \oplus \mathfrak{m}_2$, that is, $\Delta_{\mathfrak{t}}^{+} = \{\,\xi, 2\xi\,\}$. We say this case that $\mathfrak{t}$-roots system is of type $A_1(2)$. In this case we have $A_{112} \neq 0$ (cf. [3]). Thus, by Corollary 3.1, we see that $r_1 = r_2$ and the $\mathscr{B}$-metrics are Einstein.

If $\nu = 3$, then we have $\mathfrak{m} = \mathfrak{m}_1 \oplus \mathfrak{m}_2 \oplus \mathfrak{m}_3$, that is, $\Delta_{\mathfrak{t}}^{+} = \{\,\xi, 2\xi, 3\xi\,\}$. There are 7 cases and the Lie group G is always exceptional, that is, E_6, E_7, E_8, F_4 and G_2 (for E_7, E_8, there are 2 cases). In this case we have $A_{112} \neq 0, A_{123} \neq 0$ (cf. [13]). Thus, by Corollaries 3.1 and 3.2, we see that $r_1 = r_2 = r_3$ and thus the $\mathscr{B}$-metrics are Einstein.

If $\nu = 4$, then we see that $\mathfrak{t}$-roots system is of type $A_1(4)$, that is, $\Delta_{\mathfrak{t}}^{+} = \{\,\xi, 2\xi, 3\xi,\, 4\xi\,\}$. There are 4 cases and G is always exceptional Lie group, that is, E_7, E_8, F_4. In this case we have $A_{112} \neq 0$, $A_{123} \neq 0$, $A_{134} \neq 0$, $A_{224} \neq 0$ (cf. [4]). Thus, by Corollaries 3.1 and 3.2, we see that $r_1 = r_2 = r_3 = r_4$ and thus the $\mathscr{B}$-metrics are Einstein.

If $\nu = 5$, then we see that $\mathfrak{t}$-roots system is of type $A_1(5)$, that is, $\Delta_{\mathfrak{t}}^{+} = \{\,\xi, 2\xi, 3\xi,\, 4\xi,\, 5\xi\,\}$. There is only one case, $G = E_8$ and $K = \mathrm{SU}(4) \times \mathrm{SU}(5) \times \mathrm{U}(1)$. In this case we have $A_{112} \neq 0$, $A_{123} \neq 0$, $A_{134} \neq 0$, $A_{145} \neq 0$, $A_{235} \neq 0$, $A_{224} \neq 0$ (cf. [10]). Thus, by Corollaries 3.1 and 3.2, we see that $r_1 = r_2 = r_3 = r_4 = r_5$ and thus the $\mathscr{B}$-metrics are Einstein.

If $\nu = 6$, then we see that $\mathfrak{t}$-roots system $\mathfrak{t}$-roots system is of type $A_1(6)$, that is, $\Delta_{\mathfrak{t}}^{+} = \{\,\xi, 2\xi, 3\xi,\, 4\xi,\, 5\xi,\, 6\xi\}$. There is only one case, $G = E_8$ and $K = \mathrm{SU}(5) \times \mathrm{SU}(3) \times \mathrm{SU}(2) \times \mathrm{U}(1)$. In this case we have $A_{112} \neq 0$, $A_{123} \neq 0$, $A_{134} \neq 0$, $A_{145} \neq 0$, $A_{156} \neq 0$, $A_{224} \neq 0$, $A_{235} \neq 0$, $A_{246} \neq 0$, $A_{336} \neq 0$ (cf. [10]). Thus, by Corollaries 3.1 and 3.2, we see that $r_1 = r_2 = r_3 = r_4 = r_5 = r_6$ and thus the $\mathscr{B}$-metrics are Einstein. $\square$

5.2. *The case $r = 2$ and $\nu = 3$*

We have $\mathfrak{m} = \mathfrak{m}(1,0) \oplus \mathfrak{m}(0,1) \oplus \mathfrak{m}(1,1) = \mathfrak{m}_1 \oplus \mathfrak{m}_2 \oplus \mathfrak{m}_3$, up to isometry, that is, $\Delta_{\mathfrak{t}}^{+} = \{\,\xi_1, \xi_2, \xi_1 + \xi_2\,\}$ which is called of **type $\mathbf{A}_2$**.

We see there are 3 cases (cf. [2, 13]).

1) $\mathrm{SU}(p+q+r)/\mathrm{S}(\mathrm{U}(p)\times\mathrm{U}(q)\times\mathrm{U}(r))$ $(\ell = p+q+r-1)$

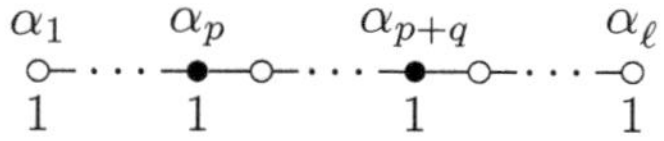

In this case we have $\xi_1 = \kappa(\alpha_p)$, $\xi_2 = \kappa(\alpha_{p+q})$, $d_1 = 2pq$, $d_2 = 2qr$, $d_3 = 2pr$ and $A_{123} = \dfrac{pqr}{p+q+r}$.

2) $\mathrm{SO}(2l)/(\mathrm{U}(l-1)\times \mathrm{U}(1))$ $(\ell \geq 4)$

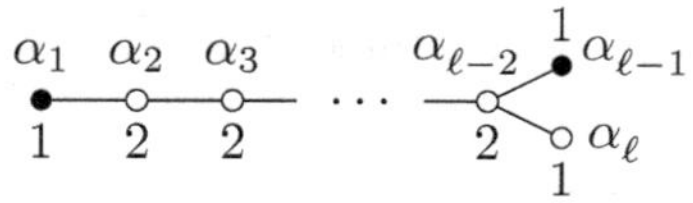

In this case we have $\xi_1 = \kappa(\alpha_1)$, $\xi_2 = \kappa(\alpha_{\ell-1})$, $d_1 = 2(\ell-1)$, $d_2 = (\ell-1)(\ell-2)$, $d_3 = 2(\ell-1)$ and $A_{123} = \dfrac{\ell-2}{2}$.

3) $\mathrm{E}_6/(\mathrm{SO}(8)\times\mathrm{U}(1)\times\mathrm{U}(1))$

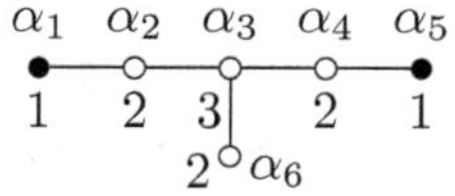

In this case we have $\xi_1 = \kappa(\alpha_1)$, $\xi_2 = \kappa(\alpha_5)$, $d_1 = d_2 = d_3 = 16$ and $A_{123} = \dfrac{8}{3}$.

We see that, for a G-invariant metric $g = x_1(-B)|_{\mathfrak{m}_1} + x_2(-B)|_{\mathfrak{m}_2} + x_3(-B)|_{\mathfrak{m}_3}$, the Ricci components r_1, r_2, r_3 for the metric g are given by

$$\begin{aligned} r_1 &= \frac{1}{2x_1} + \frac{1}{2d_1}A_{123}\left(\frac{x_1}{x_2x_3} - \frac{x_2}{x_1x_3} - \frac{x_3}{x_1x_2}\right), \\ r_2 &= \frac{1}{2x_2} + \frac{1}{2d_2}A_{123}\left(\frac{x_2}{x_1x_3} - \frac{x_1}{x_2x_3} - \frac{x_3}{x_2x_1}\right), \\ r_3 &= \frac{1}{2x_3} + \frac{1}{2d_3}A_{123}\left(\frac{x_3}{x_1x_2} - \frac{x_1}{x_3x_2} - \frac{x_2}{x_3x_1}\right). \end{aligned} \tag{6}$$

Theorem 5.2. *The $\mathscr{B}$-metrics g on a generalized flag manifold G/K of type A_2 are Einstein.*

Proof. Assume that g is a $\mathscr{B}$-metric. From Corollary 3.3, if ${x_1}^2 - 2x_1x_2 - 2x_1x_3 + {x_2}^2 - 2x_2x_3 + {x_3}^2 \neq 0$, we see that $r_1 = r_2 = r_3$ and hence, g is

Einstein in these cases. We consider the case

$$x_1{}^2 - 2x_1x_2 - 2x_1x_3 + x_2{}^2 - 2x_2x_3 + x_3{}^2 = 0.$$

We normalize the metric g by $x_1 = 1$ and put $x_2 = t^2$. Then we see that

$$1 - 2t^2 - 2x_3 + t^4 - 2t^2x_3 + x_3{}^2 = 0$$

and hence, we have $x_3 = (1+t)^2$ or $x_3 = (1-t)^2$. From Theorem 3.2, we have

$$(x_2 + x_1 - x_3)r_2 + (x_3 + x_1 - x_2)r_3 - 2x_1r_1 = 0. \tag{7}$$

By substituting $x_1 = 1$, $x_2 = t^2$, $x_3 = (1 \pm t)^2$ into (6) and (7), we obtain

$$\begin{aligned} \frac{1}{d_1d_2d_3(1 \pm t)} \Big(& 2A_{123}(2d_1d_2 + 2d_1d_3 + d_2d_3) - d_1d_2d_3 \\ & \pm \big(2A_{123}(d_1d_2 + 3d_1d_3 + d_2d_3) - d_1d_2d_3\big)t \\ & + \big(2A_{123}(d_1d_2 + 2d_1d_3 + 2d_2d_3) - d_1d_2d_3\big)t^2 \Big) = 0. \end{aligned} \tag{8}$$

We claim there are no real solutions of the equations (8).

For the case 1) $\mathrm{SU}(p+q+r)/\mathrm{S}(\mathrm{U}(p) \times \mathrm{U}(q) \times \mathrm{U}(r))$, we see that the numerator of equation (8) is given by

$$p + q \pm 2pt + (p+r)t^2 = 0.$$

Hence, there are no real solutions for t.

For the case 2) $\mathrm{SO}(2\ell)/(\mathrm{U}(\ell-1) \times \mathrm{U}(1))$ $(\ell \geq 4)$, we see that the numerator of equation (8) is given by

$$\ell \pm 4t + \ell t^2 = 0.$$

Hence, there are no real solutions for t.

For the case 3) $\mathrm{E}_6/(\mathrm{SO}(8) \times \mathrm{U}(1) \times \mathrm{U}(1))$, we see that the numerator of equation (8) is given by

$$2(1 \pm t + t^2) = 0.$$

Hence, there are no real solutions for t. □

5.3. *The case* $r = 2$ *and* $\nu = 4$

We have $\mathfrak{m} = \mathfrak{m}(1,0) \oplus \mathfrak{m}(0,1) \oplus \mathfrak{m}(1,1) \oplus \mathfrak{m}(1,2) = \mathfrak{m}_1 \oplus \mathfrak{m}_2 \oplus \mathfrak{m}_3 \oplus \mathfrak{m}_4$, up to isometry, that is, $\Delta_{\mathfrak{t}}^+ = \{\xi_1, \xi_2, \xi_1 + \xi_2,\ \xi_1 + 2\xi_2\}$ which is called of **type B**$_2$.

We see there are 6 cases (cf. [4]).

1) $\mathrm{SO}(2\ell+1)/(\mathrm{SO}(2\ell-3)\times \mathrm{U}(1)\times \mathrm{U}(1))$

α_1 α_2 α_3 $\cdots$ $\alpha_{\ell-1}$ α_ℓ

1 2 2 2 2

In this case we have $\xi_1 = \kappa(\alpha_1)$, $\xi_2 = \kappa(\alpha_2)$, $d_1 = 2$, $d_2 = 2(2\ell-3)$, $d_3 = 2(2\ell-3)$, $d_3 = 2$ and $A_{123} = \dfrac{2\ell-3}{2\ell-1}$, $A_{234} = \dfrac{2\ell-3}{2\ell-1}$.

2) $\mathrm{SO}(2\ell)/(\mathrm{SO}(2\ell-4)\times \mathrm{U}(1)\times \mathrm{U}(1))$

α_1 α_2 α_3 $\cdots$ $\alpha_{\ell-2}$ $\alpha_{\ell-1}$ α_ℓ

1 2 2 2 1 1

In this case we have $\xi_1 = \kappa(\alpha_1)$, $\xi_2 = \kappa(\alpha_2)$, $d_1 = 2$, $d_2 = 4(\ell-2)$, $d_3 = 4(\ell-2)$, $d_3 = 2$ and $A_{123} = \dfrac{\ell-2}{\ell-1}$, $A_{234} = \dfrac{\ell-2}{\ell-1}$.

3) $\mathrm{E}_6/(\mathrm{SU}(5)\times \mathrm{U}(1)\times \mathrm{U}(1))$

α_1 α_2 α_3 α_4 α_5 α_6

1 2 3 2 1 2

In this case we have $\xi_1 = \kappa(\alpha_1)$, $\xi_2 = \kappa(\alpha_2)$, $d_1 = 2$, $d_2 = 20$, $d_3 = 20$, $d_3 = 2$ and $A_{123} = \dfrac{5}{6}$, $A_{234} = \dfrac{5}{2}$.

4) $\mathrm{E}_7/(\mathrm{SO}(10))\times \mathrm{U}(1)\times \mathrm{U}(1))$

α_1 α_2 α_3 α_4 α_5 α_6 α_7

1 2 3 4 3 2 2

In this case we have $\xi_1 = \kappa(\alpha_1)$, $\xi_2 = \kappa(\alpha_2)$, $d_1 = 2$, $d_2 = 32$, $d_3 = 32$, $d_3 = 2$ and $A_{123} = \dfrac{8}{9}$, $A_{234} = \dfrac{40}{9}$.

5) $\mathrm{SO}(2\ell)/(\mathrm{U}(p)\times \mathrm{U}(\ell-p))$ $(2 \le p \le \ell-2)$

α_1 α_2 $\cdots$ α_p $\cdots$ $\alpha_{\ell-2}$ $\alpha_{\ell-1}$ α_ℓ

1 2 2 2 1 1

In this case we have $\xi_1 = \kappa(\alpha_\ell)$, $\xi_2 = \kappa(\alpha_p)$, $d_1 = (\ell-p)(\ell-p-1)$, $d_2 = 2p(\ell-p)$, $d_3 = 2p(\ell-p)$, $d_4 = p(p-1)$ and $A_{123} = \dfrac{p(\ell-p)(\ell-p-1)}{2(\ell-1)}$, $A_{234} = \dfrac{p(p-1)(\ell-p)}{2(\ell-1)}$.

6) $\mathrm{Sp}(\ell)/(\mathrm{U}(p)\times\mathrm{U}(\ell-p))\quad(1\leq p\leq\ell-1)$

$$\begin{array}{cccccc}\alpha_1 & \alpha_2 & \cdots & \alpha_p & \cdots & \alpha_{\ell-1}\ \alpha_\ell\\ \circ\!-\!\!\!-\!\!\!- & \circ\!-\!\!\!- & \cdots & -\!\!\bullet\!\!- & \cdots & -\!\!\circ\!\Leftarrow\!\bullet\\ 2 & 2 & & 2 & & 2\quad 1\end{array}$$

In this case we have $\xi_1=\kappa(\alpha_\ell)$, $\xi_2=\kappa(\alpha_p)$, $d_1=(\ell-p)(\ell-p+1)$, $d_2=2p(\ell-p)$, $d_3=2p(\ell-p)$, $d_4=p(p+1)$ and $A_{123}=\dfrac{p(\ell-p)(\ell-p+1)}{2(\ell+1)}$, $A_{234}=\dfrac{p(p+1)(\ell-p)}{2(\ell+1)}$.

We see that, for a G-invariant metric $g=x_1(-B)|_{\mathfrak{m}_1}+x_2(-B)|_{\mathfrak{m}_2}+x_3(-B)|_{\mathfrak{m}_3}+x_4(-B)|_{\mathfrak{m}_4}$, the Ricci components r_1,r_2,r_3,r_4 for the metric g are given by

$$\begin{aligned}
r_1&=\frac{1}{2x_1}+\frac{1}{2d_1}A_{123}\left(\frac{x_1}{x_2x_3}-\frac{x_2}{x_1x_3}-\frac{x_3}{x_1x_2}\right),\\
r_2&=\frac{1}{2x_2}+\frac{1}{2d_2}A_{123}\left(\frac{x_2}{x_1x_3}-\frac{x_1}{x_2x_3}-\frac{x_3}{x_2x_1}\right)\\
&\qquad+\frac{1}{2d_2}A_{234}\left(\frac{x_2}{x_3x_4}-\frac{x_3}{x_2x_4}-\frac{x_4}{x_2x_3}\right),\\
r_3&=\frac{1}{2x_3}+\frac{1}{2d_3}A_{123}\left(\frac{x_3}{x_1x_2}-\frac{x_1}{x_3x_2}-\frac{x_2}{x_3x_1}\right)\\
&\qquad+\frac{1}{2d_3}A_{234}\left(\frac{x_3}{x_2x_4}-\frac{x_2}{x_3x_4}-\frac{x_4}{x_2x_3}\right),\\
r_4&=\frac{1}{2x_4}+\frac{1}{2d_4}A_{234}\left(\frac{x_4}{x_2x_3}-\frac{x_2}{x_3x_4}-\frac{x_3}{x_2x_4}\right).
\end{aligned}$$

Theorem 5.3. *The $\mathscr{B}$-metrics g on a generalized flag manifold G/K of type B_2 are Einstein.*

Proof. Assume that g is a $\mathscr{B}$-metric. Note that and $A_{123}\neq 0$ and $A_{234}\neq 0$. If ${x_1}^2-2x_1x_2-2x_1x_3+{x_2}^2-2x_2x_3+{x_3}^2\neq 0$, we see that $r_1=r_2=r_3$ from Corollary 3.3. Now we see that $r_4=r_2=r_3$ from Corollary 3.2. Similarly, if ${x_2}^2-2x_2x_3-2x_2x_4+{x_3}^2-2x_3x_4+{x_4}^2\neq 0$, we see that $r_1=r_2=r_3=r_4$. Thus $\mathscr{B}$-metrics g are Einstein in these cases.

We consider the cases

$$
{x_1}^2-2x_1x_2-2x_1x_3+{x_2}^2-2x_2x_3+{x_3}^2=0 \tag{9}
$$
$$
{x_2}^2-2x_2x_3-2x_2x_4+{x_3}^2-2x_3x_4+{x_4}^2=0. \tag{10}
$$

We normalize the metric g by $x_1 = 1$ and put $x_2 = t^2$. Then we have $x_3 = (1+t)^2$ or $x_3 = (1-t)^2$. By substituting these values into (10) and solving the equation for x_4, we obtain

$$x_1 = 1,\ x_2 = t^2,\ x_3 = (1-t)^2,\ x_4 = (2t-1)^2, \tag{11}$$

$$x_1 = 1,\ x_2 = t^2,\ x_3 = (1-t)^2,\ x_4 = 1, \tag{12}$$

$$x_1 = 1,\ x_2 = t^2,\ x_3 = (1+t)^2,\ x_4 = (2t+1)^2, \tag{13}$$

$$x_1 = 1,\ x_2 = t^2,\ x_3 = (1+t)^2,\ x_4 = 1. \tag{14}$$

From Theorem 3.2, we have

$$(x_2 + x_1 - x_3)r_2 + (x_3 + x_1 - x_2)r_3 - 2x_1 r_1 = 0. \tag{15}$$

We claim there are no real solutions t for these cases.

For the case 1) $\mathrm{SO}(2\ell+1)/(\mathrm{SO}(2\ell-3)\times\mathrm{U}(1)\times\mathrm{U}(1))$, by substituting (11), (12), (13) and (14) into (15) and taking the numerator of the equations, we see that

$$1 + 2\ell - 6t + 6t^2 = 0, \quad -3 + 2\ell - 2t + 2t^2 = 0,$$
$$1 + 2\ell + 6t + 6t^2 = 0, \quad -3 + 2\ell + 2t + 2t^2 = 0$$

respectively. Thus there are no real solutions t.

For the case 2) $\mathrm{SO}(2\ell)/(\mathrm{SO}(2\ell-4)\times\mathrm{U}(1)\times\mathrm{U}(1))$, by substituting (11), (12), (13) and (14) into (15) and taking the numerator of the equations, we see that

$$\ell - 3t + 3t^2 = 0, \quad -2 + \ell - t + t^2 = 0,$$
$$\ell + 3t + 3t^2 = 0, \quad -2 + \ell + t + t^2 = 0$$

respectively. Thus there are no real solutions t.

For the case 3) $\mathrm{E}_6/(\mathrm{SU}(5)\times\mathrm{U}(1)\times\mathrm{U}(1))$, by substituting (11), (12), (13) and (14) into (15) and taking the numerator of the equations, we see that

$$10 - 7t + 7t^2 = 0,\ 4 - t + t^2 = 0,\ 10 + 7t + 7t^2 = 0,\ 4 + t + t^2 = 0$$

respectively. Thus there are no real solutions t.

For the case 4) $\mathrm{E}_7/(\mathrm{SO}(10))\times\mathrm{U}(1)\times\mathrm{U}(1))$, by substituting (11), (12), (13) and (14) into (15) and taking the numerator of the equations, we see that

$$16 - 11t + 11t^2 = 0,\ 6 - t + t^2 = 0,\ 16 + 11t + 11t^2 = 0,\ 6 + t + t^2 = 0$$

respectively. Thus there are no real solutions t.

For the case 5) $\mathrm{SO}(2\ell)/(\mathrm{U}(p)\times\mathrm{U}(\ell-p))$ $(2\leq p\leq \ell-2)$, by substituting (11), (12), (13) and (14) into (15) and taking the numerator of the equations, we see that

$$\begin{aligned}
-4+\ell+4p-2(-3+\ell+p)t+2(-3+\ell+p)t^2&=0,\\
\ell-2(-1+\ell-p)t+2(-1+\ell-p)t^2&=0,\\
-4+\ell+4p+2(-3+\ell+p)t+2(-3+\ell+p)t^2&=0,\\
\ell+2(-1+\ell-p)t+2(-1+\ell-p)t^2&=0,
\end{aligned}$$

respectively. Since $\ell\geq p+2\geq 4$, we see there are no real solutions t.

For the case 6) $\mathrm{Sp}(\ell)/(\mathrm{U}(p)\times\mathrm{U}(\ell-p))$ $(1\leq p\leq \ell-1)$, by substituting (11), (12), (13) and (14) into (15) and taking the numerator of the equations, we see that

$$\begin{aligned}
4+\ell+4p-2(3+\ell+p)t+2(3+\ell+p)t^2&=0,\\
\ell-2(1+\ell-p)t+2(1+\ell-p)t^2&=0,\\
4+\ell+4p+2(3+\ell+p)t+2(3+\ell+p)t^2&=0,\\
\ell+2(1+\ell-p)t+2(1+\ell-p)t^2&=0,
\end{aligned}$$

respectively. Thus there are no real solutions t. □

5.4. *The case $r=2$ and $\nu=5$*

We have $\mathfrak{m}=\mathfrak{m}(1,0)\oplus\mathfrak{m}(0,1)\oplus\mathfrak{m}(1,1)\oplus\mathfrak{m}(0,2)\oplus\mathfrak{m}(1,2)=\mathfrak{m}_1\oplus\mathfrak{m}_2\oplus\mathfrak{m}_3\oplus\mathfrak{m}_4\oplus\mathfrak{m}_5$, up to isometry, that is, $\Delta_{\mathfrak{t}}^{+}=\{\xi_1,\xi_2,\xi_1+\xi_2,\ 2\xi_2,\ \xi_1+2\xi_2\}$ which is called of **type $\mathbf{BC_{2,1}}$** (cf. [11]).

We see there are 4 cases (cf. [6]).

1) $\mathrm{SO}(2\ell+1)/(\mathrm{U}(1)\times\mathrm{U}(p)\times\mathrm{SO}(2(\ell-p-1)+1))$ $(2\leq p\leq \ell-1)$

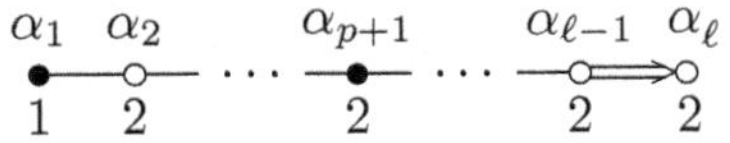

α_1 α_2 α_{p+1} $\alpha_{\ell-1}$ α_ℓ
1 2 2 2 2

In this case we have $\xi_1=\kappa(\alpha_1)$, $\xi_2=\kappa(\alpha_{p+1})$, $d_1=2p$, $d_2=2p(2\ell-2p-1)$, $d_3=2(2\ell-2p-1)$, $d_4=p(p-1)$, $d_5=2p$ and $A_{123}=\dfrac{(2\ell-2p-1)p}{2\ell-1}$, $A_{224}=\dfrac{(2\ell-2p-1)p(p-1)}{2\ell-1}$, $A_{235}=\dfrac{p(2l-2p-1)}{2\ell-1}$, $A_{145}=\dfrac{p(p-1)}{2\ell-1}$.

2) $\mathrm{SO}(2\ell)/(\mathrm{U}(1)\times\mathrm{U}(p)\times\mathrm{SO}(2(\ell-p-1)))$ $(2\leq p\leq \ell-4)$

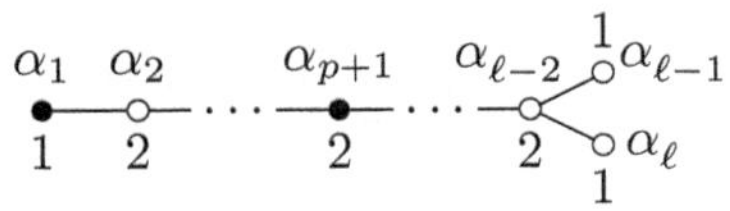

α_1 α_2 α_{p+1} $\alpha_{\ell-2}$ $\alpha_{\ell-1}$ α_ℓ
1 2 2 2 1 1

In this case we have $\xi_1 = \kappa(\alpha_1)$, $\xi_2 = \kappa(\alpha_{p+1})$, $d_1 = 2p$, $d_2 = 4p(\ell-p-1)$, $d_3 = 4(\ell-p-1)$, $d_4 = p(p-1)$, $d_5 = 2p$ and $A_{123} = \dfrac{(\ell-p-1)p}{\ell-1}$, $A_{224} = \dfrac{(\ell-p-1)p(p-1)}{\ell-1}$, $A_{235} = \dfrac{p(\ell-p-1)}{\ell-1}$, $A_{145} = \dfrac{p(p-1)}{2(\ell-1)}$.

3) $\mathrm{E}_6/(\mathrm{U}(1)\times\mathrm{SU}(4)\times\mathrm{U}(2))$

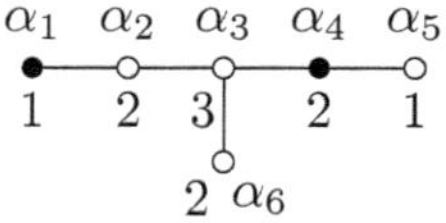

α_1 α_2 α_3 α_4 α_5
1 2 3 2 1
2 α_6

In this case we have $\xi_1 = \kappa(\alpha_1)$, $\xi_2 = \kappa(\alpha_4)$, $d_1 = 8$, $d_2 = 24$, $d_3 = 16$, $d_4 = 2$, $d_5 = 8$ and $A_{123} = 2$, $A_{224} = 1$, $A_{235} = 2$, $A_{145} = \dfrac{1}{3}$.

4) $\mathrm{E}_7/(\mathrm{SU}(6))\times\mathrm{U}(1)\times\mathrm{U}(1))$

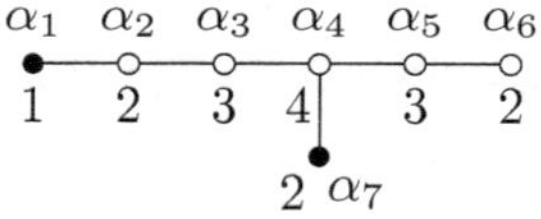

α_1 α_2 α_3 α_4 α_5 α_6
1 2 3 4 3 2
2 α_7

In this case we have $\xi_1 = \kappa(\alpha_1)$, $\xi_2 = \kappa(\alpha_7)$, $d_1 = 12$, $d_2 = 40$, $d_3 = 30$, $d_4 = 2$, $d_5 = 12$ and $A_{123} = \dfrac{10}{3}$, $A_{224} = \dfrac{10}{9}$, $A_{235} = \dfrac{10}{3}$, $A_{145} = \dfrac{1}{3}$.

We see that, for a G-invariant metric $g = x_1(-B)|_{\mathfrak{m}_1} + x_2(-B)|_{\mathfrak{m}_2} + x_3(-B)|_{\mathfrak{m}_3} + x_4(-B)|_{\mathfrak{m}_4} + x_5(-B)|_{\mathfrak{m}_5}$, the Ricci components r_1, r_2, r_3, r_4, r_5 for the metric g are given by

$$
\begin{aligned}
r_1 &= \frac{1}{2x_1} + \frac{1}{2d_1}A_{123}\left(\frac{x_1}{x_2x_3} - \frac{x_2}{x_1x_3} - \frac{x_3}{x_1x_2}\right)\\
&\quad + \frac{1}{2d_1}A_{145}\left(\frac{x_1}{x_4x_5} - \frac{x_4}{x_1x_5} - \frac{x_5}{x_1x_4}\right)\\
r_2 &= \frac{1}{2x_2} + \frac{1}{2d_2}A_{123}\left(\frac{x_2}{x_1x_3} - \frac{x_1}{x_2x_3} - \frac{x_3}{x_2x_1}\right) - \frac{1}{2d_2}A_{224}\frac{x_4}{{x_2}^2}\\
&\quad + \frac{1}{2d_2}A_{235}\left(\frac{x_2}{x_3x_5} - \frac{x_3}{x_2x_5} - \frac{x_5}{x_2x_3}\right),\\
r_3 &= \frac{1}{2x_3} + \frac{1}{2d_3}A_{123}\left(\frac{x_3}{x_1x_2} - \frac{x_1}{x_3x_2} - \frac{x_2}{x_3x_1}\right)\\
&\quad + \frac{1}{2d_3}A_{235}\left(\frac{x_3}{x_2x_5} - \frac{x_2}{x_3x_5} - \frac{x_5}{x_2x_3}\right),
\end{aligned}
$$

$$r_4 = \frac{1}{2x_4} + \frac{1}{2d_4} A_{145} \left(\frac{x_4}{x_1 x_5} - \frac{x_1}{x_4 x_5} - \frac{x_5}{x_1 x_4} \right) + \frac{1}{4d_4} A_{224} \left(-\frac{2}{x_4} + \frac{x_4}{{x_2}^2} \right),$$

$$r_5 = \frac{1}{2x_5} + \frac{1}{2d_5} A_{145} \left(\frac{x_5}{x_1 x_4} - \frac{x_1}{x_4 x_5} - \frac{x_4}{x_5 x_1} \right) + \frac{1}{2d_5} A_{235} \left(\frac{x_5}{x_2 x_3} - \frac{x_3}{x_2 x_5} - \frac{x_2}{x_3 x_5} \right).$$

Theorem 5.4. *The $\mathscr{B}$-metrics g on a generalized flag manifold G/K of type $BC_{2,1}$ are Einstein.*

Proof. Assume that g is a $\mathscr{B}$-metric. Note that $A_{224} \neq 0$. Thus, from Corollary 3.1, we have $r_2 = r_4$. Note that $A_{123} \neq 0$, $A_{145} \neq 0$ and $A_{235} \neq 0$. If ${x_1}^2 - 2x_1x_2 - 2x_1x_3 + {x_2}^2 - 2x_2x_3 + {x_3}^2 \neq 0$, we see that $r_1 = r_2 = r_3$ from Corollary 3.3. Now we see that $r_5 = r_1 = r_4$ from $A_{145} \neq 0$ and Corollary 3.2. Similarly, if ${x_1}^2 - 2x_1x_4 - 2x_1x_5 + {x_4}^2 - 2x_4x_5 + {x_5}^2 \neq 0$, or if ${x_2}^2 - 2x_2x_3 - 2x_2x_5 + {x_3}^2 - 2x_3x_5 + {x_5}^2 \neq 0$, we see that $r_1 = r_2 = r_3 = r_4 = r_5$. Thus $\mathscr{B}$-metrics g are Einstein in these cases.

We consider the cases

$${x_1}^2 - 2x_1x_2 - 2x_1x_3 + {x_2}^2 - 2x_2x_3 + {x_3}^2 = 0, \tag{16}$$

$${x_2}^2 - 2x_2x_3 - 2x_2x_5 + {x_3}^2 - 2x_3x_5 + {x_5}^2 = 0. \tag{17}$$

$${x_1}^2 - 2x_1x_4 - 2x_1x_5 + {x_4}^2 - 2x_4x_5 + {x_5}^2 = 0. \tag{18}$$

We normalize the metric g by $x_1 = 1$ and put $x_2 = t^2$. Then, from the equation (16), we obtain $x_3 = (1+t)^2$ or $x_3 = (1-t)^2$. By substituting these values into (17) and solving the equation for x_5, we obtain

$$x_1 = 1,\ x_2 = t^2,\ x_3 = (1-t)^2,\ x_5 = (2t-1)^2, \tag{19}$$

$$x_1 = 1,\ x_2 = t^2,\ x_3 = (1-t)^2,\ x_5 = 1, \tag{20}$$

$$x_1 = 1,\ x_2 = t^2,\ x_3 = (1+t)^2,\ x_5 = (2t+1)^2, \tag{21}$$

$$x_1 = 1,\ x_2 = t^2,\ x_3 = (1+t)^2,\ x_5 = 1. \tag{22}$$

By substituting these values into (18) and solving the equation for x_4, we obtain six solutions:

$$x_1 = 1,\ x_2 = t^2,\ x_3 = (1-t)^2,\ x_4 = 4t^2,\ x_5 = (2t-1)^2, \tag{23}$$

$$x_1 = 1,\ x_2 = t^2,\ x_3 = (1-t)^2,\ x_4 = 4(t-1)^2,\ x_5 = (2t-1)^2, \tag{24}$$

$$x_1 = 1,\ x_2 = t^2,\ x_3 = (1-t)^2,\ x_4 = 4,\ x_5 = 1, \tag{25}$$

$$x_1 = 1,\ x_2 = t^2,\ x_3 = (1+t)^2,\ x_4 = 4t^2,\ x_5 = (2t+1)^2, \tag{26}$$
$$x_1 = 1,\ x_2 = t^2,\ x_3 = (1+t)^2,\ x_4 = 4(t+1)^2,\ x_5 = (2t+1)^2, \tag{27}$$
$$x_1 = 1,\ x_2 = t^2,\ x_3 = (1+t)^2,\ x_4 = 4,\ x_5 = 1. \tag{28}$$

From Theorem 3.2, we have

$$(x_2 + x_1 - x_3)r_2 + (x_3 + x_1 - x_2)r_3 - 2x_1r_1 = 0, \tag{29}$$
$$(x_4 + x_1 - x_5)r_4 + (x_5 + x_1 - x_4)r_5 - 2x_1r_1 = 0, \tag{30}$$
$$(x_3 + x_2 - x_5)r_3 + (x_5 + x_2 - x_3)r_5 - 2x_2r_2 = 0, \tag{31}$$
$$r_2 - r_4 = 0. \tag{32}$$

We claim there are no real solutions t for these cases.

We consider the cases 1) and 2) together by putting $m = 2\ell + 1$ and $m = 2\ell$ respectively. By substituting the values (23) into equations (29), (30), (31), (32), and by taking numerator of the equations, we obtain

$$\begin{aligned}
f_1 &= -3 + m - p - 2pt + t^2 + pt^2 = 0 \\
f_2 &= 1 + p + 2t - 2pt + 2t^2 + 2pt^2 = 0, \\
f_3 &= -3 + m - p + 12t - 4mt + 6pt - 11t^2 + 4mt^2 - 7pt^2 = 0, \\
f_4 &= 5 - 2m + 3p = 0.
\end{aligned}$$

Now we consider f_1, f_2, f_3, f_4 as polynomials of t, m, p with integer coefficients. We denote by J_1 the ideal of the polynomial ring $\mathbb{Q}[t, m, p]$ defined by $\{f_1, f_2, f_3, f_4\}$. We take a lexicographic order $>$ with $t > m > p$. By computing a Gröbner basis of J_1, we see that the Gröbner basis is given by

$$\{(-1+p)(1+p),\ -5 + 2m - 3p,\ -1 + p + 4t\}.$$

Hence, for positive integers $p \geq 2$, there are no solutions t for a system of equations $f_1 = 0,\ f_2 = 0,\ f_3 = 0,\ f_4 = 0$.

Similarly, by substituting the values (24) into equations (29), (30), (31), (32), and by taking numerator of the equations, we obtain

$$\begin{aligned}
f_1 &= 2 - 2p - 10t + 10pt + 18t^2 - mt^2 - 14pt^2 - 14t^3 + 2mt^3 \\
&\quad + 8pt^3 - t^4 - 5pt^4 + 2t^5 + 2pt^5 = 0, \\
f_2 &= -8 + 4m - 8p + 48t - 24mt + 48pt - 112t^2 + 56mt^2 \\
&\quad - 112pt^2 + 128t^3 - 64mt^3 + 128pt^3 - 61t^4 + 32mt^4 \\
&\quad - 63pt^4 - 6t^5 - 2pt^5 + 2t^6 + 2pt^6 = 0, \\
f_3 &= -2 + 2p + 10t - 10pt - 18t^2 + mt^2 + 14pt^2 + 14t^3 \\
&\quad - 4mt^3 + 4pt^3 + t^4 + 4mt^4 - 19pt^4 - 8t^5 + 8pt^5 = 0,
\end{aligned}$$

$$
\begin{aligned}
f_4 = &-12 + 4m - 4p + 72t - 24mt + 24pt - 160t^2 + 54mt^2 \\
&- 56pt^2 + 160t^3 - 56mt^3 + 64pt^3 - 53t^4 + 26mt^4 \\
&- 35pt^4 - 22t^5 - 4mt^5 + 6pt^5 + 16t^6 = 0.
\end{aligned}
$$

As above, we consider f_1, f_2, f_3, f_4 as polynomials of t, m, p with integer coefficients. We denote by J_2 the ideal of the polynomial ring $\mathbb{Q}[t, m, p]$ defined by $\{f_1, f_2, f_3, f_4\}$. We take a lexicographic order $>$ with $t > m > p$. By computing a Gröbner basis of J_2, we see that the Gröbner basis is given by

$$\{-1 + p, \ -4 + m, \ (-1 + t)t^3\}.$$

Hence, for positive integers $p \geq 2$, there are no solutions t for a system of equations $f_1 = 0, \ f_2 = 0, \ f_3 = 0, \ f_4 = 0$.

Now, by substituting the values (25) into equations (29), (30), (31), (32), and by taking numerator of the equations, we obtain

$$
\begin{aligned}
f_1 &= -2 + 2p + 2t - 2pt + mt^2 - 6pt^3 + 2t^4 + 4pt^4 = 0, \\
f_2 &= 8 - 4m + 8p - 8t + 4mt - 8pt - 8t^3 + 4mt^3 - 8pt^3 \\
&\quad - 3t^4 - pt^4 + 3t^5 + pt^5 = 0, \\
f_3 &= 2 - 2p - 2t + 2pt - mt^2 + 6pt^3 - 2t^4 - 4pt^4 - 0, \\
f_4 &= -12 + 4m - 4p + 12t - 4mt + 4pt + 8t^2 - 2mt^2 \\
&\quad - 8t^3 + 2mt^3 + 15t^4 + pt^4 - 7t^5 - pt^5 = 0.
\end{aligned}
$$

As above, we consider f_1, f_2, f_3, f_4 as polynomials of t, m, p with integer coefficients. We denote by J_3 the ideal of the polynomial ring $\mathbb{Q}[t, m, p]$ defined by $\{f_1, f_2, f_3, f_4\}$. We take a lexicographic order $>$ with $t > m > p$. By computing a Gröbner basis of J_3, we see that the Gröbner basis is contained a polynomial $g_3(p)$ given by

$$
\begin{aligned}
g_3(p) &= (-1 + p)(-333943 - 1933145p + 134500p^2 + 11933980p^3 \\
&\qquad + 9759086p^4 + 19640434p^5 + 67008468p^6 \\
&\qquad + 72103980p^7 + 23401937p^8 + 217503p^9) \\
&= (-1 + p)\{201932800 + 1267328000(-1 + p) \\
&\qquad + 3473290240(-1 + p)^2 + 5439962048(-1 + p)^3 \\
&\qquad + 5302268544(-1 + p)^4 + 3273788672(-1 + p)^5 \\
&\qquad + 1245260816(-1 + p)^6 + 267149584(-1 + p)^7 \\
&\qquad + 25359464(-1 + p)^8 + 217503(-1 + p)^9\}.
\end{aligned}
$$

Thus we see that $g_3(p) > 0$ for $p \geq 2$, and hence, there are no solutions t for a system of equations $f_1 = 0$, $f_2 = 0$, $f_3 = 0$, $f_4 = 0$ for $p \geq 2$.

By the same way, we see there are no solutions t for a system of equations for the cases (26), (27), (28). Thus we obtain our claim for the cases 1) and 2).

We consider the case 3). By substituting the values (23) or (26) into the equations (32), we obtain

$$r_2 - r_4 = 13/(48t^2) \neq 0.$$

Thus we obtain our claim.

By substituting the values (24) into the equations (29), (31), and by taking numerator of the equations, we obtain the system of equations:

$$-2 + 10t - 22t^2 + 28t^3 - 17t^4 + 10t^5 = 0,$$
$$2 - 10t + 22t^2 - 24t^3 + 9t^4 + 8t^5 = 0.$$

By adding these equations, we have $2t^3(2 - 4t + 9t^2) = 0$, and hence, we see there are no real solutions for the system. Thus we obtain our claim. By the same way, for the values (25), (27), (28), we see that there are no real solutions for the system. Thus we obtain our claim.

We consider the case 4). By substituting the values (23) or (26) into the equations (32), we obtain

$$r_2 - r_4 = 7/(24t^2) \neq 0.$$

Thus we obtain our claim.

By substituting the values (24) into the equations (29), (31), and by taking numerator of the equations, we obtain the system of equations:

$$-2 + 10t - 26t^2 + 38t^3 - 23t^4 + 14t^5 = 0,$$
$$2 - 10t + 26t^2 - 38t^3 + 23t^4 + 8t^5 = 0.$$

By adding these equations, we have $22t^5 = 0$, and hence, we see there are no real solutions for the system. Thus we obtain our claim.

By substituting the values (25) into the equations (29), (30), and by taking numerator of the equations, we obtain the system of equations:

$$1 - t + 12t^2 - 12t^3 + 11t^4 = 0,$$
$$-40 + 40t + 40t^3 - 13t^4 + 13t^5 = 0.$$

By computing the resultant of these polynomials, we see the resultant is non-zero. Thus we see there are no real solutions for the system. Thus we obtain our claim. By the same way, for the values (27), (28), we see there are no real solutions for the system. Thus we obtain our claim. □

5.5. *The case $r = 2$, $\nu = 6$ and G_2-type*

We have $\mathfrak{m} = \mathfrak{m}(1,0) \oplus \mathfrak{m}(0,1) \oplus \mathfrak{m}(1,1) \oplus \mathfrak{m}(1,2) \oplus \mathfrak{m}(1,3) \oplus \mathfrak{m}(2,3) = \mathfrak{m}_1 \oplus \mathfrak{m}_2 \oplus \mathfrak{m}_3 \oplus \mathfrak{m}_4 \oplus \mathfrak{m}_5 \oplus \mathfrak{m}_6$, that is, $\Delta_{\mathfrak{t}}^+ = \{\,\xi_1, \xi_2,\ \xi_1 + \xi_2,\ \xi_1 + 2\xi_2,\ \xi_1 + 3\xi_2,\ 2\xi_1 + 3\xi_2\,\}$, which is called of **type $\mathbf{G}_2$**.

We see there are 5 cases (cf. [5, 7]).

1) $G_2/(\mathrm{U}(1)\times\mathrm{U}(1))$

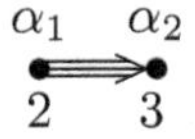

α_1 α_2
2 3

In this case we have $\xi_1 = \kappa(\alpha_1)$, $\xi_2 = \kappa(\alpha_2)$, $d_1 = d_2 = d_3 = d_4 = d_5 = d_6 = 2$ and $A_{123} = \frac{1}{4}$, $A_{156} = \frac{1}{4}$, $A_{234} = \frac{1}{3}$, $A_{245} = \frac{1}{4}$, $A_{346} = \frac{1}{4}$.

2) $\mathrm{F}_4/(U(3)\times U(1))$

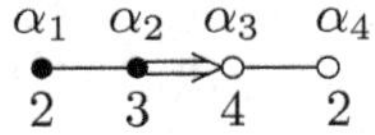

α_1 α_2 α_3 α_4
2 3 4 2

In this case we have $\xi_1 = \kappa(\alpha_1)$, $\xi_2 = \kappa(\alpha_2)$, $d_1 = 2$, $d_2 = d_3 = d_4 = 12$, $d_5 = 2$, $d_6 = 2$ and $A_{123} = \frac{2}{3}$, $A_{156} = \frac{1}{9}$, $A_{234} = 2$, $A_{245} = \frac{2}{3}$, $A_{346} = \frac{2}{3}$.

3) $\mathrm{E}_6/(\mathrm{U}(3)\times\mathrm{U}(3))$

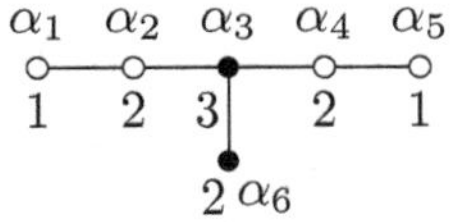

α_1 α_2 α_3 α_4 α_5
1 2 3 2 1
2 α_6

In this case we have $\xi_1 = \kappa(\alpha_6)$, $\xi_2 = \kappa(\alpha_3)$, $d_1 = 2$, $d_2 = d_3 = d_4 = 18$, $d_5 = 2$, $d_6 = 2$ and $A_{123} = \frac{3}{4}$, $A_{156} = \frac{1}{12}$, $A_{234} = 3$, $A_{245} = \frac{3}{4}$, $A_{346} = \frac{3}{4}$.

4) $\mathrm{E}_7/(\mathrm{U}(6)\times \mathrm{U}(1))$

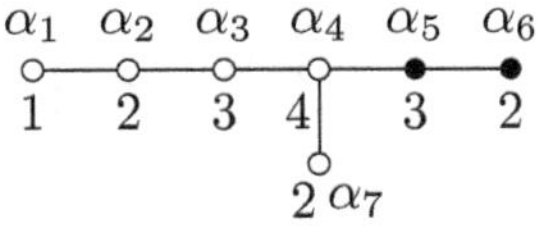

α_1 α_2 α_3 α_4 α_5 α_6
1 2 3 4 3 2
2 α_7

In this case we have $\xi_1 = \kappa(\alpha_6)$, $\xi_2 = \kappa(\alpha_5)$, $d_1 = 2$, $d_2 = d_3 = d_4 = 30$, $d_5 = 2$, $d_6 = 2$ and $A_{123} = \frac{5}{6}$, $A_{156} = \frac{1}{18}$, $A_{234} = 5$, $A_{245} = \frac{5}{6}$, $A_{346} = \frac{5}{6}$.

5) $\mathrm{E}_8/(\mathrm{E}_6\times\mathrm{U}(1)\times\mathrm{U}(1))$

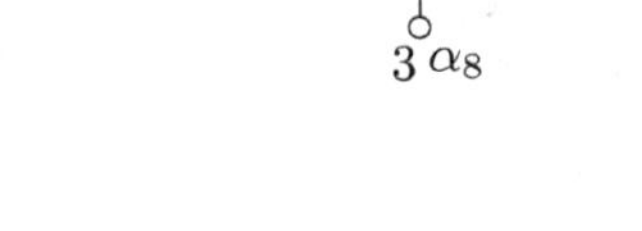

α_1 α_2 α_3 α_4 α_5 α_6 α_7
2 3 4 5 6 4 2
3 α_8

In this case we have $\xi_1 = \kappa(\alpha_1)$, $\xi_2 = \kappa(\alpha_2)$, $d_1 = 2$, $d_2 = d_3 = d_4 = 54$, $d_5 = d_6 = 2$ and $A_{123} = \dfrac{9}{10}$, $A_{156} = \dfrac{1}{30}$, $A_{234} = 9$, $A_{245} = \dfrac{9}{10}$, $A_{346} = \dfrac{9}{10}$.

By putting $a = \dfrac{1}{3}$ for the case 1), $a = 2$ for the case 2), $a = 3$ for the case 3), $a = 5$ for the case 4), $a = 9$ for the case 5), we have $d_1 = 2$, $d_2 = d_3 = d_4 = 6a$ and $A_{123} = \dfrac{a}{a+1}$, $A_{156} = \dfrac{1}{3(a+1)}$, $A_{234} = a$, $A_{245} = \dfrac{a}{a+1}$, $A_{346} = \dfrac{a}{a+1}$.

We see that, for a G-invariant metric $g = x_1(-B)|_{\mathfrak{m}_1} + x_2(-B)|_{\mathfrak{m}_2} + x_3(-B)|_{\mathfrak{m}_3} + x_4(-B)|_{\mathfrak{m}_4} + x_5(-B)|_{\mathfrak{m}_5} + x_6(-B)|_{\mathfrak{m}_6}$, the Ricci components $r_1, r_2, r_3, r_4, r_5, r_6$ for the metric g are given by

$$
\begin{aligned}
r_1 &= \frac{1}{2x_1} + \frac{1}{2d_1}A_{123}\left(\frac{x_1}{x_2x_3} - \frac{x_2}{x_1x_3} - \frac{x_3}{x_1x_2}\right)\\
&\quad + \frac{1}{2d_1}A_{156}\left(\frac{x_1}{x_5x_6} - \frac{x_5}{x_1x_6} - \frac{x_6}{x_1x_5}\right)\\
r_2 &= \frac{1}{2x_2} + \frac{1}{2d_2}A_{123}\left(\frac{x_2}{x_1x_3} - \frac{x_1}{x_2x_3} - \frac{x_3}{x_2x_1}\right)\\
&\quad + \frac{1}{2d_2}A_{234}\left(\frac{x_2}{x_3x_4} - \frac{x_3}{x_2x_5} - \frac{x_4}{x_2x_3}\right) + \frac{1}{2d_2}A_{245}\left(\frac{x_2}{x_4x_5} - \frac{x_4}{x_2x_5} - \frac{x_5}{x_2x_4}\right),\\
r_3 &= \frac{1}{2x_3} + \frac{1}{2d_3}A_{123}\left(\frac{x_3}{x_1x_2} - \frac{x_1}{x_3x_2} - \frac{x_2}{x_3x_1}\right)\\
&\quad + \frac{1}{2d_3}A_{234}\left(\frac{x_3}{x_2x_4} - \frac{x_2}{x_3x_4} - \frac{x_4}{x_2x_3}\right) + \frac{1}{2d_3}A_{346}\left(\frac{x_3}{x_4x_6} - \frac{x_6}{x_3x_4} - \frac{x_4}{x_3x_6}\right),\\
r_4 &= \frac{1}{2x_4} + \frac{1}{2d_4}A_{234}\left(\frac{x_4}{x_2x_3} - \frac{x_2}{x_4x_3} - \frac{x_3}{x_4x_2}\right)\\
&\quad + \frac{1}{4d_4}A_{245}\left(\frac{x_4}{x_2x_5} - \frac{x_2}{x_4x_5} - \frac{x_5}{x_4x_2}\right) + \frac{1}{4d_4}A_{346}\left(\frac{x_4}{x_3x_6} - \frac{x_3}{x_4x_6} - \frac{x_6}{x_4x_3}\right),\\
r_5 &= \frac{1}{2x_5} + \frac{1}{2d_5}A_{156}\left(\frac{x_5}{x_1x_6} - \frac{x_1}{x_5x_6} - \frac{x_6}{x_5x_1}\right)\\
&\quad + \frac{1}{2d_5}A_{245}\left(\frac{x_5}{x_2x_4} - \frac{x_4}{x_5x_2} - \frac{x_2}{x_5x_4}\right),\\
r_6 &= \frac{1}{2x_6} + \frac{1}{2d_6}A_{156}\left(\frac{x_6}{x_1x_5} - \frac{x_1}{x_6x_5} - \frac{x_4}{x_6x_1}\right)\\
&\quad + \frac{1}{2d_6}A_{346}\left(\frac{x_6}{x_3x_4} - \frac{x_3}{x_6x_4} - \frac{x_4}{x_6x_3}\right).
\end{aligned}
$$

Theorem 5.5. *The $\mathscr{B}$-metrics g on a generalized flag manifold G/K of type G_2 are Einstein.*

Proof. Assume that g is a $\mathscr{B}$-metric. Note that $A_{123} \neq 0$, $A_{234} \neq 0$, $A_{245} \neq 0$ and $A_{346} \neq 0$. If ${x_1}^2 - 2x_1x_2 - 2x_1x_3 + {x_2}^2 - 2x_2x_3 + {x_3}^2 \neq 0$, we see that $r_1 = r_2 = r_3$ from Corollary 3.3. Now we see that $r_4 = r_2 = r_3$ from $A_{234} \neq 0$ and Corollary 3.2. Thus we see that $r_5 = r_2 = r_4$ and $r_6 = r_3 = r_4$ from $A_{245} \neq 0$, $A_{346} \neq 0$ and Corollary 3.2. Similarly, if ${x_2}^2 - 2x_2x_3 - 2x_2x_4 + {x_3}^2 - 2x_3x_4 + {x_4}^2 \neq 0$, if ${x_2}^2 - 2x_2x_4 - 2x_2x_5 + {x_4}^2 - 2x_4x_5 + {x_5}^2 \neq 0$ or if ${x_3}^2 - 2x_3x_4 - 2x_3x_6 + {x_4}^2 - 2x_4x_6 + {x_6}^2 \neq 0$, we see that $r_1 = r_2 = r_3 = r_4 = r_5 = r_6$. Thus $\mathscr{B}$-metrics g are Einstein in these cases.

We consider the cases

$$
\begin{aligned}
&{x_1}^2 - 2x_1x_2 - 2x_1x_3 + {x_2}^2 - 2x_2x_3 + {x_3}^2 = 0, && (33)\\
&{x_2}^2 - 2x_2x_3 - 2x_2x_4 + {x_3}^2 - 2x_3x_4 + {x_4}^2 = 0, && (34)\\
&{x_2}^2 - 2x_2x_4 - 2x_2x_5 + {x_4}^2 - 2x_4x_5 + {x_5}^2 = 0, && (35)\\
&{x_3}^2 - 2x_3x_4 - 2x_3x_6 + {x_4}^2 - 2x_4x_6 + {x_6}^2 = 0. && (36)
\end{aligned}
$$

We normalize the metric g by $x_1 = 1$ and put $x_2 = t^2$. Then, from the equation (33), we obtain $x_3 = (1+t)^2$ or $x_3 = (1-t)^2$. By substituting these values into (34) and solving the equation for x_4, we obtain

$$
\begin{aligned}
&x_1 = 1,\ x_2 = t^2,\ x_3 = (1-t)^2,\ x_4 = (2t-1)^2, && (37)\\
&x_1 = 1,\ x_2 = t^2,\ x_3 = (1-t)^2,\ x_4 = 1, && (38)\\
&x_1 = 1,\ x_2 = t^2,\ x_3 = (1+t)^2,\ x_4 = (2t+1)^2, && (39)\\
&x_1 = 1,\ x_2 = t^2,\ x_3 = (1+t)^2,\ x_4 = 1. && (40)
\end{aligned}
$$

By substituting these values into (35) and (36), and solving the equations for x_5 and x_6, we obtain 16 solutions:

$$
\begin{aligned}
&x_1 = 1,\ x_2 = t^2,\ x_3 = (-1+t)^2,\ x_4 = 1,\ x_5 = (-1+t)^2,\ x_6 = (-2+t)^2,\\
&x_1 = 1,\ x_2 = t^2,\ x_3 = (-1+t)^2,\ x_4 = 1,\ x_5 = (-1+t)^2,\ x_6 = t^2,\\
&x_1 = 1,\ x_2 = t^2,\ x_3 = (-1+t)^2,\ x_4 = 1,\ x_5 = (1+t)^2,\ x_6 = (-2+t)^2,\\
&x_1 = 1,\ x_2 = t^2,\ x_3 = (-1+t)^2,\ x_4 = 1,\ x_5 = (1+t)^2,\ x_6 = t^2,\\
&x_1 = 1,\ x_2 = t^2,\ x_3 = (-1+t)^2,\ x_4 = (-1+2t)^2,\ x_5 = (-1+t)^2,\ x_6 = t^2,\\
&x_1 = 1,\ x_2 = t^2,\ x_3 = (-1+t)^2,\ x_4 = (-1+2t)^2,\ x_5 = (-1+t)^2,\\
&\qquad x_6 = (-2+3t)^2,\\
&x_1 = 1,\ x_2 = t^2,\ x_3 = (-1+t)^2,\ x_4 = (-1+2t)^2,\ x_5 = (-1+3t)^2,\ x_6 = t^2,
\end{aligned}
$$

$x_1 = 1$, $x_2 = t^2$, $x_3 = (-1+t)^2$, $x_4 = (-1+2t)^2$, $x_5 = (-1+3t)^2$,
$x_6 = (-2+3t)^2$
$x_1 = 1$, $x_2 = t^2$, $x_3 = (1+t)^2$, $x_4 = 1$, $x_5 = (-1+t)^2$, $x_6 = t^2$,
$x_1 = 1$, $x_2 = t^2$, $x_3 = (1+t)^2$, $x_4 = 1$, $x_5 = (-1+t)^2$, $x_6 = (2+t)^2$,
$x_1 = 1$, $x_2 = t^2$, $x_3 = (1+t)^2$, $x_4 = 1$, $x_5 = (1+t)^2$, $x_6 = t^2$,
$x_1 = 1$, $x_2 = t^2$, $x_3 = (1+t)^2$, $x_4 = 1$, $x_5 = (1+t)^2$, $x_6 = (2+t)^2$,
$x_1 = 1$, $x_2 = t^2$, $x_3 = (1+t)^2$, $x_4 = (1+2t)^2$, $x_5 = (1+t)^2$, $x_6 = t^2$,
$x_1 = 1$, $x_2 = t^2$, $x_3 = (1+t)^2$, $x_4 = (1+2t)^2$, $x_5 = (1+t)^2$, $x_6 = (2+3t)^2$,
$x_1 = 1$, $x_2 = t^2$, $x_3 = (1+t)^2$, $x_4 = (1+2t)^2$, $x_5 = (1+3t)^2$, $x_6 = t^2$,
$x_1 = 1$, $x_2 = t^2$, $x_3 = (1+t)^2$, $x_4 = (1+2t)^2$, $x_5 = (1+3t)^2$,
$x_6 = (2+3t)^2$.

Noting that $A_{234} \neq 0$, we have

$$(x_3 + x_2 - x_4)r_3 + (x_4 + x_2 - x_3)r_4 - 2x_2r_2 = 0, \tag{41}$$

by Theorem 3.2. We claim there are no real solutions t the solutions above.

By substituting the solutions above into the equation (41), we obtain equations:

$$\frac{2(a+4)\left(t^2-t+1\right)}{3(a+1)(t-1)} = 0, \qquad \frac{2(a+6)(t^2-t+1)}{3(a+1)(t-1)} = 0,$$

$$\frac{2(a+2)\left(t^2-t+1\right)}{3(a+1)(t-1)} = 0, \qquad \frac{2(a+4)\left(t^2-t+1\right)}{3(a+1)(t-1)} = 0,$$

$$\frac{2(a+4)\left(3t^2-3t+1\right)}{3(a+1)(t-1)(2t-1)} = 0, \qquad \frac{2(a+2)\left(3t^2-3t+1\right)}{3(a+1)(t-1)(2t-1)} = 0,$$

$$\frac{2(a+2)\left(3t^2-3t+1\right)}{3(a+1)(t-1)(2t-1)} = 0, \qquad \frac{2a\left(3t^2-3t+1\right)}{3(a+1)(t-1)(2t-1)} = 0,$$

$$\frac{2(a+4)\left(t^2+t+1\right)}{3(a+1)(t+1)} = 0, \qquad \frac{2(a+2)\left(t^2+t+1\right)}{3(a+1)(t+1)} = 0,$$

$$\frac{2(a+6)\left(t^2+t+1\right)}{3(a+1)(t+1)} = 0, \qquad \frac{2(a+4)\left(t^2+t+1\right)}{3(a+1)(t+1)} = 0,$$

$$\frac{2(a+4)\left(3t^2+3t+1\right)}{3(a+1)(t+1)(2t+1)} = 0, \qquad \frac{2(a+2)\left(3t^2+3t+1\right)}{3(a+1)(t+1)(2t+1)} = 0,$$

$$\frac{2(a+2)\left(3t^2+3t+1\right)}{3(a+1)(t+1)(2t+1)} = 0, \qquad \frac{2a\left(3t^2+3t+1\right)}{3(a+1)(t+1)(2t+1)} = 0,$$

respectively. Thus we see there are no real solutions t and obtained our claim. □

Appendix

The notion of a generalized Wallach space is introduced by Nikonorov in [15]. As in Section 2, we have a reductive decomposition $\mathfrak{g} = \mathfrak{k} \oplus \mathfrak{m}$ of $\mathfrak{g}$ with respect to negative of Killing form B.

A homogeneous space G/K is called a generalized Wallach space, if it has the following property:

The modules $\mathfrak{m}$ is decomposed as a direct sum of three $\mathrm{Ad}(K)$-invariant irreducible modules pairwise orthogonal with respect to $-B$:

$$\mathfrak{m} = \mathfrak{m}_1 \oplus \mathfrak{m}_2 \oplus \mathfrak{m}_3, \tag{42}$$

such that

$$[\mathfrak{m}_i,\ \mathfrak{m}_i] \subset \mathfrak{k} \quad \text{for} \quad i \in \{1, 2, 3\}. \tag{43}$$

Note that we have

$$[\mathfrak{m}_j,\ \mathfrak{m}_k] \subset \mathfrak{m}_i \tag{44}$$

for pairwise distinct i, j, k.

Every generalized Wallach space admits a 3-parameter family of invariant metrics:

$$g = x_1(-B)|_{\mathfrak{m}_1} + x_1(-B)|_{\mathfrak{m}_1} + x_3(-B)|_{\mathfrak{m}_3}, \tag{45}$$

for positive real numbers $(x_1, x_2, x_3) \in \mathbb{R}^3_+$. Note that, from (44), we see that the Ricci tensor r of the invariant metric (45) is also diagonal. Thus we can deal with these cases by the same method as in Section 5.2.

Due to Nikonorov [15] Table 1, we have 15 cases. F. Li, H. Chen and Z. Chen [14] proved that $\mathscr{B}$-metrics g on a generalized Wallach space of exceptional type are Einstein.

Theorem 5.6. *The $\mathscr{B}$-metrics g on a generalized Wallach space G/K are Einstein.*

Proof. Assume that g is a $\mathscr{B}$-metric. From Corollary 3.3, if ${x_1}^2 - 2x_1x_2 - 2x_1x_3 + {x_2}^2 - 2x_2x_3 + {x_3}^2 \neq 0$, we see that $r_1 = r_2 = r_3$ and hence, g is Einstein in these cases. We consider the case

$${x_1}^2 - 2x_1x_2 - 2x_1x_3 + {x_2}^2 - 2x_2x_3 + {x_3}^2 = 0.$$

We normalize the metric g by $x_1 = 1$ and put $x_2 = t^2$. Then we see that

$$x_1 = 1,\ x_2 = t^2,\ x_3 = (1+t)^2, \tag{46}$$

$$x_1 = 1,\ x_2 = t^2,\ x_3 = (1-t)^2. \tag{47}$$

From Theorem 3.2, we have

$$(x_2 + x_1 - x_3)r_2 + (x_3 + x_1 - x_2)r_3 - 2x_1 r_1 = 0. \tag{48}$$

We claim there are no real solutions t of the equations (8) by the same way as in §5.2.

1) $\mathrm{SO}(k+\ell+m)/(\mathrm{SO}(k)\times\mathrm{SO}(\ell)\times\mathrm{SO}(m))$.

In this case we have $d_1 = k\ell$, $d_2 = km$, $d_3 = \ell m$ and $A_{123} = \dfrac{k\ell m}{2(k+\ell+m-2)}$. By substituting the solutions (46), (47) into the equation (8), we see that the numerators of equations (8) are given by

$$(2+k+\ell) + 2(1+\ell)t + (2+\ell+m)t^2 = 0,$$
$$(2+k+\ell) - 2(1+\ell)t + (2+\ell+m)t^2 = 0.$$

Hence, there are no real solutions for t.

2) $\mathrm{SU}(k+\ell+m)/\mathrm{S}(\mathrm{U}(k)\times\mathrm{U}(\ell)\times\mathrm{U}(m))$.

This case is also a generalized flag manifold of type A_2. See Theorem 5.2.

3) $\mathrm{Sp}(k+\ell+m)/(\mathrm{Sp}(k)\times\mathrm{Sp}(\ell)\times\mathrm{Sp}(m))$.

In this case we have $d_1 = 4k\ell$, $d_2 = 4km$, $d_3 = 4\ell m$ and $A_{123} = \dfrac{2k\ell m}{k+\ell+m+1}$ and we see that the numerators of equations (8) are given by

$$(-1+k+\ell) + (-1+2\ell)t + (-1+\ell+m)t^2 = 0,$$
$$(-1+k+\ell) + (1-2\ell)t + (-1+\ell+m)t^2 = 0.$$

Thus we see there are no real solutions t.

4) $\mathrm{SU}(2\ell)/\mathrm{U}(\ell)$.

In this case we have $d_1 = 4k\ell$, $d_2 = 4km$, $d_3 = 4\ell m$ and $A_{123} = \dfrac{2k\ell m}{k+\ell+m+1}$ and we see that the numerators of equations (8) are given by

$$(-1+k+\ell) + (-1+2\ell)t + (-1+\ell+m)t^2 = 0,$$
$$(-1+k+\ell) + (1-2\ell)t + (-1+\ell+m)t^2 = 0.$$

Thus we see there are no real solutions t.

5) $\mathrm{SO}(2l)/(\mathrm{U}(1) \times \mathrm{U}(l-1))$ and **7)** $E_6/(\mathrm{SO}(8) \times \mathrm{U}(1)) \times \mathrm{U}(1))$.

These cases are also a generalized flag manifold of type A_2. See Theorem 5.2.

For generalized Wallach spaces of exceptional type, we just write down quadratic equations of numerators of equations (8).

6) $E_6/(\mathrm{SU}(4) \times \mathrm{Sp}(1) \times \mathrm{Sp}(1) \times \mathrm{U}(1))$.

In this case we have $d_1 = d_2 = 16$, $d_3 = 24$ and $A_{123} = 4$, thus we have $7 \pm 8t + 8t^2 = 0$.

8) $E_6/(\mathrm{Sp}(3) \times \mathrm{Sp}(1))$

In this case we have $d_1 = 14$, $d_2 = 28$, $d_3 = 12$ and $A_{123} = \dfrac{7}{2}$, thus we have $14 \pm 10t + 13t^2 = 0$.

9) $E_7/(\mathrm{SO}(8) \times \mathrm{Sp}(1) \times \mathrm{Sp}(1) \times \mathrm{Sp}(1))$.

In this case we have $d_1 = d_2 = d_3 = 32$ and $A_{123} = \dfrac{64}{9}$, thus we have $11(1 \pm t + t^2) = 0$.

10) $E_7/(\mathrm{SU}(6) \times \mathrm{Sp}(1) \times \mathrm{U}(1))$.

In this case we have $d_1 = 30$, $d_2 = 40$, $d_3 = 24$ and $A_{123} = \dfrac{20}{3}$, thus we have $11 \pm 9t + 10t^2 = 0$.

11) $E_7/\mathrm{SO}(8)$.

In this case we have $d_1 = d_2 = d_3 = 35$ and $A_{123} = \dfrac{175}{18}$, thus we have $16(1 \pm t + t^2) = 0$.

12) $E_8/(\mathrm{SO}(12) \times \mathrm{Sp}(1) \times \mathrm{Sp}(1))$.

In this case we have $d_1 = d_2 = 64$, $d_3 = 48$ and $A_{123} = \dfrac{64}{5}$, thus we have $19 \pm 17t + 17t^2 = 0$.

13) $E_8/(\mathrm{SO}(8) \times \mathrm{SO}(8))$.

In this case we have $d_1 = d_2 = 8$, $d_3 = 20$ and $A_{123} = \dfrac{20}{9}$, thus we have $5(1 \pm t + t^2) = 0$.

14) $F_4/(\mathrm{SO}(5) \times \mathrm{Sp}(1) \times \mathrm{Sp}(1))$.

In this case we have $d_1 = d_2 = d_3 = 8$ and $A_{123} = \dfrac{8}{9}$, thus we have $10 \pm 13t + 13t^2 = 0$.

15) $F_4/\mathrm{SO}(8)$.

In this case we have $d_1 = 14$, $d_2 = 28$, $d_3 = 12$ and $A_{123} = \dfrac{7}{2}$, thus we have $1 \pm t + t^2 = 0$.

Thus we obtain our claim. □

Acknowledgments

The second author was supported by JSPS KAKENHI Grant Number JP21K03224. The third author was partially supported by a grant from the Empirikeion Foundation of Athens, Greece.

References

[1] D. Alekseevsky and A. Perelomov, Invariant Kähler-Einstein metrics on compact homogeneous spaces, *Funct. Anal. Appl.* **20**(3) (1986) 171–182.

[2] A. Arvanitoyeorgos, New invariant Einstein metrics on generalized flag manifolds, *Trans. Amer. Math. Soc.* **337**(2) (1993) 981–995.

[3] A. Arvanitoyeorgos and I. Chrysikos, Motion of charged particles and homogeneous geodesics in Kähler C-spaces with two isotropy summands, *Tokyo J. Math.* **32**(2) (2009) 487–500.

[4] A. Arvanitoyeorgos and I. Chrysikos, Invariant Einstein metrics on generalized flag manifolds with four isotropy summands, *Ann. Global Anal. Geom.* **37**(2) (2010) 185–219.

[5] A. Arvanitoyeorgos, I. Chrysikos and Y. Sakane, Homogeneous Einstein metrics on G_2/T, *Proc. Amer. Math. Soc.* **141** (2013) 2485–2499.

[6] A. Arvanitoyeorgos, I. Chrysikos and Y. Sakane, Homogeneous Einstein metrics on generalized flag manifolds with five isotropy summands, *Internat. J. Math.* **24**(10) (2013), 1350077, 52 pp.

[7] A. Arvanitoyeorgos, I. Chrysikos and Y. Sakane, Homogeneous Einstein metrics on generalized flag manifolds with G_2-type $\mathfrak{t}$-roots, *Prospects of differential geometry and its related fields*, 15–38, World Scientific Publishing Co. Pte. Ltd., Hackensack, NJ, 2014.

[8] A. L. Besse, *Einstein Manifolds*, Springer-Verlag, Berlin, 1986.

[9] A. Borel and F. Hirzebruch, Characteristics classes and homogeneous spaces I, *Amer. J. Math.* **80** (1958), 458–538.

[10] I. Chrysikos and Y. Sakane, The classification of homogeneous Einstein metrics on flag manifolds with $b_2(M) = 1$, *Bull. Sci. Math.* **138**(6), (2014), 665–692.

[11] M. Graev, T-root systems of exceptional flag manifolds, *J. Geom. Phys.* **76** (2014), 217–234.

[12] A. Gray, Einstein-like manifolds which are not Einstein, *Geom. Dedicata* **7** (1978), 259–280.

[13] M. Kimura, Homogeneous Einstein metrics on certain Kähler C-spaces, *Adv. Stud. Pure Math.* **18**(1) (1990), 303–320.

[14] F. Li, H. Chen and Z. Chen, Einstein-like metrics on compact homogeneous spaces, *Manuscripta Math.* **174** (2024), 505–516.
[15] Yu. G. Nikonorov, Classification of generalized Wallach spaces, *Geom. Dedicata* **181**(1) (2016), 193–212.
[16] J-S. Park and Y. Sakane, Invariant Einstein metrics on certain homogeneous spaces, *Tokyo J. Math.* **20**(1) (1997), 51–61.
[17] C. Peng and C. Qian, Homogeneous Einstein-like metrics on spheres and projective spaces, *Differ. Geom. Appl.* **44** (2016), 63–76.
[18] C. Qian and A. Wu, Homogeneous Einstein-like metrics on symmetric spaces, *Sci. China Math.* **64**(5), (2021), 1045–1060.
[19] M. Wang and W. Ziller, Existence and non-existence of homogeneous Einstein metrics, *Invent. Math.* **84** (1986), 177–194.

Received February 14, 2024
Revised March 22, 2024

Modern Approaches to Differential Geometry
and its Related Fields 73 – 84

REPRESENTATION OF THE COMPLEX STRUCTURE OF THE T^2 FIBRE BUNDLE OVER THE HIRZEBRUCH SURFACE $\mathbb{CP}^2\#\overline{\mathbb{CP}^2}$

Hideya HASHIMOTO*

Department of Mathematics, Meijo University,
Nagoya 468-8502, Japan
E-mail:hhashi@meijo-u.ac.jp

Misa OHASHI

Department of Mathematics, Nagoya Institute of Technology,
Nagoya 466-8555, Japan
E-mail:ohashi.misa@nitech.ac.jp

The product $S^3 \times S^5$ of 3-dimensional and 5-dimensional spheres is a complex manifold, which is defined by Calabi-Eckmann ([2]). This complex structure comes from the fact that $S^3 \times S^5$ is a total space of a T^2-bundle over $\mathbb{CP}^1 \times \mathbb{CP}^2$, with a natural projection map $\pi_1 \times \pi_2 : S^3 \times S^5 \to \mathbb{CP}^1 \times \mathbb{CP}^2$. Let $W_1 = \mathbb{CP}^2\#\overline{\mathbb{CP}^2}$ be one of Hirzebruch surfaces which is obtained as a blowing up of $\mathbb{CP}^2$ at one point. This surface realizes as a complex submanifold in $\mathbb{CP}^1 \times \mathbb{CP}^2$. Then the inverse image $\widetilde{W_1} = (\pi_1 \times \pi_2)^{-1}(W_1)$ of W_1 is a complex submanifold in $S^3 \times S^5$. First we shall show that $\widetilde{W_1}$ is diffeomorphic to $S^3 \times S^3$ concretely. By using this diffeomorphism, we give a representation of the induced complex structure J of $\widetilde{W_1}$ from $S^3 \times S^5$, as a $(1,1)$ tensor field on $\widetilde{W_1}$ with respect to the left invariant vector fields on $S^3 \times S^3$. We note that the induced complex structure J is different from the (Calabi-Eckmann type) complex structure of a direct Riemannian product $S^3 \times S^3$, which is obtained as a total space of a T^2 bundle over $\mathbb{CP}^1 \times \mathbb{CP}^1$. Also we describe a blowing up of $\mathbb{RP}^2$ at one point, as a Lagrangian surface of W_1. Then this Lagrangian surface is a Klein bottle, which can be considered as a connected sum $\mathbb{RP}^2\#\mathbb{RP}^2$ in $\mathbb{CP}^2\#\overline{\mathbb{CP}^2}$.

Keywords: Hirzebruch surface; complex structure; Klein bottle.

1. Introduction

Let $W_1 = \mathbb{CP}^2\#\overline{\mathbb{CP}^2}$ be the Hirzebruch surface defined by

$$W_1 = \left\{ \left([x_0 : x_1], [y_0 : y_1 : y_2]\right) \in \mathbb{CP}^1 \times \mathbb{CP}^2 \mid x_0 y_2 - x_1 y_1 = 0 \right\}.$$

*This work is supported by JSPS KAKENHI Grant Number 19K03482.

Then W_1 is a complex submanifold of $\mathbb{CP}^1 \times \mathbb{CP}^2$. By using the two Hopf fibrations

$$\pi_1 : S^3 \to \mathbb{CP}^1 \quad \text{and} \quad \pi_2 : S^5 \to \mathbb{CP}^2,$$

with S^1 fibre, we obtain a principal $T^2 (= S^1 \times S^1)$ fibre bundle over $\mathbb{CP}^1 \times \mathbb{CP}^2$. That is, we have the following map

$$\pi_1 \times \pi_2 : S^3 \times S^5 \to \mathbb{CP}^1 \times \mathbb{CP}^2$$

and its inverse image at each point of $\mathbb{CP}^1 \times \mathbb{CP}^2$ coincides with T^2. From this, we obtain the inverse image $\widetilde{W_1} = (\pi_1 \times \pi_2)^{-1}(W_1)$ of the Hirzebruch surface W_1. Then the inverse image $\widetilde{W_1}$ is a total space of a T^2 fibre bundle over the Hirzebruch surface, which is a complex submanifold of $S^3 \times S^5$ with respect to the complex structure defined by Calabi and Eckmann ([2]).

The purpose of this paper is to describe a complex structure of $\widetilde{W_1}$ as a tensor field on $\widetilde{W_1}$ by making use of the left invariant vector fields on $S^3 \times S^3$. To do this, first we describe the following embedding

$$\Psi : S^3 \times S^3 \to \widetilde{W_1} \subset S^3 \times S^5,$$

explicitly. Then we have the following diagram.

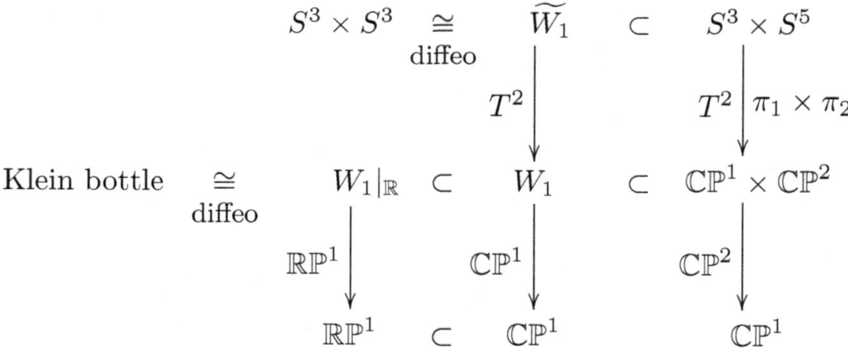

By using this diagram, we give a representation of the complex structure on $\widetilde{W_1}$. Also we obtain $W_1|_{\mathbb{R}} = W_1 \cap (\mathbb{RP}^1 \times \mathbb{RP}^2)$ as a Lagrangian surface in W_1.

2. Sasakian structure on an odd dimensional sphere

In this section we recall fundamental facts on a contact structure of an odd dimensional sphere, and the complex structure on a product of two odd dimensional spheres which is defined in [2].

First we define a contact structure on an odd dimensional sphere S^{2n-1} in $\mathbb{C}^n$ by using the canonical complex structure $J^{\mathbb{C}^n}$ (which is the right multiplication of the purely imaginary unit $i = \sqrt{-1}$). Let S^{2n-1} be a $(2n-1)$ dimensional unit sphere in $\mathbb{C}^n$ centered at the origin of $\mathbb{C}^n$ defined by

$$S^{2n-1} = \Big\{(z_1, \dots, z_n) \in \mathbb{C}^n \;\Big|\; \sum_{\alpha=1}^{n} |z_\alpha|^2 = 1\Big\}.$$

We denote by x as a position vector of S^{2n-1} in $\mathbb{C}^n$, and $T_x S^{2n-1}$ is a tangent space at x. For all $X \in T_x S^{2n-1}$, we set $(1,1)$ tensor field $\varphi^{S^{2n-1}}$ as

$$\varphi^{S^{2n-1}}(X) = J^{\mathbb{C}^n}(X) - \langle J^{\mathbb{C}^n}(X), x\rangle_{\mathbb{C}^n} x = J^{\mathbb{C}^n}(X) + \langle X, J^{\mathbb{C}^n} x\rangle x,$$

where $\langle\ ,\ \rangle$ denote the metric on S^{2n-1} which is induced from that on $\mathbb{R}^{2n} \simeq \mathbb{C}^n$. Also we set $\eta^{S^{2n-1}}$ as a dual 1-form of the vector field $J^{\mathbb{C}^n} x$ on S^{2n-1} by

$$\eta^{S^{2n-1}}(X) = \langle X, J^{\mathbb{C}^n} x\rangle.$$

Then the quartet $(\varphi^{S^{2n-1}},\ \eta^{S^{2n-1}},\ J^{\mathbb{C}^n}x,\ \langle\ ,\ \rangle)$ satisfies the following conditions

$$\begin{aligned}
&(1) \quad (\varphi^{S^{2n-1}})^2 = -id_{T_x S^{2n-1}} + \eta^{S^{2n-1}} \otimes J^{\mathbb{C}^n} x, \quad \varphi^{S^{2n-1}}(J^{\mathbb{C}^n} x) = 0\\
&(2) \quad \eta^{S^{2n-1}} \circ \varphi^{S^{2n-1}} = 0, \quad \eta^{S^{2n-1}}(J^{\mathbb{C}^n} x) = 1\\
&(3) \quad \langle \varphi^{S^{2n-1}}(X),\ \varphi^{S^{2n-1}}(Y)\rangle = \langle X,\ Y\rangle - \eta^{S^{2n-1}}(X)\eta^{S^{2n-1}}(Y)\\
&(4) \quad \eta^{S^{2n-1}} \wedge (d\eta^{S^{2n-1}})^{n-1} \neq 0\\
&(5) \quad (\nabla_X \varphi^{S^{2n-1}})Y = -\langle X,\ Y\rangle J^{\mathbb{C}^n} x + \eta^{S^{2n-1}}(Y)X
\end{aligned}$$

where $X, Y \in T_x S^{2n-1}$ and ∇ is a Levi-Civita connection of the Riemannian metric on S^{2n-1}. The condition (5) implies that the quartet $(\varphi^{S^{2n-1}}, \eta^{S^{2n-1}}, J^{\mathbb{C}^n} x, \langle\ ,\ \rangle)$ is a Sasakian structure.

By using this Sasakian structure, we define the complex structure $J_{S^{2n-1}\times S^{2m-1}}$ on a product of two spheres $S^{2n-1} \times S^{2m-1}$ by

$$\begin{aligned}
&J_{S^{2n-1}\times S^{2m-1}}(X, Y)\\
&= (\varphi^{S^{2n-1}}(X) - \eta^{S^{2m-1}}(Y)\, J^{\mathbb{C}^n} x,\ \varphi^{S^{2m-1}}(Y) + \eta^{S^{2n-1}}(X) J^{\mathbb{C}^m} y),
\end{aligned}$$

for $(X, Y) \in T_{(x,y)}(S^{2n-1} \times S^{2m-1}) \simeq T_x S^{2n-1} \times T_y S^{2m-1}$ and $(x, y) \in S^{2n-1} \times S^{2m-1} \subset \mathbb{C}^n \times \mathbb{C}^m$. Here, $(\varphi^{S^{2n-1}}, \eta^{S^{2n-1}}, J^{\mathbb{C}^n} x, \langle\ ,\ \rangle_{S^{2n-1}})$ and $(\varphi^{S^{2m-1}}, \eta^{S^{2m-1}}, J^{\mathbb{C}^m} y, \langle\ ,\ \rangle_{S^{2m-1}})$ are the Sasakian structures of the unit $(2n-1)$-sphere S^{2n-1} in $\mathbb{C}^n$ and of the unit $(2m-1)$-sphere S^{2m-1} in $\mathbb{C}^m$

both of which are centered at the origins of $\mathbb{C}^n$ and $\mathbb{C}^m$, respectively. See also Morimoto [4] for more details.

3. Geometrical properties of $\widetilde{W_1}$

We set $U_0 = \mathbb{CP}^1 \setminus \{[0:1]\}$ and $U_1 = \mathbb{CP}^1 \setminus \{[1:0]\}$. Since the Hirzebruch surface W_1 is a non-trivial $\mathbb{CP}^1$ fibre bundle over $\mathbb{CP}^1$, it can be realized as the image of the following map

$$\psi : U_0 \times \mathbb{CP}^1 \sqcup U_1 \times \mathbb{CP}^1 \to \mathbb{CP}^1 \times \mathbb{CP}^2,$$

which is defined as follows. We define two maps $\psi_\alpha : U_\alpha \times \mathbb{C} \to \mathbb{CP}^1 \times \mathbb{CP}^2$, for $\alpha \in \{0,1\}$, by

$$\psi_0\big([z_0 : z_1], [\nu_0 : \nu_1]\big) = \big([z_0 : z_1], [z_0\nu_0 : z_0\nu_1 : z_1\nu_1]\big),$$

for $([z_0 : z_1], [\nu_0 : \nu_1]) \in U_0 \times \mathbb{CP}^1$, and

$$\psi_1\big([z_0 : z_1], [\mu_0 : \mu_1]\big) = \big([z_0 : z_1], [z_1\mu_0 : z_0\mu_1 : z_1\mu_1]\big),$$

for $([z_0 : z_1], [\mu_0 : \mu_1]) \in U_1 \times \mathbb{CP}^1$. We see that two maps ψ_0, ψ_1 are well-defined. For $[z_0 : z_1] \in U_0 \cap U_1$, we have

$$\begin{aligned} \psi_0\big([z_0 : z_1], [\nu_0 : \nu_1]\big) &= \psi_1\big([z_0 : z_1], [z_0\nu_0 : z_1\nu_1]\big), \\ \psi_1\big([z_0 : z_1], [\mu_0 : \mu_1]\big) &= \psi_0\left([z_0 : z_1], \left[\frac{\mu_0}{z_0} : \frac{\mu_1}{z_1}\right]\right). \end{aligned}$$

Therefore we can define the map ψ such that $\psi|_{U_0\times\mathbb{CP}^1} = \psi_0$ and $\psi|_{U_1\times\mathbb{CP}^1} = \psi_1$, then ψ define a map whole on $U_0 \times \mathbb{CP}^1 \sqcup U_1 \times \mathbb{CP}^1$. Also we see that the image of the map ψ coincides with W_1. It is known that W_1 is not diffeomorphic to a product of two complex projective spaces $\mathbb{CP}^1 \times \mathbb{CP}^1$ in [3].

From the representation of the map ψ, to describe an inverse image $\widetilde{W_1}$, we use the algebra of quaternions $\mathbb{H}$ as a 4-dimensional vector space $\mathbb{H} = \mathrm{span}_{\mathbb{R}}\{1, i, j, k\}$. Then the basis of $\mathbb{H}$ satisfies the following algebraic property

$$i^2 = j^2 = k^2 = -1, ij = -ji = k, \text{and } ijk = -1.$$

The algebra of quaternions $\mathbb{H}$ is a non-commutative, associative division algebra. From the definition of W_1 and the above map ψ, we define the map Ψ from $S^3 \times S^3$ to $S^3 \times S^5$ as:

$$\begin{aligned} \Psi(q, \lambda_0, \lambda_1) = (q, \lambda_0, q\lambda_1) \ &\in \mathbb{H} \times \mathbb{C} \times \mathbb{H} \\ &\Big(\simeq (z_0, z_1, \lambda_0, z_0\lambda_1, z_1\lambda_1) \in \mathbb{C}^2 \times \mathbb{C}^3\Big), \end{aligned}$$

for $q = z_0 + jz_1 \in S^3(\simeq Sp(1)) \subset \mathbb{H} = \mathbb{C} \oplus j\mathbb{C}$ and $(\lambda_0, \lambda_1) \in \mathbb{C}^2$ with $|\lambda_0|^2 + |\lambda_1|^2 = 1$. Then we shall show that the inverse image $\widetilde{W_1}$ coincides with the image $\Psi(S^3 \times S^3)$, and that the map Ψ implies a diffeomorphism between $S^3 \times S^3$ and $\widetilde{W_1}$.

To compute the almost complex structure as a tensor field on $\widetilde{W_1}$, we need the differential Ψ_* of the map Ψ. To do this, we put $q = q_0 + q_1 i + q_2 j + q_3 k \in \mathbb{H}$ and $p = \lambda_0 + j\lambda_1 = p_0 + p_1 i + p_2 j + p_3 k \in \mathbb{H}$, where $q_\alpha, p_\beta \in \mathbb{R}$ for $\alpha, \beta \in \{0, 1, 2, 3\}$, then we have $\lambda_0 = p_0 + p_1 i$ and $\lambda_1 = p_2 - p_3 i$. From these relations, the left invariant vector fields qi, qj, qk on S^3 with respect to a position vector q are given by

$$\begin{aligned} qi &= -q_1 + q_0 i + q_3 j - q_2 k \simeq -q_1 \frac{\partial}{\partial q_0} + q_0 \frac{\partial}{\partial q_1} + q_3 \frac{\partial}{\partial q_2} - q_2 \frac{\partial}{\partial q_3}, \\ qj &= -q_2 - q_3 i + q_0 j + q_1 k \simeq -q_2 \frac{\partial}{\partial q_0} - q_3 \frac{\partial}{\partial q_1} + q_0 \frac{\partial}{\partial q_2} + q_1 \frac{\partial}{\partial q_3}, \\ qk &= -q_3 + q_2 i - q_1 j + q_0 k \simeq -q_3 \frac{\partial}{\partial q_0} + q_2 \frac{\partial}{\partial q_1} - q_1 \frac{\partial}{\partial q_2} + q_0 \frac{\partial}{\partial q_3}. \end{aligned}$$

In the same way, the left invariant vector fields pi, pj, pk on S^3 with respect to a position vector p are given by

$$\begin{aligned} pi &= -p_1 + p_0 i + p_3 j - p_2 k \simeq -p_1 \frac{\partial}{\partial p_0} + p_0 \frac{\partial}{\partial p_1} + p_3 \frac{\partial}{\partial p_2} - p_2 \frac{\partial}{\partial p_3}, \\ pj &= -p_2 - q_3 i + q_0 j + q_1 k \simeq -q_2 \frac{\partial}{\partial p_0} - p_3 \frac{\partial}{\partial p_1} + p_0 \frac{\partial}{\partial p_2} + p_1 \frac{\partial}{\partial p_3}, \\ pk &= -p_3 + p_2 i - p_1 j + p_0 k \simeq -p_3 \frac{\partial}{\partial p_0} + p_2 \frac{\partial}{\partial p_1} - p_1 \frac{\partial}{\partial p_2} + p_0 \frac{\partial}{\partial p_3}. \end{aligned}$$

From the definition of Ψ, we have

$$\begin{aligned} \mathfrak{q}_1 &= \Psi_*\big((qi, O_{\mathbb{H}})\big) = (qi, O_{\mathbb{C}}, qi\lambda_1), \\ \mathfrak{q}_2 &= \Psi_*\big((qj, O_{\mathbb{H}})\big) = (qj, O_{\mathbb{C}}, qj\lambda_1), \\ \mathfrak{q}_3 &= \Psi_*\big((qk, O_{\mathbb{H}})\big) = (qk, O_{\mathbb{C}}, qk\lambda_1), \\ \mathfrak{p}_1 &= \Psi_*\big((O_{\mathbb{H}}, pi)\big) = (O_{\mathbb{H}}, \lambda_0 i, qi\lambda_1), \\ \mathfrak{p}_2 &= \Psi_*\big((O_{\mathbb{H}}, pj)\big) = (O_{\mathbb{H}}, -\overline{\lambda_1}, q\overline{\lambda_0}), \\ \mathfrak{p}_3 &= \Psi_*\big((O_{\mathbb{H}}, pk)\big) = (O_{\mathbb{H}}, \overline{\lambda_1} i, -qi\overline{\lambda_0}), \end{aligned}$$

where $O_{\mathbb{C}} = 0 + 0i$ and $O_{\mathbb{H}} = O_{\mathbb{C}} + jO_{\mathbb{C}}$. Since S^3 is parallelizable, the above 6-vector fields are defined whole on $S^3 \times S^3$. Therefore, we obtain the induced metric on $S^3 \times S^3$ as follows:

Proposition 3.1. *The matrix G corresponding to the induced metric g of Ψ is given by*

$$G = \left(\begin{array}{c|cc|ccc} \langle \mathfrak{q}_1, \mathfrak{q}_1\rangle & \langle \mathfrak{q}_1, \mathfrak{q}_2\rangle & \langle \mathfrak{q}_1, \mathfrak{q}_3\rangle & \langle \mathfrak{q}_1, \mathfrak{p}_1\rangle & \langle \mathfrak{q}_1, \mathfrak{p}_2\rangle & \langle \mathfrak{q}_1, \mathfrak{p}_3\rangle \\ \hline \langle \mathfrak{q}_2, \mathfrak{q}_1\rangle & \langle \mathfrak{q}_2, \mathfrak{q}_2\rangle & \langle \mathfrak{q}_2, \mathfrak{q}_3\rangle & \langle \mathfrak{q}_2, \mathfrak{p}_1\rangle & \langle \mathfrak{q}_2, \mathfrak{p}_2\rangle & \langle \mathfrak{q}_2, \mathfrak{p}_3\rangle \\ \langle \mathfrak{q}_3, \mathfrak{q}_1\rangle & \langle \mathfrak{q}_3, \mathfrak{q}_2\rangle & \langle \mathfrak{q}_3, \mathfrak{q}_3\rangle & \langle \mathfrak{q}_3, \mathfrak{p}_1\rangle & \langle \mathfrak{q}_3, \mathfrak{p}_2\rangle & \langle \mathfrak{q}_3, \mathfrak{p}_3\rangle \\ \hline \langle \mathfrak{p}_1, \mathfrak{q}_1\rangle & \langle \mathfrak{p}_1, \mathfrak{q}_2\rangle & \langle \mathfrak{p}_1, \mathfrak{q}_3\rangle & \langle \mathfrak{p}_1, \mathfrak{p}_1\rangle & \langle \mathfrak{p}_1, \mathfrak{p}_2\rangle & \langle \mathfrak{p}_1, \mathfrak{p}_3\rangle \\ \langle \mathfrak{p}_2, \mathfrak{q}_1\rangle & \langle \mathfrak{p}_2, \mathfrak{q}_2\rangle & \langle \mathfrak{p}_2, \mathfrak{q}_3\rangle & \langle \mathfrak{p}_2, \mathfrak{p}_1\rangle & \langle \mathfrak{p}_2, \mathfrak{p}_2\rangle & \langle \mathfrak{p}_2, \mathfrak{p}_3\rangle \\ \langle \mathfrak{p}_3, \mathfrak{q}_1\rangle & \langle \mathfrak{p}_3, \mathfrak{q}_2\rangle & \langle \mathfrak{p}_3, \mathfrak{q}_3\rangle & \langle \mathfrak{p}_3, \mathfrak{p}_1\rangle & \langle \mathfrak{p}_3, \mathfrak{p}_2\rangle & \langle \mathfrak{p}_3, \mathfrak{p}_3\rangle \end{array}\right)$$

$$= \left(\begin{array}{c|c|ccc} 1+|\lambda_1|^2 & O_{1\times 2} & |\lambda_1|^2 & p_0p_3-p_1p_2 & -(p_0p_2+p_1p_3) \\ \hline O_{2\times 1} & (1+|\lambda_1|^2)I_2 & O_{2\times 1} & O_{2\times 1} & O_{2\times 1} \\ \hline |\lambda_1|^2 & O_{1\times 2} & 1 & 0 & 0 \\ p_0p_3-p_1p_2 & O_{1\times 2} & 0 & 1 & 0 \\ -(p_0p_2+p_1p_3) & O_{1\times 2} & 0 & 0 & 1 \end{array}\right),$$

where $|\lambda_1|^2 = p_2^2 + p_3^2$, and $\langle\ ,\ \rangle$ denote the metric on $S^3 \times S^5$. The matrix G is positive definite, therefore the map Ψ is an immersion. Moreover, the map Ψ is an embedding.

We note that the following equality

$$|\lambda_1|^4 + (p_0p_3 - p_1p_2)^2 + (p_0p_2 + p_1p_3)^2 = |\lambda_1|^2$$

holds, since $|\lambda_0|^2 + |\lambda_1|^2 = 1$.

Proof of Proposition 3.1. In order to calculate the metric, frequently, we identify $(q, \lambda_0, q\lambda_1) \in \mathbb{H} \times \mathbb{C} \times \mathbb{H}$ with $(q, p_0, p_1, q\lambda_1) \in \mathbb{H} \times \mathbb{R}^2 \times \mathbb{H}$, for any $\lambda_0 = p_0 + p_1 i \in \mathbb{C}$. We recall

$$\begin{aligned} \mathfrak{q}_1 &= (qi, 0, 0, p_3\, q + p_2\, qi), \\ \mathfrak{q}_2 &= (qj, 0, 0, p_2\, qj + p_3\, qk), \\ \mathfrak{q}_3 &= (qk, 0, 0, -p_3\, qj + p_2\, qk), \\ \mathfrak{p}_1 &= (O_{\mathbb{H}}, -p_1,\ p_0,\ p_3\, q + p_2\, qi), \\ \mathfrak{p}_2 &= (O_{\mathbb{H}}, -p_2, -p_3,\ p_0\, q - p_1\, qi), \\ \mathfrak{p}_3 &= (O_{\mathbb{H}}, -p_3,\ p_2, -p_1\, q - p_0\, qi). \end{aligned}$$

Then we can easily see that, for example,

$$\begin{aligned} \langle \mathfrak{q}_1, \mathfrak{p}_1\rangle &= \big\langle (qi, 0, 0, p_3\, q + p_2\, qi),\ \ (O_{\mathbb{H}}, -p_1,\ p_0,\ p_3\, q + p_2\, qi)\big\rangle \\ &= p_2^2 + p_3^2 = |\lambda_1|^2, \end{aligned}$$

and

$$\begin{aligned}\langle \mathfrak{q}_1, \mathfrak{p}_2\rangle &= \big\langle (qi, 0, 0, p_3\, q + p_2\, qi),\ \ (O_{\mathbb{H}}, -p_2, -p_3,\ p_0\, q - p_1\, qi)\big\rangle \\ &= p_0p_3 - p_1p_2.\end{aligned}$$

In the same way, we obtain another element of the matrix G of the induced metric g. For brevity, we set $\alpha = |\lambda_1|^2$. Then the eigenvalues of G are given by

$$\left\{1+\alpha,\ 1,\ 1+\frac{\alpha \pm \sqrt{\alpha(\alpha+4)}}{2}\right\}.$$

Since $0 \le \alpha \le 1$, we see that $\alpha - \sqrt{\alpha(\alpha+4)} \ge 1 - \sqrt{5}$. Therefore, we have

$$1 + \frac{\alpha - \sqrt{\alpha(\alpha+4)}}{2} \ge \frac{3-\sqrt{5}}{2} > 0.$$

Thus, all the eigenvalues of G are positive, hence the map Ψ is an immersion. Moreover, if we assume $\Psi(q, \lambda_0, \lambda_1) = \Psi(q', \lambda_0', \lambda_1')$, then we can easily check that $(q, \lambda_0, \lambda_1) = (q', \lambda_0', \lambda_1')$. Therefore Ψ is an embedding. □

Since the image $\Psi(S^3 \times S^3)$ coincides with $\widetilde{W_1}$ and the map Ψ is injective, we find that Ψ is a diffeomorphism. Hence, we obtain

Corollary 3.1. *The inverse image $\widetilde{W_1}$ is diffeomorphic to $S^3 \times S^3$.*

To give a representation of the complex structure of $\widetilde{W_1}$, we put

$$\begin{aligned}u_1 &= (qi, O_{\mathbb{H}}),\ u_2 = (qj, O_{\mathbb{H}}),\ u_3 = (qk, O_{\mathbb{H}}),\\ v_1 &= (O_{\mathbb{H}}, pi),\ v_2 = (O_{\mathbb{H}}, pj),\ v_3 = (O_{\mathbb{H}}, pk).\end{aligned}$$

From this, we obtain

Theorem 3.1. *The induced complex structure $J = \Psi^* J_{S^3\times S^5}$ of $\widetilde{W_1}$ is represented as a $(1,1)$ tensor field by*

$$\begin{aligned}&(Ju_2, Ju_3, Jv_2, Jv_3, Jv_1, Ju_1)\\ &= (u_2, u_3, v_2, v_3, v_1, u_1)\begin{pmatrix} 0 & 1 & 0 & 0 & 0 & 0\\ -1 & 0 & 0 & 0 & 0 & 0\\ 0 & 0 & 0 & 1 & m_1 & \alpha m_1 + m_2\\ 0 & 0 & -1 & 0 & m_2 & \alpha m_2 - m_1\\ 0 & 0 & 0 & 0 & \alpha & 1+\alpha^2\\ 0 & 0 & 0 & 0 & -1 & -\alpha \end{pmatrix},\end{aligned}$$

where $\alpha = |\lambda_1|^2$, $m_1 = p_0p_3 - p_1p_2$ and $m_2 = -(p_0p_2 + p_1p_3)$.

Proof. The complex structure $J_{S^3\times S^5}$ of the product of two spheres $S^3\times S^5$ is given by

$$J_{S^3\times S^5}(X,Y)=\left(\varphi^{S^3}(X)-\eta^{S^5}(Y)\ J^{\mathbb{C}^2}x,\ \varphi^{S^5}(Y)+\eta^{S^3}(X)J^{\mathbb{C}^3}y\right),$$

for $(X,Y)\in T_{(x,y)}(S^3\times S^5)$ and $(x,y)\in S^3\times S^5\subset\mathbb{C}^2\times\mathbb{C}^3$. From this, we get

$$\begin{aligned}&J_{S^3\times S^5}(\mathfrak{q}_2)=J_{S^3\times S^5}(\Psi_*(u_2))=J_{S^3\times S^5}(qj,O_{\mathbb{C}},qj\lambda_1)\\&=\left(\varphi^{S^3}(qj)-\eta^{S^5}(O_{\mathbb{C}},qj\lambda_1)\ J^{\mathbb{C}^2}q,\ \ \varphi^{S^5}(O_{\mathbb{C}},qj\lambda_1)+\eta^{S^3}(qj)\ J^{\mathbb{C}^3}(\lambda_0,q\lambda_1)\right).\end{aligned}$$

Since $\eta^{S^3}(qj)=0$ and $\eta^{S^5}(O_{\mathbb{C}},qj\lambda_1)=0$, we have

$$J_{S^3\times S^5}(\mathfrak{q}_2)=-(qk,O_{\mathbb{C}},qk\lambda_1)=-\Psi_*(u_3)=-\mathfrak{q}_3.$$

In the same way, we obtain the followings

$$J_{S^3\times S^5}(\mathfrak{p}_2)=J_{S^3\times S^5}(\Psi_*(v_2))=(O_{\mathbb{H}},-\overline{\lambda_1}i,qi\overline{\lambda_0})=-\Psi_*(v_3)=-\mathfrak{p}_3$$

and

$$\begin{aligned}&J_{S^3\times S^5}(\mathfrak{p}_1)=J_{S^3\times S^5}(\Psi_*(v_1))=(-qi,O_{\mathbb{C}},O_{\mathbb{H}})\\&\quad=\Psi_*(-qi,\alpha pi+m_1pj\ +\ m_2pk)=-\mathfrak{q}_1+\alpha\mathfrak{p}_1+m_1\mathfrak{p}_2+m_2\mathfrak{p}_3.\end{aligned}$$

Hence we obtain the orthonormal (unitary) frame on $T_qS^3\times T_pS^3$ as follows:

$$e_1=\frac{1}{\sqrt{1+\alpha}}u_2,\quad Je_1=-\frac{1}{\sqrt{1+\alpha}}u_3,\quad e_2=v_2,\quad Je_2=-v_3,$$
$$e_3=v_1,\quad Je_3=(-qi,\alpha pi+m_1pj\ +\ m_2pk)=-u_1+\alpha v_1+m_1v_2\ +\ m_2v_3.$$

Note that the above base satisfies the following

$$\begin{aligned}\Psi_*(e_1)&=\frac{1}{\sqrt{1+\alpha}}(qj,O_{\mathbb{C}},p_2\,qj+p_3\,qk),\\ \Psi_*(Je_1)&=-\frac{1}{\sqrt{1+\alpha}}(qk,O_{\mathbb{C}},-p_3\,qj+p_2\,qk),\\ \Psi_*(e_2)&=(O_{\mathbb{H}},-\overline{\lambda_1},\ p_0\,q-p_1\,qi),\\ \Psi_*(Je_2)&=-(O_{\mathbb{H}},-\overline{\lambda_1}i,-p_1\,q-p_0\,qi),\\ \Psi_*(e_3)&=(O_{\mathbb{H}},\lambda_0 i,p_3\,q+p_2\,qi),\\ \Psi_*(Je_3)&=(-qi,O_{\mathbb{C}},O_{\mathbb{H}}).\end{aligned}$$

Then $\{\Psi_*(e_1),\Psi_*(Je_1),\dots,\Psi_*(Je_3)\}$ is an orthonormal basis of a tangent space $T_{\Psi(q,\lambda_0,\lambda_1)}\Psi(S^3\times S^3)$ at $\Psi(q,\lambda_0,\lambda_1)=(q,\lambda_0,q\lambda_1)$. The matrix $P\in$

$M_{6\times 6}(\mathbb{R})$ of the above two basis is defined by the solution of the following equation

$$(e_1, Je_1, e_2, Je_2, e_3, Je_3) = (u_2, u_3, v_2, v_3, v_1, u_1)P.$$

Then P and P^{-1} are given by, respectively,

$$P = \begin{pmatrix} \frac{1}{\sqrt{1+\alpha}} & 0 & 0 & 0 & 0 & 0 \\ 0 & -\frac{1}{\sqrt{1+\alpha}} & 0 & 0 & 0 & 0 \\ 0 & 0 & 1 & 0 & 0 & m_1 \\ 0 & 0 & 0 & -1 & 0 & m_2 \\ 0 & 0 & 0 & 0 & 1 & \alpha \\ 0 & 0 & 0 & 0 & 0 & -1 \end{pmatrix},$$

$$P^{-1} = \begin{pmatrix} \sqrt{1+\alpha} & 0 & 0 & 0 & 0 & 0 \\ 0 & -\sqrt{1+\alpha} & 0 & 0 & 0 & 0 \\ 0 & 0 & 1 & 0 & 0 & m_1 \\ 0 & 0 & 0 & -1 & 0 & -m_2 \\ 0 & 0 & 0 & 0 & 1 & \alpha \\ 0 & 0 & 0 & 0 & 0 & -1 \end{pmatrix}.$$

Hence, the complex structure J with respect to the basis (u_2, u_3, v_2, v_3, v_1, u_1) is given by

$$\begin{aligned}(Ju_2, Ju_3, Jv_2, Jv_3, Jv_1, Ju_1) &= (Je_1, J^2e_1, Je_2, J^2e_2, Je_3, J^2e_3)P^{-1} \\ &= (e_1, Je_1, e_2, Je_2, e_3, Je_3)J_0P^{-1} = (u_2, u_3, v_2, v_3, v_1, u_1)PJ_0P^{-1}\end{aligned}$$

where

$$J_0 = \begin{pmatrix} 0 & -1 & 0 & 0 & 0 & 0 \\ 1 & 0 & 0 & 0 & 0 & 0 \\ 0 & 0 & 0 & -1 & 0 & 0 \\ 0 & 0 & 1 & 0 & 0 & 0 \\ 0 & 0 & 0 & 0 & 0 & -1 \\ 0 & 0 & 0 & 0 & 1 & 0 \end{pmatrix}.$$

Therefore we obtain the matrix of representation of J with respect to the fixed basis $(u_2, u_3, v_2, v_3, v_1, u_1)$ by PJ_0P^{-1}. By simple calculation, we get the desired result. □

Let $\{\omega_1, \omega_2, \omega_3, \eta_1, \eta_2, \eta_3\}$ be the left invariant dual 1-forms on $S^3 \times S^3 (\simeq Sp(1) \times Sp(1))$ which satisfy

$$\omega_\alpha(u_\beta) = \delta_{\alpha\beta},\ \omega_\alpha(v_\beta) = 0,\ \eta_\alpha(u_\beta) = 0,\ \eta_\alpha(v_\beta) = \delta_{\alpha\beta},$$

for $\alpha, \beta \in \{1, 2, 3\}$. Then we obtain

Corollary 3.2. *The canonical 2-form* Ω *with respect to the induced metric* g *and complex structure* J *on* $S^3 \times S^3 (\simeq \widetilde{W_1})$ *is given by*

$$\Omega = -\{(1 + |\lambda_1|^2)\,\omega_2 \wedge \omega_3 + \eta_2 \wedge \eta_3 + \eta_1 \wedge \omega_1\},$$

where $\Omega(X,Y) = g(JX,Y)$ *for smooth vector fields* X, Y *on* $S^3 \times S^3$.

Proof. Vector fields X, Y are represented by

$$X = \sum_{\alpha=1}^{3} u_\alpha \omega_\alpha(X) + \sum_{\alpha=1}^{3} v_\alpha \eta_\alpha(X), \quad Y = \sum_{\alpha=1}^{3} u_\alpha \omega_\alpha(Y) + \sum_{\alpha=1}^{3} v_\alpha \eta_\alpha(Y).$$

We substitute these into $g(JX,Y)$, and obtain the desired result. □

Remark 3.1. Geometrical structures of the Calabi-Eckmann manifold of a product of two 3-dimensional sphere $S^3 \times S^3$ is different from the above triplet (g, J, Ω) on $S^3 \times S^3 (\simeq \widetilde{W_1})$. In fact the pair of the metric $\langle\ ,\ \rangle_{S^3\times S^3}$ and the complex structure $J_{S^3\times S^3}$ on $S^3 \times S^3$ corresponds to the matrix form (I_6, J_0) with respect to the left invariant frame field. Here I_6 is a 6×6 identity matrix. The canonical 2-form Ω_0 is given by

$$\Omega_0 = -\{\omega_2 \wedge \omega_3 + \eta_2 \wedge \eta_3 + \eta_1 \wedge \omega_1\},$$

where $\Omega_0(X,Y) = \langle J_{S^3\times S^3} X, Y\rangle_{S^3\times S^3}$ for vector fields X, Y on $S^3 \times S^3$. Since $|\lambda_1|^2$ is a non-constant function on $S^3 \times S^3 (\simeq \widetilde{W_1})$, we see that Ω_0 is different from Ω.

4. Realization of Klein bottle of W_1

In this section, we shall show that there exists a Klein bottle of W_1 in the natural way. Let $W_1|_{\mathbb{R}} = W_1 \cap (\mathbb{RP}^1 \times \mathbb{RP}^2)$ be the intersection defined by

$$W_1|_{\mathbb{R}} = \big\{([x_0 : x_1], [y_0 : y_1 : y_2]) \in \mathbb{RP}^1 \times \mathbb{RP}^2 \;\big|\; x_0 y_2 - x_1 y_1 = 0\big\}.$$

By using the map Ψ, we see that the inverse image $\widetilde{W_1}|_{\mathbb{R}} = (\pi_1 \times \pi_2)^{-1}(W_1|_{\mathbb{R}})$ of $W_1|_{\mathbb{R}}$ coincides with the image

$$\big\{\Psi(e^{j\theta_1}, e^{j\theta_2}) \;\big|\; (e^{j\theta_1}, e^{j\theta_2}) \in S^1 \times S^1\big\}$$

where $e^{j\theta_\alpha} = \cos\theta_\alpha + j \sin\theta_\alpha \in \mathbb{H}$ for $\alpha = 1, 2$. In fact,

$$\Psi(e^{j\theta_1}, e^{j\theta_2}) = \Big(\cos\theta_1 + j\sin\theta_1, \cos\theta_2, (\cos\theta_1 + j\sin\theta_1)\sin\theta_2\Big).$$

Therefore, if we put

$$x_0 = \cos\theta_1,\ x_1 = \sin\theta_1,\ y_0 = \cos\theta_2,\ y_1 = \cos\theta_1\sin\theta_2,\ y_2 = \sin\theta_1\sin\theta_2,$$

then we have

$$\begin{aligned}&\pi_1 \times \pi_2\big(\Psi(e^{j\theta_1}, e^{j\theta_2})\big)\\ &\quad = \big([\cos\theta_1 : \sin\theta_1], [\cos\theta_2 : \cos\theta_1 \sin\theta_2 : \sin\theta_1 \sin\theta_2]\big) \in W_1|_{\mathbb{R}}.\end{aligned}$$

Since the dimension of $W_1|_{\mathbb{R}}$ is two, the map Ψ is surjective. From this we shall show that

Proposition 4.1. *The surface* $W_1|_{\mathbb{R}} = W_1 \cap (\mathbb{RP}^1 \times \mathbb{RP}^2)$ *in* $\mathbb{CP}^1 \times \mathbb{CP}^2$ *is diffeomorphic to a Klein bottle.*

Proof. If $(e^{j\theta_1}, e^{j\theta_2})$, $(e^{j\theta_1'}, e^{j\theta_2'}) \in S^1 \times S^1$ satisfy

$$(\pi_1 \times \pi_2 \circ \Psi)(e^{j\theta_1}, e^{j\theta_2}) = (\pi_1 \times \pi_2 \circ \Psi)(e^{j\theta_1'}, e^{j\theta_2'}),$$

then there exists $(n, \ell) \in \mathbb{Z} \times \mathbb{Z}$ such that $(\theta_1', \theta_2') = (\theta_1 + n\pi, (-1)^n\theta_2 + \ell\pi))$. That is,

$$(\pi_1 \times \pi_2 \circ \Psi)(e^{j(\theta_1+n\pi)}, e^{j((-1)^n\theta_2+\ell\pi)}) = (\pi_1 \times \pi_2 \circ \Psi)(e^{j\theta_1}, e^{j\theta_2}) \qquad (1)$$

holds. In fact, we see that

$$\begin{aligned}e^{j(\theta_1+n\pi)} &= \cos(\theta_1 + n\pi) + j\sin(\theta_1 + n\pi)\\ &= (-1)^n(\cos\theta_1 + j\sin\theta_1) = (-1)^n e^{j\theta_1},\\ e^{j((-1)^n\theta_2+\ell\pi)} &= \cos((-1)^n\theta_2 + \ell\pi) + j\sin((-1)^n\theta_2 + \ell\pi)\\ &= (-1)^\ell\cos\theta_2 + j(-1)^{n+\ell}\sin\theta_2,\end{aligned}$$

therefore, we have

$$\begin{aligned}&\Psi(e^{j(\theta_1+n\pi)}, e^{j((-1)^n\theta_2+\ell\pi)})\\ &= \Big((-1)^n e^{j\theta_1},\ (-1)^\ell\cos\theta_2,\ (-1)^n e^{j\theta_1}(-1)^{n+\ell}\sin\theta_2\Big)\\ &= \Big((-1)^n e^{j\theta_1},\ (-1)^\ell\cos\theta_2,\ (-1)^\ell e^{j\theta_1}\sin\theta_2\Big).\end{aligned}$$

By (1), the orbit space of this action $\mathbb{Z} \times \mathbb{Z}$ on $\mathbb{R}^2$ corresponds to a Klein bottle. We get the desired result. □

References

[1] M. Audin. *Torus actions on symplectic manifolds* (Second revised edition). Progress in Mathematics, 93. Birkhäuser Verlag, Basel, 2004.

[2] E. Calabi and B. Eckmann. A class of compact complex manifolds which are not algebraic. *Ann. Math,* **58** (1953), 494–500.

[3] K. Kodaira. *Complex manifolds and deformation of complex structures.* Grundlehren der mathematischen Wissenschaften 283, Springer-Verlag, New York, 1986.

[4] A. Morimoto. On normal almost contact structures. *J. Math. Soc. Japan* **15** (4) (1963), 420–436.

Received January 12, 2024
Revised April 2, 2024

Modern Approaches to Differential Geometry
and its Related Fields 85 – 96

NON-HOROCYCLIC UNBOUNDED TRAJECTORIES ON A COMPLEX HYPERBOLIC SPACE ARE NOT EXPRESSED BY THOSE ON TUBES OF TYPE (A)

Yusei AOKI

Division of Mathematics and Mathematical Science,
Nagoya Institute of Technology,
Nagoya 466-8555, Japan
E-mail: yusei11291@outlook.jp

Toshiaki ADACHI*

Department of Mathematics, Nagoya Institute of Technology,
Nagoya 466-8555, Japan
E-mail: adachi@nitech.ac.jp

We study whether trajectories for Kähler magnetic fields on a complex hyperbolic space are obtained as extrinsic shapes of trajectories on some real hypersurface of type (A). We show that bounded trajectories are expressed by geodesics and by trajectories on some real hypersurfaces but unbounded trajectories are not expressed by them if strengths of Kähler magnetic fields are small compared with the absolute value of the holomorphic sectional curvature.

Keywords: Trajectories; Kähler and Sasakian magnetic fields; horocyclic; real hypersurfaces of type (A); extrinsic shapes.

1. Introduction

We shall start with quite an elementary fact. We take a circle of positive geodesic curvature in a Euclidean 3-space $\mathbb{R}^3$. Clearly, there is a unique sphere where this circle rides on and is seen as a geodesic. In this paper, we study whether a corresponding property holds on a complex hyperbolic space $\mathbb{C}H^n$ or not. Being different from circles on a Euclidean space, on a complex hyperbolic space, we have many circles which are not congruent to each other. In this paper, we restrict ourselves to special circles whose velocity and acceleration vectors form complex lines. Such circles are

*The second author is partially supported by Grant-in-Aid for Scientific Research (C) (No. 20K03581), Japan Society for the Promotion of Science.

interpreted as trajectories for Kähler magnetic fields on $\mathbb{C}H^n$. Though spheres in a Euclidean space are totally umbilic, there are no totally umbilic real hypersurfaces in a complex hyperbolic space. We therefore consider totally η-umbilic hypersurfaces, which are horospheres, geodesic spheres and tubes around totally geodesic complex hypersurfaces $\mathbb{C}H^{n-1}$, and also consider tubes of type (A_2), which are tubes around totally geodesic submanifolds $\mathbb{C}H^{\ell}$ $(1 \leq \ell \leq n-2)$. In the preceding paper [4], we showed that all non-geodesic trajectories for Kähler magnetic fields on a complex projective space are expressed by geodesics on geodesic spheres. It is natural to consider the same property holds for complex hyperbolic space. But the situation is not the same. We show that bounded trajectories for Kähler magnetic fields are expressed in many ways by trajectories for Sasakian magnetic fields on real hypersurface of type (A), but unbounded non-horocyclic trajectories are not expressed by them.

2. Trajectories for Kähler magnetic fields

Generally, a closed 2-form on a Riemannian manifold is said to be a *magnetic field* (see [8], for example). Regarding the $(1,1)$-tensor field $\Omega_{\mathbb{B}}$ defined by $\mathbb{B}(v,w) = \langle v, \Omega_{\mathbb{B}}(w)\rangle$ for each $v, w \in T_pM$ at an arbitrary point $p \in M$ as the Lorentz force associated with $\mathbb{B}$, we say that a smooth curve γ parameterized by its arclength is a *trajectory* for $\mathbb{B}$ if it satisfies $\nabla_{\dot\gamma}\dot\gamma = \Omega_{\mathbb{B}}(\dot\gamma)$. When $\mathbb{B}$ is the null 2-form, trajectories are geodesics. Hence we may consider that trajectories are natural generalizations of geodesics.

On a complex hyperbolic space $\mathbb{C}H^n(c)$ of constant holomorphic sectional curvature c, we have a natural closed 2-form $\mathbb{B}_J$ called the Kähler form which is defined by $\mathbb{B}_J(v,w) = \langle v, Jw\rangle$ with complex structure J on $\mathbb{C}H^n(c)$. We call its constant multiple $\mathbb{B}_\kappa = \kappa\mathbb{B}_J$ $(\kappa \in \mathbb{R})$ a *Kähler magnetic field.* We say that $|\kappa|$ is the *strength* of $\mathbb{B}_\kappa$. A smooth curve γ parameterized by its arclength is hence a trajectory for $\mathbb{B}_\kappa$ if it satisfies $\nabla_{\dot\gamma}\dot\gamma = \kappa J\dot\gamma$. Clearly, every trajectory for a Kähler magnetic field lies on a totally geodesic $\mathbb{C}H^1$. In view of the equation of trajectories for Kähler magnetic fields, we may say that they form a natural family of curves associated with complex structure J on $\mathbb{C}H^n$. From the viewpoint of the Frenet-Serre formula, trajectories are most elementary curves next to geodesics. We say a smooth curve γ on a Riemannian manifold is said to be a circle if there is a nonnegative constant k and a field Y of unit vectors along γ satisfying $\nabla_{\dot\gamma}\dot\gamma = kY$ and $\nabla_{\dot\gamma}Y = -k\dot\gamma$. This constant k is called the geodesic curvature of a circle γ. Since J is parallel, for a trajectory γ for $\mathbb{B}_\kappa$, we have $\nabla_{\dot\gamma}(J\dot\gamma) = -\kappa\dot\gamma$, hence we find that it is a circle.

Since $\mathbb{C}H^n$ is a symmetric space of rank 1, for arbitrary unit tangent vectors $v, w \in U\mathbb{C}H^n$, there are a holomorphic isometry φ_+ and an anti-holomorphic isometry φ_- of $\mathbb{C}H^n$ satisfying $d\varphi_\pm(v) = w$ and $d\varphi_\pm(Jv) = \pm Jw$. Therefore, we can conclude congruency and homogeneity of trajectories. We say that two smooth curves γ_1, γ_2 on a Riemannian manifold M parameterized by their arclength are *congruent* to each other if and only if there is an isometry φ of M satisfying $\gamma_2(t) = \varphi \circ \gamma_2(t)$ for all t. We call a smooth curve γ homogeneous if it is an orbit of one parameter family $\{\varphi_t\}$ of isometries, that is, $\gamma(t) = \varphi_t(\gamma(0))$. With these terminologies, trajectories for Kähler magnetic fields are homogeneous, and two trajectories γ_i for $\mathbb{B}_{\kappa_i}$ $(i = 1, 2)$ are congruent to each other if and only if $|\kappa_1| = |\kappa_2|$. Under such consideration, we take trajectories as corresponding curves to circles in a Euclidean 3-space.

In [1], we studied some properties of trajectories for a Kähler magnetic field $\mathbb{B}_\kappa$ on $\mathbb{C}H^n(c)$. Since $\mathbb{C}H^n(c)$ is a typical example of Hadamard manifolds, we can define its ideal boundary $\partial\mathbb{C}H^n(c)$ as the set of asymptotic classes of geodesic rays, and compactify it with this boundary (see [6]).

1) When $|\kappa| > \sqrt{|c|}$, trajectories are closed and have length $2\pi/\sqrt{\kappa^2 + c}$.
2) When $|\kappa| \leq \sqrt{|c|}$, trajectories are unbounded and do not have self-intersections.
3) When $|\kappa| < \sqrt{|c|}$, each trajectory γ for $\mathbb{B}_\kappa$ has two points at infinity, that is, $\gamma(\pm\infty) = \lim_{t\to\pm\infty} \gamma(t) \in \partial\mathbb{C}H^n(c)$ exist and do not coincide with each other.
4) Each trajectory γ for $\mathbb{B}_{\pm\sqrt{|c|}}$ has single limit point, i.e. $\gamma(\infty) = \gamma(-\infty)$. If it crosses with a geodesic σ with $\sigma(\infty) = \gamma(\infty)$, then they cross orthogonally. We hence call this trajectory *horocyclic*.

3. Real hypersurfaces of type (A)

A real hypersurface is said to be of type (A) in $\mathbb{C}H^n$ if it is one of a geodesic sphere, a horosphere and a tube around totally geodesic and totally complex $\mathbb{C}H^\ell$ with $1 \leq \ell \leq n-1$. In this paper, we take these real hypersurfaces as correspondence of spheres in a Euclidean space.

On a real hypersurface M in $\mathbb{C}H^n$, we have an almost contact metric structure, which is a quartet $(\xi, \eta, \phi, \langle\ ,\ \rangle)$ of a vector field ξ, a 1-form η, a $(1,1)$-tensor field ϕ and a Riemannian metric, induced by complex structure J on $\mathbb{C}H^n$. By taking a unit normal $\mathcal{N}$ and the induced metric on M, we set $\xi = -J\mathcal{N}$, and define η and ϕ by $\eta(v) = \langle v, \xi\rangle$ and $\phi v = Jv - \eta(v)\mathcal{N}$, respectively. We call ξ the characteristic vector field and ϕ the structure tensor field. With the Riemannian connection of $\mathbb{C}H^n(c)$ being denoted by $\widetilde{\nabla}$, for vector fields X, Y tangent to M, the Gauss and Weingarten formulae are given as $\widetilde{\nabla}_X Y = \nabla_X Y + \langle A_M X, Y\rangle \mathcal{N}$ and $\widetilde{\nabla}_X \mathcal{N} = -A_M X$ with the shape operator A_M associated with $\mathcal{N}$. These show $\nabla_X \xi = \phi A_M X$ and $(\nabla_X \phi)Y = \eta(Y)A_M X - \langle A_M X, Y\rangle\xi$. If we set the fundamental form $\mathbb{F}_\phi$ on M by $\mathbb{F}_\phi(v, w) = \langle v, \phi w\rangle$, the latter guarantees that it is a closed 2-form (see [5]). We call a constant multiple $\mathbb{F}_\kappa = \kappa\mathbb{F}_\phi$ ($\kappa \in \mathbb{R}$) a *Sasakian magnetic field* or a contact magnetic field. A trajectory σ for this magnetic field is hence a smooth curve on M parameterized by its arclength which satisfies the equation $\nabla_{\dot\sigma}\dot\sigma = \kappa\phi\dot\sigma$. We define a function ρ_σ along σ by $\rho_\sigma = \langle\dot\sigma, \xi\rangle$, and call it the *structure torsion* of σ. The strength of a Sasakian magnetic field $\mathbb{F}_\kappa$ acting on σ is $\|\nabla_{\dot\sigma}\dot\sigma\| = |\kappa|\sqrt{1-\rho_\sigma^2}$. Thus it depends on σ. In this sense, trajectories for Kähler and Sasakian magnetic fields are not so similar to each other. The derivative of structure torsion is given as

$$\frac{d}{dt}\rho_\sigma = \kappa\langle\phi\dot\sigma, \xi\rangle + \langle\dot\sigma, \phi A_M\dot\sigma\rangle = -\langle A_M\phi\dot\sigma, \dot\sigma\rangle,$$

hence

$$\frac{d}{dt}\rho_\sigma = \langle\dot\sigma, (\phi A_M - A_M\phi)\dot\sigma\rangle. \tag{3.1}$$

We now restrict ourselves to real hypersurfaces of type (A). A horosphere is denoted by HS, a geodesic sphere of radius r by $G(r)$, a tube of radius r around $\mathbb{C}H^{n-1}$ by $T(r)$, and a tube of radius r around $\mathbb{C}H^\ell$ ($1 \le \ell \le n-2$) by $T_\ell(r)$. They are so called Hopf hypersurfaces, that is, their characteristic vectors ξ are principal curvature vectors i.e. $A_M\xi = \delta_M\xi$. For $HS, G(r)$ and $T(r)$, tangent vectors orthogonal to ξ are principal. These real hypersurfaces are called totally η-umbilic. For $T_\ell(r)$, its tangent bundle splits into three subbundles of principal curvature vectors as $TT_\ell(r) = V_\lambda \oplus V_\mu \oplus \mathbb{R}\xi$. This real hypersurface is called type (A_2). Correspondingly, horospheres are called type (A_0), and geodesic spheres and tubes around $\mathbb{C}H^{n-1}$ are called type (A_1). Their principal curvatures are as in the following table. It is known that a real hypersurface M is of type (A) if and only if one of the following conditions holds (see [7], for example):

i) $A_M\phi = \phi A_M$,
ii) $(\nabla_X A_M)Y = -\frac{c}{4}\{\langle\phi X, Y\rangle\xi + \eta(Y)\phi X\}$ for arbitrary vector fields X, Y.

Therefore, the structure torsion of each trajectory σ on a real hypersurface of type (A) is constant along σ. In particular, when $\rho_\sigma = \pm 1$, we have $\nabla_{\dot\sigma}\dot\sigma = \pm\kappa\phi\xi = 0$, hence the trajectory σ is a geodesic.

	λ_M	μ_M	δ_M
HS	$\frac{\sqrt{\lvert c\rvert}}{2}$	—	$\sqrt{\lvert c\rvert}$
$G(r)$	$\frac{\sqrt{\lvert c\rvert}}{2}\coth(\frac{\sqrt{\lvert c\rvert}}{2}r)$	—	$\sqrt{\lvert c\rvert}\coth(\sqrt{\lvert c\rvert}r)$
$T(r)$	$\frac{\sqrt{\lvert c\rvert}}{2}\tanh(\frac{\sqrt{\lvert c\rvert}}{2}r)$	—	$\sqrt{\lvert c\rvert}\coth(\sqrt{\lvert c\rvert}r)$
$T_\ell(r)$	$\frac{\sqrt{\lvert c\rvert}}{2}\coth(\frac{\sqrt{\lvert c\rvert}}{2}r)$	$\frac{\sqrt{\lvert c\rvert}}{2}\tanh(\frac{\sqrt{\lvert c\rvert}}{2}r)$	$\sqrt{\lvert c\rvert}\coth(\sqrt{\lvert c\rvert}r)$

Table. Principal curvatures of real hypersurfaces of type (A)

For a trajectory σ on a real hypersurface M of type (A_2), by using the projection $\mathrm{Proj}_\lambda : TM \to V_\lambda$ onto the subbundle of principal curvature vectors associated with λ_M, we set $\omega_\sigma = \|\mathrm{Proj}_\lambda(\dot\sigma)\|$ and call it its *principal torsion*. Since we have $\langle A_M\dot\sigma, \dot\sigma\rangle = \lambda_M\omega_\sigma^2 + \mu_M(1-\omega_\sigma^2-\rho_\sigma^2) + \delta_M\rho_\sigma^2$ and

$$\begin{aligned}\frac{d}{dt}\langle A_M\dot\sigma,\dot\sigma\rangle &= -\frac{c\rho_\sigma}{2}\langle\phi\dot\sigma,\dot\sigma\rangle + \kappa\langle A_M\phi\dot\sigma,\dot\sigma\rangle + \kappa\langle A_M\dot\sigma,\phi\dot\sigma\rangle\\ &= \kappa\big\langle(A_M\phi-\phi A_M)\dot\sigma,\dot\sigma\big\rangle = 0,\end{aligned}$$

we find that ω_σ is also constant along σ.

Structure and principal torsions are important invariants of trajectories on real hypersurfaces of type (A). When M is one of *HS*, $G(r)$ and $T(r)$, for arbitrary unit tangent vectors $v, w \in TM$, if they satisfy $\langle v,\xi\rangle = \pm\langle w,\xi\rangle$, there are isometries ψ_+,ψ_- of M satisfying $d\psi_\pm(v-\eta(v)\xi) = w-\eta(w)\xi$ and $d\psi_\pm(\xi) = \pm\xi$. Since we have $\phi\xi = 0$, we find that every trajectory for a Sasakian magnetic field on M is an orbit of a one-parameter family of isometries of M, and that two trajectories σ_i for $\mathbb{F}_{\kappa_i}$ $(i = 1,2)$ on M are congruent to each other if and only if they satisfy one of the following conditions (see [2]):

i) $|\rho_{\sigma_1}| = |\rho_{\sigma_2}| = 1$,
ii) $|\rho_{\sigma_1}| = |\rho_{\sigma_2}| < 1$, $|\kappa_1| = |\kappa_2|$ and $\kappa_1\rho_{\sigma_1} = \kappa_2\rho_{\sigma_2}$.

When $M = T_\ell(r)$, for arbitrary unit tangent vectors $v, w \in TM$, if they satisfy $\langle v,\xi\rangle = \pm\langle w,\xi\rangle$ and $\|\mathrm{Proj}_\lambda(v)\| = \|\mathrm{Proj}_\lambda(w)\|$, there are isometries ψ_+,ψ_- of M satisfying $d\psi_\pm(v-\eta(v)\xi) = w-\eta(w)\xi$ and $d\psi_\pm(\xi) = \pm\xi$ (see [3]). We therefore find that every trajectory for a Sasakian magnetic field on M is an orbit of a one-parameter family of isometries of M, and that two trajectories σ_i for $\mathbb{F}_{\kappa_i}$ $(i = 1, 2)$ on M are congruent to each other if and only if they satisfy one of the following conditions (see [5]):

i) $|\rho_{\sigma_1}| = |\rho_{\sigma_2}| = 1$,
ii) $|\rho_{\sigma_1}| = |\rho_{\sigma_2}| < 1$, $|\kappa_1| = |\kappa_2|, \omega_{\sigma_1} = \omega_{\sigma_2}$ and $\kappa_1\rho_{\sigma_1} = \kappa_2\rho_{\sigma_2}$.

4. Congruency of expressions

Let γ be a trajectory on $\mathbb{C}H^n(c)$. If there is a real hypersurface M in $\mathbb{C}H^n(c)$ and a trajectory σ for a Sasakian magnetic field on M satisfying $\gamma(t) = \iota \circ \sigma(t)$ for all t with an isometric immersion $\iota : M \to \mathbb{C}H^n(c)$, we say that γ is *expressed* by (M, σ). When both (M_1, σ_1) and (M_2, σ_2) are expressions of γ, we say that they are congruent to each other if there is an isometry φ of $\mathbb{C}H^n(c)$ with $\varphi(M_1) = M_2$ which either preserves γ or reverses γ, that is, either $\varphi\circ\gamma(t) = \gamma(t)$ for all t or $\varphi\circ\gamma(t) = \gamma(-t)$ for all t. Though it seems a bit strange, as $\mathbb{C}H^n$ is a symmetric space of rank 1, one can understand that this definition is natural by considering the following example. Given a circle γ of positive geodesic curvature in $\mathbb{R}^3$, we take a sphere S^2 so that γ lies on it and can be seen as a small circle on S^2 showing a latitude line in the northern hemisphere. Also, we take another sphere S^2 of the same radius so that γ lies on it and can be seen as a small circle on S^2 showing a latitude line in the southern hemisphere. If we turn the latter S^2 upside down and make the north pole to the south pole, then γ turns to the small circle in the northern hemisphere of reversed orientation. We consider these two expressions to be congruent to each other.

Given a smooth curve σ on a real hypersurface M, we call the curve $\iota\circ\sigma$ on $\mathbb{C}H^n(c)$ with an isometric immersion $\iota : M \to \mathbb{C}H^n(c)$ its *extrinsic shape*. For the sake of simplicity, we usually also denote it by σ. Here we study conditions that extrinsic shapes of trajectories for Sasakian magnetic fields on real hypersurfaces of type (A) to be those for Kähler magnetic fields.

Lemma 4.1. *Let σ be a trajectory for a Sasakian magnetic field $\mathbb{F}_\kappa$ on a real hypersurface M of type* (A). *Its extrinsic shape is a trajectory for some Kähler magnetic field if and only if one of the following conditions holds*:

i) $\rho_\sigma = \pm 1$,
ii) $|\rho_\sigma| < 1$ *and* $\kappa\rho_\sigma = \langle A_M\dot\sigma, \dot\sigma\rangle$.

In these cases, the extrinsic shapes are trajectories for $\mathbb{B}_{\rho_\sigma\delta_M}$ and for $\mathbb{B}_\kappa$ $(\kappa \neq 0)$, respectively.

Proof. By Gauss formula, we have

$$\widetilde{\nabla}_{\dot\sigma}\dot\sigma = \kappa\phi\dot\sigma + \langle A_M\dot\sigma, \dot\sigma\rangle\mathcal{N} = \kappa J\dot\sigma + \big(\langle A_M\dot\sigma, \dot\sigma\rangle - \kappa\rho_\sigma\big)\mathcal{N}.$$

When $\rho_\sigma = \pm 1$ or when $\kappa = 0$, this becomes $\widetilde{\nabla}_{\dot\sigma}\dot\sigma = \langle A_M\dot\sigma, \dot\sigma\rangle\mathcal{N}$. Hence the extrinsic shape is a trajectory for a Kähler magnetic field if $\rho_\sigma = \pm 1$. In this case we have $\widetilde{\nabla}_{\dot\sigma}\dot\sigma = \pm\delta_M J\dot\sigma$. When $|\rho_\sigma| < 1$, since $J\dot\sigma$ is not parallel to $\mathcal{N}$, the extrinsic shape of σ is a trajectory for some Kähler magnetic field if and only if $\kappa\rho_\sigma = \langle A_M\dot\sigma, \dot\sigma\rangle$ under the assumption $|\rho_\sigma| < 1$. In this case the extrinsic shape is a trajectory for $\mathbb{B}_\kappa$. Considering principal curvatures of real hypersurfaces of type (A), we see $\kappa \neq 0$. $\square$

To conduct our study, we investigate the relationship between congruency of two expressions and congruency of trajectories on real hypersurfaces. Let (M_1, σ_1) and (M_2, σ_2) be two expressions of a trajectory γ for a Kähler magnetic field $\mathbb{B}_\kappa$ by trajectories for $\mathbb{F}\kappa_i$ on M_i $(i = 1, 2)$.

First, we suppose that they are congruent to each other. For the sake of simplicity, we consider both M_1 and M_2 as subsets of $\mathbb{C}H^n(c)$, and drop the notation of isometric embeddings ι_i. Then there is an isometry $\tilde\varphi$ of $\mathbb{C}H^n(c)$ with $\tilde\varphi(M_1) = M_2$ which satisfies either $\tilde\varphi \circ \sigma_1(t) = \sigma_2(t)$ or $\tilde\varphi \circ \sigma_1(t) = \sigma_2(-t)$. We may suppose $d\tilde\varphi(\mathcal{N}_{M_1}) = \mathcal{N}_{M_2}$. We note that if $d\tilde\varphi(\mathcal{N}_{M_1}) = -\mathcal{N}_{M_2}$ we need to change the signatures of principal curvatures and those of structure torsions. We therefore consider only this case. We set

$$\epsilon = \begin{cases} 1, & \text{when } \tilde\varphi \circ \sigma_1(t) = \sigma_2(t) \text{ holds,} \\ -1, & \text{when } \tilde\varphi \circ \sigma_1(t) = \sigma_2(-t) \text{ holds.} \end{cases}$$

We then have $(d\tilde\varphi \circ \dot\sigma_1)(t) = \epsilon\dot\sigma_2(\epsilon t)$. By Lemma 4.1, we have $\kappa_1 = \kappa_2 = \kappa$ and $\kappa \neq 0$. Since $\tilde\varphi$ is $\pm$-holomorphic, that is, it is either holomorphic or anti-holomorphic, we have

$$\begin{aligned} \kappa J\dot\sigma_2 &= \widetilde{\nabla}_{\dot\sigma_2}\dot\sigma_2 = \widetilde{\nabla}_{d\tilde\varphi\circ\dot\sigma_1}(d\tilde\varphi \circ \dot\sigma_1) = d\tilde\varphi\big(\widetilde{\nabla}_{\dot\sigma_1}\dot\sigma_1\big) \\ &= d\tilde\varphi(\kappa J\dot\sigma_1) = \pm\kappa_1 J(d\tilde\varphi \circ \dot\sigma_1) = \pm\epsilon\kappa J\dot\sigma_2. \end{aligned}$$

Hence, we find that $\epsilon = 1$ when $\tilde\varphi$ is holomorphic and $\epsilon = -1$ when $\tilde\varphi$ is anti-holomorphic. Therefore, we obtain

$$\begin{aligned} \rho_{\sigma_2} &= \langle \dot\sigma_2, -J\mathcal{N}_{M_2}\rangle = \epsilon\langle d\tilde\varphi(\dot\sigma_1), -Jd\tilde\varphi(\mathcal{N}_{M_1})\rangle \\ &= \langle d\tilde\varphi(\dot\sigma_1), d\tilde\varphi(-J\mathcal{N}_{M_1})\rangle = \langle \dot\sigma_1, \xi_{M_1}\rangle = \rho_{\sigma_1}. \end{aligned}$$

In particular, we have $\kappa_1\rho_{\sigma_1} = \kappa_2\rho_{\sigma_2}$. Moreover, if M_1, M_2 are real hypersurfaces of type (A_2), denoting by $TM_i = V_\lambda^{(i)} \oplus V_\mu^{(i)} \oplus \mathbb{R}\xi_{M_i}$ $(i = 1, 2)$, the decompositions into subbundles of principal curvature vectors, we have $d\varphi(V_\lambda^{(1)}) = V_\lambda^{(2)}$ and $d\varphi(V_\mu^{(1)}) = V_\mu^{(2)}$. Thus we find $\omega_{\sigma_1} = \omega_{\sigma_2}$.

On the other hand, we suppose that M_1, M_2 are isometric to each other. Also we suppose that $\rho_{\sigma_1} = \rho_{\sigma_2}$ when these real hypersurfaces M_1, M_2 are totally η-umbilic and that $\rho_{\sigma_1} = \rho_{\sigma_2}$ and $\omega_{\sigma_1} = \omega_{\sigma_2}$ when they are real hypersurfaces of type (A_2). We take an isometry $\tilde{\varphi}$ of $\mathbb{C}H^n(c)$ with $\tilde{\varphi}(M_1) = M_2$. Since it is $\pm$-holomorphic, the curves $\tilde{\varphi}|_{M_1} \circ \sigma_1$ and σ_2 satisfy the congruency conditions in §3. Hence, there is an isometry ψ of M_2 satisfying $\psi \circ (\tilde{\varphi}|_{M_1} \circ \sigma_1)(t) = \sigma_2(t)$ for all t. It is well known that isometries on real hypersurfaces of type (A) are equivariant. We therefore have an isometry $\widetilde{\psi}$ of $\mathbb{C}H^n(c)$ satisfying $\widetilde{\psi}\big|_{M_2} = \psi$. Thus, the isometry $\widetilde{\psi} \circ \tilde{\varphi}$ of $\mathbb{C}H^n(c)$ satisfies $(\widetilde{\psi} \circ \tilde{\varphi})(M_1) = M_2$ and $(\widetilde{\psi} \circ \tilde{\varphi}) \circ \sigma_1(t) = \sigma_2(t)$. Therefore, expressions (M_1, σ_1) and (M_2, σ_2) are congruent to each other.

Our discussion guarantees that if there are two congruent expressions (M_1, σ_1) and (M_2, σ_2) of a trajectory γ for a Kähler magnetic field $\mathbb{B}_\kappa$ by trajectories for Sasakian magnetic fields, then there exists a holomorphic isometry $\tilde{\varphi}$ of $\mathbb{C}H^n(c)$ satisfying $\tilde{\varphi}(M_1) = M_2$ and $\tilde{\varphi} \circ \sigma_1(t) = \sigma_2(t)$ for all t.

Lemma 4.2. *Let (M_1, σ_1) and (M_2, σ_2) be two expressions of a trajectory for a non-null Kähler magnetic field on $\mathbb{C}H^n(c)$ by trajectories for Sasakian magnetic fields $\mathbb{F}_{\kappa_1}, \mathbb{F}_{\kappa_2}$ on real hypersurfaces of type (A). They are congruent to each other if and only if they satisfy one of the following conditions:*

1) $\kappa_1 = \kappa_2$, $\rho_{\sigma_1} = \rho_{\sigma_2}$ *and M_1, M_2 are isometric and are totally η-umbilic,*
2) $\kappa_1 = \kappa_2$, $\rho_{\sigma_1} = \rho_{\sigma_2}$, $\omega_{\sigma_1} = \omega_{\sigma_2}$ *and M_1, M_2 are isometric and are of type* (A_2).

In this case, there is a holomorphic isometry $\tilde{\varphi}$ of $\mathbb{C}H^n(c)$ satisfying $\tilde{\varphi}(M_1) = M_2$ and $\tilde{\varphi} \circ \sigma_1(t) = \sigma_2(t)$ for all t.

We note here that if γ is a trajectory for $\mathbb{B}_\kappa$ on $\mathbb{C}H^n(c)$ the curve τ defined by $\tau(t) = \gamma(-t)$ is a trajectory for $\mathbb{B}_{-\kappa}$. If (M, σ) is an expression of γ, then (M, ς) defined by $\varsigma(t) = \sigma(-t)$ is an expression of τ. Therefore, we only need to study the case $\kappa \geq 0$.

5. Expressions of trajectories on $\mathbb{C}H^n$ by geodesics

First we study expressions by geodesics on real hypersurfaces of type (A). Since we have

$$\langle A_M\dot\sigma,\dot\sigma\rangle = \begin{cases} \lambda_M(1-\rho_\sigma^2)+\delta_M\rho_\sigma^2, & \text{when } M=HS, G(r), T(r), \\ \lambda_M\omega_\sigma^2+\mu_M(1-\omega_\sigma^2-\rho_\sigma^2)+\delta_M\rho_\sigma^2, & \text{when } M=T_\ell(r), \end{cases}$$

by checking principal curvatures in §3, we find that $\langle A_M\dot\sigma,\dot\sigma\rangle$ is positive. Thus, for geodesics on real hypersurfaces of type (A), by Lemma 4.1, only in the case that their structure torsions are ± 1, they are seen as trajectories on $\mathbb{C}H^n(c)$. Considering principal curvatures associated with characteristic vector fields, we find that extrinsic shapes of geodesics on HS are trajectories for Kähler magnetic fields of strength $\sqrt{|c|}$ and those on other real hypersurfaces of type (A) whose radii are r are for Kähler magnetic fields of strength $\sqrt{|c|}\coth(\sqrt{|c|}\,r)$. If we vary the radius r, the function $\sqrt{|c|}\coth(\sqrt{|c|}\,r)$ is monotone decreasing with respect to r and takes all values in the interval $(\sqrt{|c|},\infty)$. Therefore, we obtain the following results.

Theorem 5.1. *Trajectories for $\mathbb{B}_\kappa$ ($|\kappa|<\sqrt{|c|}$) on $\mathbb{C}H^n(c)$ are not expressed by geodesics on real hypersurfaces of type* (A).

Theorem 5.2. *Let γ be a trajectory for $\mathbb{B}_\kappa$ on $\mathbb{C}H^n(c)$.*

(1) *When $|\kappa|=\sqrt{|c|}$, it is expressed by a geodesic on some horosphere.*
(2) *When $|\kappa|>\sqrt{|c|}$, it is expressed by a geodesic on some geodesic sphere and by a geodesic on some tube around $\mathbb{C}H^\ell$ for each given ℓ ($1\le\ell\le n-1$).*

Remark 5.1. The expressions in Theorem 5.2 are unique up to congruency if we restrict the type of real hypersurfaces. For example, if a trajectory on $\mathbb{C}H^n$ is expressed two ways by geodesics on geodesic spheres, these expressions are congruent to each other.

6. Expressions by non-geodesic trajectories

In this section we study expressions of trajectories for Kähler magnetic fields on $\mathbb{C}H^n(c)$ by non-geodesic trajectories for Sasakian magnetic fields on real hypersurfaces of type (A). Hence we study the case of the second condition in Lemma 4.1.

First we study on totally η-umbilic real hypersurfaces. As we see in §4, we have $\langle A_M\dot\sigma,\dot\sigma\rangle$ is positive. Hence $\rho_\sigma\neq 0$. By congruency of trajectories for Kähler magnetic fields, we may suppose $\kappa > 0$. We consider the function $\kappa(\rho) = \lambda_M \rho^{-1} + (\delta_M - \lambda_M)\rho$ on the interval $(0,1)$. When $M = HS$, it is monotone decreasing and takes all values in the interval $(\sqrt{|c|}, \infty)$. When $M = G(r)$, it is also monotone decreasing and takes all values in the interval $(\sqrt{|c|}\coth(\sqrt{|c|}\,r), \infty)$. When $M = T(r)$, it is monotone decreasing in the interval $\big(0, \tanh(\sqrt{|c|}\,r/2)\big)$ and is monotone increasing in the interval $\big(\tanh(\sqrt{|c|}\,r/2), 1\big)$. Since we have

$$\kappa\big(\tanh(\sqrt{|c|}\,r/2)\big) = \sqrt{|c|}, \quad \lim_{\rho\uparrow 1}\kappa(\rho) = \sqrt{|c|}\coth(\sqrt{|c|}\,r), \quad \lim_{\rho\downarrow 0}\kappa(\rho) = \infty,$$

we obtain the following results.

Theorem 6.1. *Trajectories for a Kähler magnetic field* $\mathbb{B}_\kappa$ *on* $\mathbb{C}H^n(c)$ *with* $|\kappa| < \sqrt{|c|}$ *are not expressed by non-geodesic trajectories on totally* η*-umbilic real hypersurfaces.*

Theorem 6.2. *Every trajectory for a Kähler magnetic field* $\mathbb{B}_\kappa$ *on* $\mathbb{C}H^n(c)$ *with* $|\kappa| > \sqrt{|c|}$ *is expressed by a non-geodesic trajectory on a horosphere. This trajectory is for* $\mathbb{F}_\kappa$*, and such an expression is unique up to congruency.*

Theorem 6.3. *We take a trajectory for a Kähler magnetic field* $\mathbb{B}_\kappa$ *on* $\mathbb{C}H^n(c)$ *with* $|\kappa| > \sqrt{|c|}$.

(1) *It is expressed by a non-geodesic trajectory on some geodesic sphere. This trajectory is for* $\mathbb{F}_\kappa$.
(2) *If we consider only geodesic spheres of radii* r*, when* $|\kappa| > \sqrt{c}\coth(\sqrt{c}\,r)$ *it is uniquely expressed by a non-geodesic trajectory on such a geodesic sphere up to congruency.*

Theorem 6.4. *We take a trajectory for a Kähler magnetic field* $\mathbb{B}_\kappa$ *on* $\mathbb{C}H^n(c)$ *with* $|\kappa| \geq \sqrt{|c|}$.

(1) *It is expressed by a non-geodesic trajectory on some tube around* $\mathbb{C}H^{n-1}$*. This trajectory is for* $\mathbb{F}_\kappa$.
(2) *Even if we fix the radius* r *of such a tube, its expressions by non-geodesic trajectories are not unique when* $\sqrt{|c|} < |\kappa| < \sqrt{|c|}\coth(\sqrt{|c|}\,r)$*. When* $|\kappa| = \sqrt{c}$ *or when* $|\kappa| \geq \sqrt{|c|}\coth(\sqrt{|c|}\,r)$*, it is uniquely expressed by a non-geodesic trajectory on such a tube up to congruency.*

Next we consider a real hypersurface M of type (A_2). The second condition in Lemma 4.1 on a trajectory σ for $\mathbb{F}_\kappa$ on $T_\ell(r)$ in $\mathbb{C}H^n(c)$ turns into $\kappa\rho_\sigma = \lambda_M(\rho_\sigma^2 + \omega_\sigma^2) + \mu_M(1-\omega_\sigma^2)$ because we have $\delta_M = \lambda_M + \mu_M$.

We therefore need to find a pair (ρ, ω) satisfying

$$\begin{cases} (\lambda_M - \mu_M)\omega^2 = \kappa\rho - \lambda_M\rho^2 - \mu_M, \\ \rho^2 + \omega^2 \le 1, \quad |\rho| < 1. \end{cases} \tag{6.1}$$

Since $\lambda_M > \mu_M > 0$, considering the inequalities $0 \le \omega^2 \le 1 - \rho^2$, we find that if we have ρ which satisfies the following system of inequalities

$$\begin{cases} \lambda_M\rho^2 - \kappa\rho + \mu_M \le 0, \\ \mu_M\rho^2 - \kappa\rho + \lambda_M \ge 0, \\ |\rho| < 1, \end{cases} \tag{6.2}$$

then we can obtain a solution of (6.1). The first inequality in (6.2) shows that it does not have solutions when $|\kappa| < \sqrt{|c|}$, and that when $|\kappa| \ge \sqrt{|c|}$ its solutions satisfy

$$\Big(\kappa - \sqrt{\kappa^2 + c}\Big)/(2\lambda_M) \le \rho \le \Big(\kappa + \sqrt{\kappa^2 + c}\Big)/(2\lambda_M).$$

The second inequality in (6.2) shows that when $|\kappa| \ge \sqrt{|c|}$ its solutions satisfy

$$\rho \le \Big(\kappa - \sqrt{\kappa^2 + c}\Big)/(2\mu_M) \quad \text{or} \quad \rho \ge \Big(\kappa + \sqrt{\kappa^2 + c}\Big)/(2\mu_M).$$

As the above study, we only make mention of the case $\kappa > 0$. The solutions of (6.2) hence satisfy $\rho < 1$ and

$$\frac{1}{2\lambda_M}\Big(\kappa - \sqrt{\kappa^2 + c}\Big) \le \rho \le \min\Big\{\frac{1}{2\mu_M}\Big(\kappa - \sqrt{\kappa^2 + c}\Big), \frac{1}{2\lambda_M}\Big(\kappa + \sqrt{\kappa^2 + c}\Big)\Big\}.$$

And, if we take ρ satisfying the above, we get the solution of (6.1). Thus, we find infinitely many pairs (ρ, ω) satisfying (6.1), and get the following.

Theorem 6.5. *Each trajectory for $\mathbb{B}_\kappa$ with $|\kappa| \ge \sqrt{|c|}$ on $\mathbb{C}H^n(c)$ is expressed by a non-geodesic trajectory on a tube of type* (A_2)*. It is formed by a trajectory for $\mathbb{F}_\kappa$. When $|\kappa| > \sqrt{|c|}$, even if we fix the underlying hypersurface as $T_\ell(r)$, there are infinitely many non-congruent expressions.*

Theorem 6.6. *Trajectory for $\mathbb{B}_\kappa$ with $|\kappa| < \sqrt{|c|}$ on $\mathbb{C}H^n(c)$ are not expressed by non-geodesic trajectories on tubes of type* (A_2).

Combining our results in §5 and §6, we have

Corollary 6.1.

(1) *Non-horocyclic unbounded trajectories for Kähler magnetic fields on* $\mathbb{C}H^n$ *are not expressed by any trajectories for Sasakian magnetic fields on real hypersurfaces of type* (A).
(2) *Every horocyclic trajectory for Kähler magnetic field on* $\mathbb{C}H^n$ *is expressed by a geodesic on some horosphere and by a non-geodesic trajectory for a Sasakian magnetic field on some tube around* $\mathbb{C}H^{n-1}$.
(3) *Every bounded trajectory for a Kähler magnetic field on* $\mathbb{C}H^n$ *is expressed by some trajectory for a Sasakian magnetic field on some real hypersurface of type* (A).

The authors are hence interested in unbounded non-horocyclic trajectories for Kähler magnetic fields on $\mathbb{C}H^n$. Of course, just like how every geodesic on $\mathbb{R}^3$ lies on a totally geodesic $\mathbb{R}^2$, each of these trajectory lies on some totally geodesic complex hypersurface. For such a trajectory, is there a homogeneous real hypersurface containing it as a "nice" curve?

References

[1] T. Adachi, Kähler magnetic flows on a manifold of constant holomorphic sectional curvature, *Tokyo J. Math.* **18** (1995), 473–483.
[2] ———, Trajectories on geodesic spheres in a non-flat complex space form, *J. Geom.* **90** (2008), 1–29.
[3] Y. Aoki and T. Adachi, Moduli space of extrinsic circular trajectories on real hypersurfaces of type (A_2) in a complex hyperbolic space, *Note Math.* **42** (2022), 77-94.
[4] ———, Expressions of circles on a complex projective space by geodesics and by non-geodesic trajectories on geodeisc spheres, to appear in Osaka J. Math.
[5] T. Bao and T. Adachi, Circular trajectories on real hypersurfaces in a nonflat complex space form, *J. Geom.* **96** (2009), 41–55.
[6] P. Eberlein and R. O'Neill, Visibility manifolds, *Pasific J. Math.* **46** (1973), 45–109.
[7] R. Niebergall and P.J. Ryan, Real hypersurfaces in complex space forms, in *Tight and taut submanifolds, MSRI Publ.* **32** (1997), 233–305.
[8] T. Sunada, Magnetic flows on a Riemann surface, *Proc. KAIST Math. Workshop* **8** (1993), 93–108.

Received November 24, 2023

Modern Approaches to Differential Geometry
and its Related Fields 97 – 109

ON INSTABILITY OF F-YANG-MILLS CONNECTIONS

Kurando BABA

Department of Mathematics, Faculty of Science and Technology,
Tokyo University of Science, Noda, Chiba 278-8510, Japan
E-mail: kurando.baba@rs.tus.ac.jp

F-Yang-Mills connections are critical points of the F-Yang-Mills functional defined on the space of connections of a principal fiber bundle. Here, F is a nonnegative strictly increasing C^2-function. In this paper, we study the instability of the F-Yang-Mills connections. We derive a sufficient condition that any non-flat, F-Yang-Mills connection over the standard sphere is unstable.

This is an announcement of a not-yet-published paper entitled "A Simons type condition for instability of F-Yang-Mills connections" jointwork with Kazuto Shintani (arXiv:2301.0429).

Keywords: F-Yang-Mills connection; flat connection; instability; degree.

1. Introduction

A Yang-Mills connection is a critical point (connection) of the Yang-Mills functional defined on the space of connections of any principal fiber bundle over a Riemannian manifold. Several Yang-Mills type functionals were introduced, and critical connections of such functionals have been studied. In this paper, we will study the F-Yang-Mills functional and its critical connections, which were introduced by Dong-Wei [6]. Here, F is a nonnegative strictly increasing C^2-function on the interval $[0, \infty)$. The F-Yang-Mills functional is known as a generalization of the Yang-Mills functional, the p-Yang-Mills functional ([5]) and the exponential Yang-Mills functional ([14]). Critical connections of the F-Yang-Mills functional are called F-Yang-Mills connections.

There are many developments in the theory of the usual Yang-Mills connections. The study of the F-Yang-Mills connections has progressed by extending the results on Yang-Mills connections such as instability theorems (Simons [16], Kobayashi-Ohnita-Takeuchi [10]) and some types of vanishing theorems (Bourguignon-Lawson [4], Kobayashi-Ohnita-Takeuchi [10]). The purpose of this paper is to study the instability of F-Yang-Mills connections. Our motivation for studying the instability of F-Yang-Mills connections comes from the theory of characteristic classes. The flat connections on a principal fiber bundle P or its associated vector bundles are crucial in the computation of characteristic classes. By Chern-Weil theory, if P has a flat connection, then the Chern classes vanish. It is shown that any flat connection is a weakly stable F-Yang-Mills connection. In general, the converse does not hold. Our concern is to give a sufficient condition that any non-flat, F-Yang-Mills connection is unstable. We can make use of such a condition to study the existence of a flat connection of P. In the case of $F(t) = t$, the corresponding F-Yang-Mills connections are nothing but Yang-Mills connections. In this case, Simons ([16]) proved that, if $n > 4$, then any non-flat, Yang-Mills connection over the standard sphere S^n is unstable.

In this paper, we extend Simons' theorem for the instability of Yang-Mills connections to F-Yang-Mills connections as follows:

Theorem 1.1 ([3, Corollary 4.12]). *Assume that the degree of F' defined by*

$$d_{F'} = \sup_{t\in[0,\infty)} \frac{tF''(t)}{F'(t)}$$

is finite. If

$$n > 4d_{F'} + 4, \tag{1}$$

then any non-flat F-Yang-Mills connection over the n-dimensional standard sphere S^n is unstable.

Our theorem clarifies the importance of the finiteness of the degree $d_{F'}$ in order to derive the Simons type theorem for the instability of F-Yang-Mills connections. In the case of Yang-Mills connections ($F(t) = t$), we have $d_{F'} = 0$. Hence, the inequality (1) becomes $n > 4$, which coincides with Simons' theorem. Our theorem also gives a generalization of the instability theorem for the p-Yang-Mills connections due to Chen-Zhou ([5]). On the other hand, we cannot apply our theorem to exponential Yang-Mills connections ($F(t) = \exp(t)$), since the corresponding degree $d_{F'}$ is infinite. We have an alternative type instability theorem for exponential Yang-Mills connections:

Theorem 1.2 ([3, Proposition 4.14]). *Assume that $n > 4$. Let ∇ be a non-flat exponential Yang-Mills connection over S^n. If the norm of the* *curvature* R^∇ *of* ∇ *satisfies*

$$\|R^\nabla\| < \sqrt{\frac{n-4}{2}},$$

then ∇ *is unstable.*

The organization of this paper is as follows: In Section 2, we review basics of the F-Yang-Mills connections. We recall the definitions of the F-Yang-Mills functional and the F-Yang-Mills connections. We also give the first and second variational formulas for the F-Yang-Mills functional. In Section 3, we give a brief proof of Theorems 1.1 and 1.2. In Section 4, we give problems for further direction to extend Theorem 1.1, and explain ongoing studies.

2. Basics of F-Yang-Mills connection

2.1. *Preliminaries*

Let M be an oriented connected closed Riemannian manifold. Let G be a compact Lie group with Lie algebra $\mathfrak{g}$. We denote by $\mathrm{Ad} : G \to \mathrm{GL}(\mathfrak{g})$ the adjoint representation of G on $\mathfrak{g}$. Let P be a principal fiber bundle over M with structure group G. A $\mathfrak{g}$-valued 1-form A on P is called a connection (or a connection 1-form) if A is of type Ad and $A(\tilde{X}) = X$ holds for all $X \in \mathfrak{g}$, where $\tilde{X}$ denotes the fundamental vector field on P associated with X. Let $\mathfrak{g}_P = P \times_{\mathrm{Ad}} \mathfrak{g}$ denote the adjoint bundle of P. We set $\Omega^k(\mathfrak{g}_P) = \Lambda^k T^*M \otimes \mathfrak{g}_P$ for a nonnegative integer k. Each connection on P induces a connection on $\mathfrak{g}_P$. This also induces a connection on $\Omega^k(\mathfrak{g}_P)$. The space of connections on $\mathfrak{g}_P$, which we write as $\mathcal{C}$, becomes an (infinite dimensional) affine space whose associated vector space is $\Omega^1(\mathfrak{g}_P)$. Let ∇ be a connection on $\mathfrak{g}_P$. The tangent space of $\mathcal{C}$ at ∇ is naturally isomorphic to $\Omega^1(\mathfrak{g}_P)$. We denote by R^∇ its curvature 2-form. Then R^∇ is an element of $\Omega^2(\mathfrak{g}_P)$. The covariant exterior derivative $d^\nabla : \Omega^k(\mathfrak{g}_P) \to \Omega^{k+1}(\mathfrak{g}_P)$ relative to ∇ is given by

$$\begin{aligned} d^\nabla\varphi(X_1,\dots,X_{k+1}) &= \sum_{i=1}^{k+1}(-1)^{i+1}\nabla_{X_i}(\varphi(X_1,\dots,\hat{X}_i,\dots,X_{k+1})) \\ &\quad + \sum_{i<j}(-1)^{i+j}\varphi([X_i,X_j],X_1,\dots,\hat{X}_i,\dots,\hat{X}_j,\dots,X_{k+1}), \end{aligned}$$

where the symbols $\hat{X}_i$ and $\hat{X}_j$ are omitted. Then the following equation holds:

$$d^\nabla R^\nabla = 0,$$

which is called the *Bianchi identity*. In general, $d^\nabla \circ d^\nabla$ does not vanish. It is shown that $R^\nabla = 0$ yields $d^\nabla \circ d^\nabla = 0$. A connection is said to be *flat* if its curvature 2-form vanishes.

We give a fiber metric on $\mathfrak{g}_P$ which is compatible with connections on $\mathfrak{g}_P$. Such a fiber metric is induced from an adjoint invariant inner product $\langle \cdot, \cdot \rangle$ on $\mathfrak{g}$ (cf. [7, Proposition 5.9.7]). In addition, $\langle \cdot, \cdot \rangle$ also induces a pointwise inner product on the vector space $\Omega^k(\mathfrak{g}_P)$, which is denoted by the same symbol $\langle \cdot, \cdot \rangle$. We set $\|\varphi\|^2 = \langle \varphi, \varphi \rangle$ for $\varphi \in \Omega^k(\mathfrak{g}_P)$. By integrating the pointwise inner product over the base manifold M, we get an inner product on $\Omega^k(\mathfrak{g}_P)$ as follows:

$$(\varphi, \psi) = \int_M \langle \varphi, \psi \rangle dv, \quad \varphi, \psi \in \Omega^k(\mathfrak{g}_P), \tag{2}$$

where dv denotes the Riemannian volume form.

2.2. *F-Yang-Mills functional and F-Yang-Mills connection*

Definition 2.1 ([6, Definitions 3.1 and 3.2]). *Let $F : [0, \infty) \to \mathbb{R}$ be a nonnegative strictly increasing C^2-function. The F-*Yang-Mills functional *$\mathcal{YM}_F : \mathcal{C} \to \mathbb{R}$ is defined by*

$$\mathcal{YM}_F(\nabla) = \int_M F\left(\frac{1}{2}\|R^\nabla\|^2\right) dv.$$

*A critical point (connection) ∇ of the functional $\mathcal{YM}_F$ is called an F-*Yang-Mills connection.

We give examples of three types of F-Yang-Mills connections.

Example 2.1. (i) In the case when $F(t) = t$, the F-Yang-Mills functional is nothing but the usual Yang-Mills functional $\mathcal{YM} : \mathcal{C} \to \mathbb{R}$:

$$\mathcal{YM}(\nabla) = \frac{1}{2}\int_M \|R^\nabla\|^2 \, dv,$$

which indicates the total curvature of ∇. Thus, in this case, F-Yang-Mills connections are Yang-Mills connections. (ii) Let $p \geq 2$. In the case when $F = F_p : t \mapsto (1/p)(2t)^{p/2}$, F_p-Yang-Mills connections are known as p-Yang-Mills connections (cf. [5]). Clearly, 2-Yang-Mills connections are Yang-Mills connections. (iii) In the case when $F = F_e : t \mapsto \exp(t)$, F_e-Yang-Mills connections are also known as exponential Yang-Mills connections (cf. [14]).

2.3. *First variational formula for F-Yang-Mills functional*

We will recall the first variational formula for $\mathcal{YM}_F$. Let δ^∇ denote the formal adjoint operator of d^∇ relative to the inner product (2). Namely, $(d^\nabla\varphi, \psi) = (\varphi, \delta^\nabla\psi)$ holds for $\varphi \in \Omega^k(\mathfrak{g}_P)$ and $\psi \in \Omega^{k+1}(\mathfrak{g}_P)$.

Theorem 2.1 (First variational formula, [6, Lemma 3.1]). *Let $\nabla \in \mathcal{C}$. Let ∇^t be a C^∞-curve in $\mathcal{C}$ with $\nabla^0 = \nabla$. We put*

$$\alpha = \left.\frac{d}{dt}\right|_{t=0} \nabla^t \in \Omega^1(\mathfrak{g}_P).$$

Then we have

$$\left.\frac{d}{dt}\right|_{t=0} \mathcal{YM}_F(\nabla^t) = \int_M \left\langle \delta^\nabla \left(F'\left(\frac{1}{2}\|R^\nabla\|^2\right) R^\nabla \right), \alpha \right\rangle dv. \tag{3}$$

It follows from (3) that $\nabla \in \mathcal{C}$ is an F-Yang-Mills connection if and only if it satisfies the following equation:

$$\delta^\nabla \left(F'\left(\frac{1}{2}\|R^\nabla\|^2\right) R^\nabla \right) = 0. \tag{4}$$

This equation is called the *F-Yang-Mills equation*. For example, if we put $F(t) = t$, (4) is rewritten as $\delta^\nabla R^\nabla = 0$. This is known as the usual Yang-Mills equation.

2.4. *Definition of instability of F-Yang-Mills connection*

Definition 2.2. An F-Yang-Mills connection ∇ is said to be *weakly stable* if the following inequality holds:

$$\left.\frac{d^2}{dt^2}\right|_{t=0} \mathcal{YM}_F(\nabla^t) \geq 0,$$

for any C^∞-curve ∇^t in $\mathcal{C}$ with $\nabla^0 = \nabla$. An F-Yang-Mills connection ∇ is said to be *unstable* if ∇ is not weakly stable.

Example 2.2. Any flat connection ∇ is a weakly stable F-Yang-Mills connection.

It is a fundamental problem to find weakly stable F-Yang-Mills connections other than flat connections. However, our concern is to study the instability of F-Yang-Mills connections. As mentioned in Theorem 1.1, our main result means that the flat connections are the only weakly stable F-Yang-Mills connections in the case when the base space of P is the standard sphere with higher dimension.

2.5. *Second variational formula for F-Yang-Mills functional*

We give the second fundamental variational formula for F-Yang-Mills functional. Let $\mathfrak{R}^\nabla : \Omega^1(\mathfrak{g}_P) \to \Omega^1(\mathfrak{g}_P)$ denote the Weitzenböck curvature of ∇ (see [4, (3.1)] for the definition). It is shown that $\mathfrak{R}^\nabla$ satisfies the following relation:

$$\langle \mathfrak{R}^\nabla(\alpha), \alpha \rangle = \langle [\alpha \wedge \alpha], R^\nabla \rangle,$$

where $[\cdot \wedge \cdot] : \Omega^1(\mathfrak{g}_P) \times \Omega^1(\mathfrak{g}_P) \to \Omega^2(\mathfrak{g}_P)$ is defined by

$$[\alpha \wedge \beta](X, Y) = [\alpha(X), \beta(Y)] - [\alpha(Y), \beta(X)],$$

for all $X, Y \in \mathfrak{X}(M) = \Gamma(TM)$. We note that the Weitzenböck curvature was introduced by describing the difference between the Hodge Laplacian $\Delta^\nabla = \delta^\nabla \circ d^\nabla + d^\nabla \circ \delta^\nabla$ and the rough Laplacian ([4, (3.2) Theorem]).

Theorem 2.2 (Second variational formula, [3, Proposition 3.7]). *Let $\nabla \in \mathcal{C}$ be an F-Yang-Mills connection and ∇^t be a C^∞-curve in $\mathcal{C}$ with $\nabla^0 = \nabla$. We put*

$$\alpha = \left.\frac{d}{dt}\right|_{t=0} \nabla^t \in \Omega^1(\mathfrak{g}_P).$$

Then we have

$$\left.\frac{d^2}{dt^2}\right|_{t=0} \mathcal{YM}_F(\nabla^t) = \int_M F''\left(\frac{1}{2}\|R^\nabla\|^2\right) \langle d^\nabla \alpha, R^\nabla \rangle^2 \, dv + \int_M F'\left(\frac{1}{2}\|R^\nabla\|^2\right) \{\langle \mathfrak{R}^\nabla(\alpha), \alpha \rangle + \|d^\nabla \alpha\|^2\} \, dv.$$

3. Instability of *F*-Yang-Mills connection over the standard sphere

The aim of this paper is to generalize the following Simons' theorem for Yang-Mills connections to F-Yang-Mills connections.

Theorem 3.1 ([16]). *If $n > 4$, then any non-flat Yang-Mills connection over the n-dimensional standard sphere S^n is unstable.*

We will give a historical remark about the proof of Theorem 3.1. This theorem was originally announced in Symposium at Tokyo by Simons ([16]). After him, Bourguignon-Lawson ([4]) gave its proof. In order to generalize

Simons' theorem, we observe their proof. Their method is based on the so-called *average method.*

Based on Theorem 2.2, we define the *index form* $\mathcal{I}_\nabla$ for an F-Yang-Mills connection ∇ as follows:

$$\mathcal{I}_\nabla(\alpha) = \int_M F''\left(\frac{1}{2}\|R^\nabla\|^2\right)\langle d^\nabla \alpha, R^\nabla\rangle^2 \, dv + \int_M F'\left(\frac{1}{2}\|R^\nabla\|^2\right)\{\langle \mathfrak{R}^\nabla(\alpha),\alpha\rangle + \|d^\nabla\alpha\|^2\}\, dv,$$

for any $\alpha \in \Omega^1(\mathfrak{g}_P)$. We note that $\mathcal{I}_\nabla(\alpha)$ is quadratic for $\alpha \in \Omega^1(\mathfrak{g}_P)$. By the second variational formula, ∇ is weakly stable if and only if $\mathcal{I}_\nabla(\alpha) \geq 0$ holds for all $\alpha \in \Omega^1(\mathfrak{g}_P)$.

Assume that M^n is isometrically immersed in the Euclidean space $\mathbb{R}^N$ $(n < N)$. Let $\{E_A\}_{A=1}^N$ denote the standard basis of $\mathbb{R}^N$. For each $1 \leq A \leq N$, we write V_A as the tangent component of E_A with respect to $M \subset \mathbb{R}^N$. For each $1 \leq A \leq N$, we obtain the $\mathfrak{g}_P$-valued 1-form $\iota_{V_A}R^\nabla$ which is defined by

$$\iota_{V_A}R^\nabla(X) = R^\nabla(V_A, X), \quad X \in \mathfrak{X}(M).$$

Remark 3.1. Following Nayatani-Urakawa ([17]), we can give a geometrical description of $\iota_V R^\nabla \in \Omega^1(\mathfrak{g}_P)$ for any tangent vector field V on M. We denote by $\tilde{V}$ the horizontal lift of V with respect to ∇. The one-parameter group of transformations corresponding to $\tilde{V}$ gives a curve in $\mathcal{C}$ through ∇ whose velocity vector at ∇ coincides with $\iota_V R^\nabla$.

We consider the average of $\mathcal{I}_\nabla(\iota_{V_A}R^\nabla)$ for $1 \leq A \leq N$:

$$\sum_{A=1}^N \mathcal{I}_\nabla(\iota_{V_A}R^\nabla), \tag{5}$$

which is also called the trace of R^∇. The following lemma plays a fundamental role in our argument.

Lemma 3.1. *For any F-Yang-Mills connection ∇, if*

$$\sum_{A=1}^N \mathcal{I}_\nabla(\iota_{V_A}R^\nabla) < 0, \tag{6}$$

then ∇ is unstable.

Proof. We prove this lemma by contraposition. Let ∇ be a weakly stable F-Yang-Mills connection. Then, for any $\alpha \in \Omega^1(\mathfrak{g}_P)$, we have $\mathcal{I}_\nabla(\alpha) \geq 0$. In particular, by substituting $\alpha = \iota_{V_A} R^\nabla$, we have $\mathcal{I}_\nabla(\iota_{V_A} R^\nabla) \geq 0$, which yields

$$\sum_{A=1}^{N} \mathcal{I}_\nabla(\iota_{V_A} R^\nabla) \geq 0.$$

We have completed the proof. □

This lemma indicates that, if (6) holds then, for any non-flat F-Yang-Mills connection ∇, there exists at least one $A \in \{1, \dots, N\}$ such that the F-Yang-Mills functional $\mathcal{YM}_F$ decreases along the direction $\iota_{V_A} R^\nabla$.

We are ready to show the proof of Theorem 3.1:

Proof of Theorem 3.1 (Sketch). In this theorem, we have considered the case when $F(t) = t$. Let S^n be the unit standard sphere in $\mathbb{R}^{n+1}$. Bourguignon-Lawson ([4, (7.7) Theorem]) proved that

$$\sum_{A=1}^{n+1} \mathcal{I}_\nabla(\iota_{V_A} R^\nabla) = 2(4-n) \int_{S^n} \|R^\nabla\|^2 \, dv. \tag{7}$$

Hence $n > 4$ implies that the right-hand side of (7) becomes negative, so that Lemma 3.1 yields the assertion. □

In order to generalize Simons' theorem, for any non-flat F-Yang-Mills connection ∇ over S^n, we need to evaluate the average (5) and to derive a (necessary and) sufficient condition for ∇ satisfying (6). We have the following proposition.

Proposition 3.1. *Let ∇ be an F-Yang-Mills connection over S^n. Then the average* (5) *has the following description:*

$$\begin{aligned} \sum_{A=1}^{n+1} \mathcal{I}_\nabla(\iota_{V_A} R^\nabla) &= 4 \int_{S^n} F''\left(\frac{1}{2}\|R^\nabla\|^2\right) \|R^\nabla\|^4 \, dv \\ &\quad + 2(4-n) \int_{S^n} F'\left(\frac{1}{2}\|R^\nabla\|^2\right) \|R^\nabla\|^2 \, dv. \end{aligned} \tag{8}$$

This proposition is a special case of [3, Theorem 4.3]. In order to derive (8) from this theorem, we have used the fact that the Riemann curvature R of S^n and the second fundamental form h of $S^n \subset \mathbb{R}^{n+1}$ are described as follows:

$$R(X,Y)Z = \langle Z, Y\rangle X - \langle Z, X\rangle Y, \quad h(X,Y) = \langle X, Y\rangle \xi,$$

for tangent vectors X, Y, Z of S^n, where ξ is the unit normal vector field on $S^n \subset \mathbb{R}^{n+1}$ pointed toward the center.

For further calculation, we introduce a *degree* of F', which we write $d_{F'}$ ([3, Definition 4.8]):

$$d_{F'} = \sup_{t \in [0,\infty)} \frac{tF''(t)}{F'(t)}. \tag{9}$$

Example 3.1. We obtain the degree $d_{F'}$ for each function F given in Example 2.1. (i) In the case when $F(t) = t$, we have $d_{F'} = 0$. (ii) Let $p \geq 2$ and $F = F_p$. We get $d_{F'_p} = (p-2)/2$. (iii) In the case when $F = F_e$, we have $d_{F'_e} = \infty$.

We are ready to show the main result as stated in Introduction.

Theorem 3.2 ([3, Corollary 4.12], Theorem 1.1). *Assume that $d_{F'}$ is finite. If*

$$n > 4d_{F'} + 4, \tag{10}$$

then any non-flat F-Yang-Mills connection over the n-dimensional standard sphere S^n is unstable.

Proof. Let ∇ be a non-flat F-Yang-Mills connection over S^n. By the definition of $d_{F'}$, we have:

$$d_{F'} \geq \frac{\|R^\nabla\|^2}{2} \frac{F''(\|R^\nabla\|^2/2)}{F'(\|R^\nabla\|^2/2)}.$$

By using this inequality, (8) yields

$$\sum_{A=1}^{N} \mathcal{I}_\nabla(\iota_{V_A} R^\nabla) \leq 2(4d_{F'} + 4 - n) \int_{S^n} \|R^\nabla\|^2 \, dv.$$

Thus, we have the assertion by a similar argument in the proof of Theorem 3.1. □

Clearly, we can observe that our theorem is an extension of Simons' theorem because $d_{F'} = 0$ holds in the case when $F(t) = t$. Moreover, our result is also an extension of Chen-Zhou's theorem for the instability for p-Yang-Mills connections over S^n ([5]). Indeed, in the case when $F = F_p$ as in Example 2.1, (ii), the inequality (10) becomes $n > 2p$. Then any non-flat p-Yang-Mills connection is unstable.

On the other hand, we can find an alternative formula of the instability theorem for F-Yang-Mills connections over S^n by Jia-Zhou [8]. The advantage of Theorem 3.2 compared to them is that we have introduced the notion of the degree $d_{F'}$ and clarified the finiteness of $d_{F'}$ in order to derive the Simons type condition for the instability of F-Yang-Mills connections over S^n.

In the case when $d_{F'}$ has infinite value, we can observe an alternative type instability theorem for F-Yang-Mills connections. Let us consider the case of $F = F_e$ as in Example 2.1, (iii). For any exponential Yang-Mills connection ∇, a direct calculation shows that (8) is rewritten as follows:

$$\sum_{A=1}^{n+1} \mathcal{I}_{\nabla}(\iota_{V_A} R^{\nabla}) = 2 \int_{S^n} \exp\left(\frac{1}{2}\|R^{\nabla}\|^2\right) \|R^{\nabla}\|^2 \left\{2\|R^{\nabla}\|^2 - (n-4)\right\} dv.$$

Then we have the other main result as stated in Introduction.

Theorem 3.3 ([3, Proposition 4.14], Theorem 1.2). *Assume that $n > 4$. Let ∇ be a non-flat exponential Yang-Mills connection over S^n. If the norm of R^{∇} satisfies*

$$\|R^{\nabla}\| < \sqrt{\frac{n-4}{2}},$$

then ∇ is unstable.

Now, it is known that there are strong similarities between Yang-Mills theory and harmonic map theory. We will discuss counterpart of our results in harmonic map theory. The counterpart of the notion of F-Yang-Mills connections is known as F-harmonic map, which was introduced by Ara ([1]). This notion is a generalization of harmonic maps, p-harmonic maps, exponentially harmonic maps and so on. Moreover, he derived the instability theorem for F-harmonic map as follows.

Theorem 3.4 ([1, Theorem 7.1]). *Let M be a closed Riemannian manifold. For any F-harmonic map $\phi : M \to S^n$, if*

$$\int_M \|d\phi\|^2 \left\{\|d\phi\|^2 F''\left(\frac{1}{2}\|d\phi\|^2\right) + (2-n)F'\left(\frac{1}{2}\|d\phi\|^2\right)\right\} dv < 0, \quad (11)$$

then ϕ is unstable.

Here, we note that the left-hand side of inequality (11) corresponds to the average for the F-harmonic map setting (see [1, (7.5)]). On the other hand, in the case when our degree $d_{F'}$ defined in (9) is finite, we have the following inequality:

$$\int_M \|d\phi\|^2 \left\{ \|d\phi\|^2 F''\left(\frac{1}{2}\|d\phi\|^2\right) + (2-n)F'\left(\frac{1}{2}\|d\phi\|^2\right) \right\} dv$$
$$\leq \int_M \|d\phi\|^2 F'\left(\frac{1}{2}\|d\phi\|^2\right) \{2d_{F'} + (2-n)\}\, dv.$$

Therefore, we obtain the following theorem as a counterpart of Theorem 3.2:

Theorem 3.5. *Assume that $d_{F'}$ is finite. Then, if*

$$n > 2d_{F'} + 2, \tag{12}$$

then any non-constant F-harmonic map $\phi : M \to S^n$ is unstable.

The inequality (12) is an extension of Leung's theorem for the instability of harmonic maps ([13]). We also find a counterpart of Theorem 3.3 in Koh's theorem for the instability of exponentially harmonic maps ([11]).

4. Problems and ongoing studies

We give some further direction to extend Theorem 3.2. It is a natural problem to find an instability theorem of F-Yang-Mills connections over other base spaces. Kobayashi-Ohnita-Takeuchi ([10]) developed the average method to study the instability of Yang-Mills connections in the case when the base space is an irreducible compact Riemannian symmetric space.

Let M be an irreducible compact Riemannian symmetric space with the normal homogeneous Riemannian metric. We denote by λ_1 the first eigenvalue of the Laplacian acting on the space of the functions on M, and by μ the maximum eigenvalue of the Riemann curvature operator on $\Lambda^2 T_x M$ $(x \in M)$. Kobayashi-Ohnita-Takeuchi proved:

Theorem 4.1 ([10, (7.10) Theorem]). *If*

$$\lambda_1 - 1 + 2\mu < 0, \tag{13}$$

then any non-flat Yang-Mills connection over M is unstable.

After them, Shintani ([15]) extended the inequality (13) for Yang-Mills connections to F-Yang-Mills connections as follows.

Theorem 4.2 ([15]). *Let γ denote the maximum eigenvalue of the shape operator for the first imbedding from M into $S^{N-1}(r)$ with $r =$*

$\sqrt{\lambda_1/\dim(M)}$. *Assume that* $0 \leq d_{F'} < \infty$. *If*

$$\lambda_1 - 1 + 2\mu + 4d_{F'}\left\{(\operatorname{codim}(M)-1)\gamma + \frac{\lambda_1}{\dim(M)}\right\} < 0, \tag{14}$$

then any non-flat F*-Yang-Mills connection is unstable.*

A similar type inequality for p-Yang-Mills connections is found in Kawagoe ([9]).

From the above argument, we propose the following two problems.

Problem 1. Classify irreducible compact Riemannian symmetric spaces M satisfying Shintani's inequality (14).

Problem 2. Find weakly stable F-Yang-Mills connections over M (if it exists) in the case when it does not satisfy his inequality.

Shintani's inequality coincides with Kobayashi-Ohnita-Takeuchi's one in the case when $F(t) = t$. In the case of $F(t) = t$, the above two problems were solved by Kobayashi-Ohnita-Takeuchi ([10, (7.11) Theorem]). Following their result, S^n $(n > 4)$, the Cayley projective plane $F_4/Spin(9)$ and E_6/F_4 are the only compact Riemannian symmetric spaces satisfying (13). In addition, Laquer [12] proved that the canonical connection is a weakly stable Yang-Mills connection over the irreducible compact Riemannian symmetric space except for S^n $(n > 4)$, $F_4/Spin(9)$ and E_6/F_4. Therefore, in the case when $F(t) = t$, our problems are solved completely. Our concern is to solve the problems in the case when the function F is general.

Finally, we explain ongoing studies to solve Problem 1. We note that (14) gives a sufficient condition for (13). Hence we would like to derive the instability theorem for $F_4/Spin(9)$ and E_6/F_4. One of the difficulties in solving Problem 1 is to determine the exact value of γ in (14). At this moment, we have a partial answer to Problem 1 as follows.

Theorem 4.3 ([2]). *If* $0 \leq d_{F'} < 1/6$, *then any non-flat* F*-Yang-Mills connection over* $F_4/Spin(9)$ *is unstable.*

References

[1] M. Ara, Geometry of F-Harmonic Maps, *Kodai. Math. J.* **22** (1999), 243–263.

[2] K. Baba, On instability of F-Yang-Mills connections over irreducible symmetric R-spaces, submitted.

[3] K. Baba and K. Shintani, A Simons type condition for instability of F-Yang-Mills connections, arXiv:2301.0429.

[4] J. P. Bourguignon and H. B. Lawson, Jr., Stability and Isolation Phenomena for Yang-Mills Fields, *Commun. Math. Phys.* **79** (1981), 189–230.

[5] Q. Chen and Z.-R. Zhou, On Gap Properties and Instabilities of p-Yang-Mills Fields, *Cand. J. Math.* **59** (2007), 1245–1259.

[6] Y. Dong and S. W. Wei, On Vanishing Theorems for Vector Bundle Valued p-Forms and their Applications, *Commun. Math. Phys.* **304** (2011), 329–368.

[7] M. Hamilton, *Mathematical Gauge Theory*, Springer Cham, 2017.

[8] G.-Y. Jia and Z.-R. Zhou, Stabilities of F-Yang-Mills Fields on Submanifolds, *Arch. Math. (Brno)* **49** (2013), 125–139.

[9] J. Kawagoe, Instability of p-Yang-Mills fields on submanifolds, *Master thesis* (supervisor: M. S. Tanaka), Tokyo University of Science, March 2015.

[10] S. Kobayashi, Y. Ohnita and M. Takeuchi, On instability of Yang-Mills connections, *Math. Z.* **193** (1986), 165–189.

[11] S. E. Koh, A nonexistence theorem for stable exponentially harmonic maps, *Bull. Korean Math. Soc.* **32** (1995), 211–214.

[12] H. T. Laquer, Stability properties of the Yang-Mills functional neart the canonical connection, *Mich. Math. J.* **31** (1984), 139–159.

[13] P. F. Leung, On the stability of harmonic maps, in *Harmonic Maps, Lecture Notes in Math.* **949**, Springer Verlag, 1982, 122–129.

[14] F. Matsuura and H. Urakawa, On exponential Yang-Mills connections, *J. Geom. Phys.* **17** (1995), 73–89.

[15] K. Shintani, Weakly stability of F-Yang-Mills connections, *Master thesis* (supervisor: K. Baba), Tokyo University of Science, March 2022.

[16] J. Simons, Gauge Fields, *Tokyo Symposium on Minimal Submanifolds and Geodesics*, 1977.

[17] S. Nayatani and H. Urakawa, Morse indices of Yang-Mills connections over the unit sphere, *Compositio Math.* **98** (1995), 177–192.

Received December 20, 2023
Revised February 6, 2024

Modern Approaches to Differential Geometry
and its Related Fields 111 – 121

TWO WEIGHT PROJECTIVE CODES AND SOME COMBINATORIAL OBJECTS

Paskal PIPERKOV* and Mariya DZHUMALIEVA-STOEVA†

*Faculty of Mathematics and Informatics, Veliko Tarnovo University,
2 Teodosii Tarnovski Str., 5000 Veliko Tarnovo, Bulgaria
E-mail: *p.piperkov@ts.uni-vt.bg, †m.dzhumalieva@ts.uni-vt.bg*

In this paper, we present an overview of some combinatorial objects having interesting structure, namely, association schemes and strongly regular graphs. We propose an approach to obtain a certain type of association schemes from two weight linear codes. We use the projective dual transform of linear codes, and its basic concept is also described in the article.

Keywords: Bose-Mesner algebra; strongly regular graphs; coding theory; dual transform; projective codes.

1. Introduction

Association schemes originally come from the field of statistics and experimental design, but they are also very useful in algebra, combinatorics, graph theory and coding theory. In these areas, the association schemes are first introduced by Bose and Shimamoto [4] in 1952. Several years later, Bose and Mesner [3] developed an unitary commutative associative algebra on the base of the adjacency matrices of an association scheme. The concept of strongly regular graphs as association schemes with 2 classes and the relation between them and partial geometries are due to Bose [2]. Later, Delsarte [10] proves some bounds in coding theory using association schemes. A survey on the results for association schemes in coding theory is also presented in Chapter 21 of the famous book of McWilliams and Sloane [12].

In this paper, we present a brief review of these objects and how they can be used to obtain some new structures. The paper is organized as follows. In the second section, the basic concepts of the association schemes, the Bose-Mesner algebra as well as the basic definition of a strongly regular graph are given. The third section is dedicated to coding theory. There are some basic definitions about linear codes, characteristic vector of a linear code and projective dual codes. In Section 4 we give an approach for possible construction of a strongly regular graph, using the projective dual transform of a special type linear codes.

2. Association schemes

2.1. *Definition of association scheme*

Let Ω be a set with n elements. We denote by I_n the $n \times n$ identity matrix, by O_n the $n \times n$ matrix all of whose entries are 0, and by J_n the $n \times n$ matrix all of whose entries are 1.

Definition 2.1 (Third definition in [1]). *An association scheme with s associate classes of a finite set Ω with n elements is a set of $n \times n$ matrices $A_0, A_1, \ldots, A_s$, all of whose entries are equal to 0 or 1, such that*

(1) $A_0 = I_n$;
(2) A_i *is symmetric for* $i = 1, \ldots, s$;
(3) *for all* $i, j \in \{1, \ldots, s\}$, *the product* A_iA_j *is a linear combination of* $A_0, A_1, \ldots, A_s$, *namely*

$$A_iA_j = \sum_{k=0}^{s} p_{i,j}^{(k)} A_k,$$

where the scalars $p_{i,j}^{(k)}$ *are real numbers;*
(4) *none of the* A_i *is equal to* O_n *and* $\sum_{i=0}^{s} A_i = J_n$.

In this paper, we study only *symmetric* association schemes. There exists a general definition for association schemes [10], where the transpose of a matrix A_i should be among the matrices $A_0, A_1, \ldots, A_s$. Obviously, if we get the sum of A_i and its transposed matrix A_i^T instead, we will obtain a symmetric association scheme.

One of the most popular examples of association schemes is the *Hamming* scheme [1]. Let Γ be an n elements set and $\Omega = \Gamma^m$. Let the $n^m \times n^m$ matrices A_i for $i = 0, 1, \ldots, n$ be defined as follows: For every m-tuples x and y in Ω

$$A_i(x, y) = \begin{cases} 1, & \text{if } x \text{ and } y \text{ differ in } i \text{ positions;} \\ 0, & \text{otherwise.} \end{cases}$$

In other words, the (x, y)-entry is 1 only in the matrix A_i, where i is the Hamming distance between x and y.

To illustrate Hamming schemes we can consider the matrix $H_{m,n} = \sum_{i=1}^{m} iA_i$. The matrices $H_{3,2}$ and $H_{2,3}$ obtained by the Boolean cube $\Omega =$

$\mathbb{F}_2^3$ and the ternary square $\Omega = \mathbb{F}_3^2$, respectively, are given bellow:

	000	001	010	011	100	101	110	111
000	0	1	1	2	1	2	2	3
001	1	0	2	1	2	1	3	2
010	1	2	0	1	2	3	1	2
011	2	1	1	0	3	2	2	1
100	1	2	2	3	0	1	1	2
101	2	1	3	2	1	0	2	1
110	2	3	1	2	1	2	0	1
111	3	2	2	1	2	1	1	0

	00	01	02	10	11	12	20	21	22
00	0	1	1	1	2	2	1	2	2
01	1	0	1	2	1	2	2	1	2
02	1	1	0	2	2	1	2	2	1
10	1	2	2	0	1	1	1	2	2
11	2	1	2	1	0	1	2	1	2
12	2	2	1	1	1	0	2	2	1
20	1	2	2	1	2	2	0	1	1
21	2	1	2	2	1	2	1	0	1
22	2	2	1	2	2	1	1	1	0

If the elements of Ω are lexicographically ordered, it is easy to prove by induction that

$$H_{0,n} = (0) \quad \text{and} \quad H_{m,n} = J_n \otimes H_{m-1,n} + (J_n - I_n) \otimes J_{n^{m-1}}, \quad m \geq 1,$$

where $\otimes$ means the Kroneker product.

A recurrent relation between the matrices A_i of a Hamming scheme is shown in the following lemma.

Lemma 2.1. *Let the integer n (> 1) be fixed and Γ be an ordered set of n elements. For some positive integer m and $\Omega = \Gamma^m$ let the $n^m \times n^m$ matrices $A_{0,m}, A_{1,m}, \ldots, A_{m,m}$ be defined as follows: The rows and the columns of $A_{i,m}$ are labeled lexicographically with the elements of Ω and for each two elements $x, y \in \Omega$ let (x, y)-entry of $A_{i,m}$ be 1, if the Hamming distance between x and y is i, and 0, otherwise, $0 \leq i \leq m$. Then*

$$\begin{aligned}
A_{0,m} &= I_{n^m}, \quad A_{1,1} = J_n - I_n, \\
A_{i,m} &= I_n \otimes A_{i,m-1} + (J_n - I_n) \otimes A_{i-1,m-1}, \quad 1 \leq i < m, \\
A_{m,m} &= (J_n - I_n) \otimes A_{m-1,m-1} = \otimes^m (J_n - I_n), \quad m > 1.
\end{aligned}$$

Proof. The equation $A_{0,m} = I_{n^m}$ is true because of the fixed order of the elements of Ω. Only equal elements have distance 0, so in $A_{0,m}$ only the entries of the main diagonal must be different from 0. For $m = 1$, every two different elements of $\Omega = \Gamma$ have distance 1. So $A_{1,1} = J_n - I_n$.

Let $\gamma_1, \ldots, \gamma_n$ be the ordered elements of Γ. Let the matrix $A_{i,m}$ be split into square matrices $A_{i,m}^{(r,s)}$, $r, s = 1, \ldots, n$ of order n^{m-1}.

$$A_{i,m} = \left(\begin{array}{c|c|c|c} A_{i,m}^{(1,1)} & A_{i,m}^{(1,2)} & \cdots & A_{i,m}^{(1,n)} \\ \hline \vdots & \vdots & & \vdots \\ \hline A_{i,m}^{(n,1)} & A_{i,m}^{(n,2)} & \cdots & A_{i,m}^{(n,n)} \end{array} \right).$$

Because of the lexicographical order in Ω, the rows of $A_{i,m}^{(r,s)}$ correspond to those rows of $A_{i,m}$ whose labels have γ_r as first component. So the labels of the rows of $A_{i,m}^{(r,s)}$ have the same first component. Analogously, the columns of $A_{i,m}^{(r,s)}$ correspond to those columns of $A_{i,m}$ whose labels have γ_s as first component and the labels of those columns have the same first component.

Let $x = (\gamma_r|x')$ and $y = (\gamma_s|y')$ be elements of Ω with Hamming distance i. If $r = s$, then the Hamming distance between x' and y' is also i. Therefore $A_{i,m}^{(r,r)} = A_{i,m-1}$. If $r \neq s$, the Hamming distance between x' and y' is equal to $i-1$, so $A_{i,m}^{(r,s)} = A_{i-1,m-1}$. Hence, the recurrent relations are satisfied. $\square$

From Lemma 2.1, it follows that the matrices $A_{0,m}, A_{1,m}, \ldots, A_{m,m}$ satisfy Conditions (1), (2) and (4) of Definition 2.1. The meaning of Condition (3) is that for every two elements $x, y \in \Omega$ which have Hamming distance equal to k there are exactly $p_{i,j}^{(k)}$ elements $z \in \Omega$ such that the Hamming distance between x and z is i and the Hamming distance between z and y is j. Indeed, the number of elements z depends only on n, m, i, j, k, but does not depend on the choice of $x, y \in \Omega$.

2.2. *Bose-Mesner algebra*

Consider now the linear closure

$$\mathcal{BM} = \left\{ \sum_{i=0}^{s} r_i A_i \;\middle|\; r_i \in \mathbb{R}, i = 0, 1, \ldots, s \right\}$$

of the set of the matrices $A_0, A_1, \ldots, A_s$ from an association scheme which are given in Definition 2.1 over the field of real numbers $\mathbb{R}$. Because of Condition (3) in Definition 2.1, the given set is an algebra with respect to matrix addition, scalar multiplication and ordinary matrix multiplication. Such algebra is called Bose-Mesner algebra. Because of Condition (4) of Definition 2.1 and the fact that all entries of the matrices A_i have values

0 or 1, the set $\mathcal{BM}$ is an algebra with respect to matrix addition, scalar multiplication and Schur (entry-wise) multiplication [13].

Here, we give basic properties of $\mathcal{BM}$ [3, 10].

- The dimension of $\mathcal{BM}$ is $s+1$ because of the linear independence of the matrices A_i.
- The matrices A_i commute under matrix multiplication because the numbers $p_{i,j}^{(k)}$ are constants. Hence $\mathcal{BM}$ is a commutative algebra.
- The matrices A_i are simultaneously diagonalizable according to the spectral theorem.
- There exists an unique basis of mutually orthogonal minimal idempotents $E_0 = \frac{1}{n}J_n, E_1, \dots, E_s$, where n is the number of the elements of Ω.

2.3. *Strongly regular graphs*

The matrices $A_0, A_1, \dots, A_s$ from an association scheme given in Definition 2.1 can be considered as adjacency matrices of a colored complete undirected graph. If $s = 2$, i.e. $A_0 = I$, $A_2 = J - I - A_1$, the matrix A_1 can be considered also as an adjacency matrix of a relation R_1. In this case, the corresponding graph (Ω, R_1) is a strongly regular graph.

More precisely, a graph (Ω, R_1) is called *regular* of valency v if and only if every vertex has exactly v neighbors. Regular graph, for which every two neighbors have exactly λ common neighbors and every two non-neighbors have exactly μ common neighbors, is called *strongly regular graph* with parameters (n, v, λ, μ), where $n = |\Omega|$.

Some known strongly regular graphs with their parameters are described in [9].

3. Some concepts in coding theory

Let $\mathbb{F}_q^n$ be the n-dimensional vector space over the finite field $\mathbb{F}_q = \{0, \alpha_1, \alpha_2, \dots, \alpha_{q-1}\}$, where q is a power of a prime number. Each k-dimensional linear subspace $C \subset \mathbb{F}_q^n$ is called *linear code*. More precisely, it is said to be $[n,k]_q$ code with length n and dimension k. Every vector $v \in C$ is a *codeword* of C. A *generator matrix* of the code is any $k \times n$ matrix whose rows form a basis of C. The (*Hamming*) *weight* $wt(v)$ of a vector $v \in \mathbb{F}_q^n$ is the number of the nonzero coordinates of v. The *weight distribution* of the code C is the finite sequence $(A_0, A_1, \dots, A_n) \in \mathbb{Z}^{n+1}$, where A_u is the number of the codewords with weight $u, u = 0, 1, \dots, n$, and the *weight enumerator* of the code C is the polynomial $W_C(y) = A_0y^0 + A_1y^1 + \dots + A_ny^n$.

If the code C has exactly t different nonzero weights, then the code is called *t-weight code.* We are interested in two weight codes over $\mathbb{F}_2$.

3.1. *Simplex code and characteristic vector of a linear code*

Now consider the matrix, whose columns form a maximal set of pairwise non-proportional vectors in $\mathbb{F}_q^k$. Putting $\theta(q,k) = (q^k - 1)/(q-1)$, we find that this matrix generates a $[\theta(q,k), k]_q$ linear code, called a *simplex code.* It is denoted by $S(q,k)$.

We use the following special type of generator matrices of the simplex codes $S(q,k)$:

$$G_1 = (1),$$

$$G_{k+1} = \left(\begin{array}{c|c|c|c|c} 0\,0\ldots0 & \alpha_1\,\alpha_1\ldots\alpha_1 & \ldots & \alpha_{q-1}\,\alpha_{q-1}\ldots\alpha_{q-1} & 1 \\ \hline & & & & 0 \\ G_k & G_k & \ldots & G_k & 0 \\ & & & & \vdots \\ & & & & 0 \end{array}\right), \quad k \in \mathbb{N},$$

where the number q and the order of the elements of $\mathbb{F}_q$ are fixed.

Example 3.1. For $q = 4$, $\mathbb{F}_4 = \{0, x, \bar{x}, 1\}$ and $k = 3$, we have

$$G_3 = \left(\begin{array}{c|c|c|c|c} 0\,0\,0\,0\,0 & x\,x\,x\,x\,x & \bar{x}\,\bar{x}\,\bar{x}\,\bar{x}\,\bar{x} & 1\,1\,1\,1\,1 & 1 \\ \hline 0\,x\,\bar{x}\,1\,1 & 0\,x\,\bar{x}\,1\,1 & 0\,x\,\bar{x}\,1\,1 & 0\,x\,\bar{x}\,1\,1 & 0 \\ 1\,1\,1\,1\,0 & 1\,1\,1\,1\,0 & 1\,1\,1\,1\,0 & 1\,1\,1\,1\,0 & 0 \end{array}\right).$$

The extended matrix $\overline{G}_k$ consists of all vectors in $\mathbb{F}_q^k$ as columns, i.e.

$$\overline{G}_k = \big(0 \,|\, \alpha_1 G_k \,|\, \alpha_2 G_k \,|\, \ldots \,|\, \alpha_{q-1} G_k\big).$$

The standard extended matrix $\overline{G}_k^{(st)}$ is the matrix with the same set of columns but arranged in lexicographic order. Hence the rows of the matrices $\overline{G}_k^T \cdot G_k$ and $\overline{G}_k^{(st)T} \cdot G_k$ represent all codewords of the simplex code $S(q,k)$.

Furthermore, for any $[n,k]_q$ linear code C with a generator matrix G, the matrices $\overline{G}_k^T \cdot G$ and $\overline{G}_k^{(st)T} \cdot G$ have all codewords of the code C in their rows. The rows of the matrix $G_k^T \cdot G$ are the elements of a maximal set of pairwise non-proportional codewords of C.

Let C be a linear $[n,k]_q$ code. We suppose that an $m \times n$ matrix G with $m \geq k$ generates the code. That is, the matrix G has k linearly independent rows. The characteristic vector of the code C with respect to the matrix G and the corresponding simplex code with generator matrix G_m is

$$\chi(G) = (\chi_1, \chi_2, \dots, \chi_{\theta(q,m)}) \in \mathbb{Z}^{\theta(q,m)},$$

where χ_u $(u = 1, 2, \dots, \theta(q, m))$ is the number of the columns of G which are equal or proportional to the u-th column of the matrix G_m. Obviously, $\chi_1 + \chi_2 + \cdots + \chi_{\theta(q,m)} = n$. This procedure is invertible if $m = k$. In this case, every characteristic vector gives a generator matrix G_χ of a code which is equivalent to C.

Example 3.2. For $q = 4$, $\mathbb{F}_4 = \{0, x, \bar{x}, 1\}$, $n = 10$, $k = 2$, $\theta(q, k) = 5$ and

$$G = \begin{pmatrix} \bar{x} & 1 & 0 & \bar{x} & x & 0 & \bar{x} & 1 & x & 0 \\ 1 & \bar{x} & x & 0 & 1 & \bar{x} & 0 & \bar{x} & x & \bar{x} \end{pmatrix},$$

we have

$$G_2 = \begin{pmatrix} 0 & x & \bar{x} & 1 & 1 \\ 1 & 1 & 1 & 1 & 0 \end{pmatrix}, \quad \chi = (3, 3, 1, 1, 2), \quad G_\chi = \begin{pmatrix} 0 & 0 & 0 & x & x & x & \bar{x} & 1 & 1 & 1 \\ 1 & 1 & 1 & 1 & 1 & 1 & 1 & 1 & 0 & 0 \end{pmatrix}.$$

3.2. *Projective dual transform of a linear code*

The projective dual transform is well studied in coding theory (see [6, 8, 11]). It is used for obtaining construction and classification results. It can naturally be presented in terms of incidence structures (see [7]). An incidence structure $S = (P, B, I)$ is *self-dual* if it is isomorphic to its dual (B, P, I^T). That is, it is self-dual if there are bijections $f : P \to B$ and $g : B \to P$ such that $(g(b), f(p)) \in I$ if and only if $(p, b) \in I$. It is *self-polar* if we can choose $g = f^{-1}$.

An incidence matrix is defined for each incidence structure, which is a matrix $N = (a_{ij})_{|P|\times|B|}$ such that $a_{ij} = 1$ if $p_i \in B_j$ and $a_{ij} = 0$ if $p_i \notin B_j$. Then, the incidence structure $S = (P, B, I)$ is self-dual if there exist permutation matrices P_1 and P_2 such that $P_1 N = N^T P_2$ and $S = (P, B, I)$ is self-polar if there exists a permutation matrix P_3 such that $P_3 N = N^T P_3^T$.

A square matrix A is said to be *symmetrizable* if a symmetric matrix can be obtained from A by a sequence of rows and columns permutations. It is easy to prove that the incidence matrix of a self-polar incidence structure is a symmetrizable matrix. Let $A = P_3 N$. Then

$$A^T = (P_3 N)^T = N^T P_3^T = P_3 N = A.$$

We can apply similar properties to linear codes. Let C be a linear $[n, k]_q$ code with generator matrix G. We denote by $W = \{w_1, w_2, \dots, w_s\}$ the set which consists of all nonzero weights of its codewords. Let $\alpha, \beta \in \mathbb{Q}$ be rational numbers such that all the numbers $\alpha w_i + \beta$ are non-negative integers for $i = 1, \ldots, s$. We denote by P the set which consists of all vectors that are columns of the generator matrix G_k of $S(q,k)$. The *projective dual code* $D_{\alpha,\beta}(C)$ of the code C is a code, whose generator matrix $D_{\alpha,\beta}(G)$ has all vectors $v \in P$ taken $\alpha.wt(vG) + \beta$ times as columns. A linear code is formally projective self-dual if it has the same length, dimension and weight distribution as its projective dual for some α and β. The code is projective self-dual (resp. projective self-polar) if it is equivalent to (resp. coincides with) its projective dual for some α and β.

The projective dual transform can be also given in terms of characteristic vectors (see [6]). The projective dual code $D_{\alpha,\beta}(C)$ of a code C is the linear code with the characteristic vector $\chi_{\alpha,\beta} = \alpha \cdot \chi \cdot \mathcal{N}(M_k) + \beta \cdot \mathbf{1}$, where $\mathbf{1}$ is the vector of the corresponding length all of whose entries are 1, and the matrix $\mathcal{N}(M_k)$ is obtained by replacing the nonzero elements of $M_k = G_k^T \cdot G_k$ by 1. If a code C is projective self-polar, then it has a generator matrix such that C and its dual code $D_{\alpha,\beta}(C)$ have the same characteristic vector for some α and β. Let χ_0 be such characteristic vector which is both for the code and for its dual, and G_0 be the corresponding generator matrix. The matrix $G_0^T \cdot G_0$ is a symmetric matrix and it has constant number of $1's$ in each row and column. If it has only $0's$ in the main diagonal, it may be an adjacency matrix of a strongly regular graph.

4. Our approach

We consider projective linear codes over the field $\mathbb{F}_2$. In general, a linear $[n,k]_q$ code is called *projective* if no two columns of a generator matrix G are linearly dependent. In other words, the code is projective if and only if its characteristic vector consists of $0's$ and $1's$. It is proven in [6] that if C is a projective linear code then its projective dual code is a two weight code. Moreover, if such a code is formally projective self-dual, then its projective dual has weight distribution $(1, A_{w_1}, A_{w_2})$, for which either $A_{w_1} = n$ or $A_{w_2} = n$ (see [5]).

Let C be a two weight projective self-polar $[n,k]_2$ code with weight enumerator $W_C(y) = 1 + n.y^r + m.y^s$, $r < s$, i.e. the code has exactly n codewords with the smaller weight r. Taking these codewords as rows, we obtain a square $n \times n$ matrix M, which is a symmetrizable matrix because the code is projective self-polar. The corresponding symmetric matrix M' could be also an adjacency matrix of a strongly regular graph. In this way a strongly regular graph with parameters $(28, 12, 6, 4)$ is obtained.

Example 4.1. We consider the $[28, 6]_2$ linear code. It has weight enumerator $W_C(y) = 1y^0 + 28y^{12} + 35y^{16}$ and the following generator matrix:

$$G = \begin{pmatrix}
0&0&0&0&0&0&0&0&0&0&0&0&0&0&0&0&1&1&1&1&1&1&1&1&1&1&1&1\\
0&0&0&0&0&1&1&1&1&0&0&0&1&1&1&1&1&1&1&1&0&0&0&0&0&0&0&0\\
0&0&0&0&1&0&1&0&1&1&1&1&1&0&1&0&0&0&0&0&1&0&1&0&1&0&1&0\\
0&0&1&1&0&0&0&0&0&0&1&1&1&1&1&1&0&0&0&0&0&0&0&0&1&1&1&1\\
0&1&0&1&0&0&0&1&1&1&1&0&0&0&1&1&0&1&1&0&0&1&0&1&0&0&0&0\\
1&1&0&0&0&0&1&1&0&0&0&0&1&1&0&0&0&0&1&1&1&1&0&0&1&1&0&0
\end{pmatrix}.$$

We then have

$$M' = \begin{pmatrix}
0&0&0&0&0&0&0&0&0&0&0&0&0&0&0&0&1&1&1&1&1&1&1&1&1&1&1&1\\
0&0&0&0&0&1&1&1&1&0&0&0&1&1&1&1&1&1&1&1&0&0&0&0&0&0&0&0\\
0&0&0&0&1&0&1&0&1&1&1&1&1&0&1&0&0&0&0&0&1&0&1&0&1&0&1&0\\
0&0&0&0&1&1&0&1&0&1&1&1&0&1&0&1&0&0&0&0&0&1&0&1&0&1&0&1\\
0&0&1&1&0&0&0&0&0&0&1&1&1&1&1&1&0&0&0&0&0&0&0&0&1&1&1&1\\
0&1&0&1&0&0&0&1&1&1&1&0&0&0&1&1&0&1&1&0&0&1&0&1&0&0&0&0\\
0&1&1&0&0&0&0&1&1&1&0&1&1&1&0&0&1&0&0&1&1&0&1&0&0&0&0&0\\
0&1&0&1&0&1&1&0&0&1&1&0&1&1&0&0&1&0&0&1&0&1&0&1&0&0&0&0\\
0&1&1&0&0&1&1&0&0&1&0&1&0&0&1&1&0&1&1&0&1&0&1&0&0&0&0&0\\
0&0&1&1&0&1&1&1&1&0&1&1&0&0&0&0&0&0&0&0&1&1&1&1&0&0&0&0\\
0&0&1&1&1&1&0&1&0&1&0&0&1&0&1&0&0&0&0&0&0&1&0&1&1&0&1&0\\
0&0&1&1&1&0&1&0&1&1&0&0&0&1&0&1&0&0&0&0&1&0&1&0&0&1&0&1\\
0&1&1&0&1&0&1&1&0&0&1&0&0&1&1&0&1&0&0&1&0&0&0&0&1&0&1&0\\
0&1&0&1&1&0&1&1&0&0&0&1&1&0&0&1&1&0&0&1&0&0&0&0&0&1&0&1\\
0&1&1&0&1&1&0&0&1&0&1&0&1&0&0&1&0&1&1&0&0&0&0&0&1&0&1&0\\
0&1&0&1&1&1&0&0&1&0&0&1&0&1&1&0&0&1&1&0&0&0&0&0&0&1&0&1\\
1&1&0&0&0&0&1&1&0&0&0&0&1&1&0&0&0&0&1&1&1&1&0&0&1&1&0&0\\
1&1&0&0&0&1&0&0&1&0&0&0&0&0&1&1&0&0&1&1&0&0&1&1&0&0&1&1\\
1&1&0&0&0&1&0&0&1&0&0&0&0&0&1&1&1&1&0&0&1&1&0&0&1&1&0&0\\
1&1&0&0&0&0&1&1&0&0&0&0&1&1&0&0&1&1&0&0&0&0&1&1&0&0&1&1\\
1&0&1&0&0&0&1&0&1&1&0&1&0&0&0&0&1&0&1&0&0&1&1&0&1&1&0&0\\
1&0&0&1&0&1&0&1&0&1&1&0&0&0&0&0&1&0&1&0&1&0&0&1&1&1&0&0\\
1&0&1&0&0&0&1&0&1&1&0&1&0&0&0&0&0&1&0&1&1&0&0&1&0&0&1&1\\
1&0&0&1&0&1&0&1&0&1&1&0&0&0&0&0&0&1&0&1&0&1&1&0&0&0&1&1\\
1&0&1&0&1&0&0&0&0&0&1&0&1&0&1&0&1&0&1&0&1&1&0&0&0&1&1&0\\
1&0&0&1&1&0&0&0&0&0&0&1&0&1&0&1&1&0&1&0&1&1&0&0&1&0&0&1\\
1&0&1&0&1&0&0&0&0&0&1&0&1&0&1&0&0&1&0&1&0&0&1&1&1&0&0&1\\
1&0&0&1&1&0&0&0&0&0&0&1&0&1&0&1&0&1&0&1&0&0&1&1&0&1&1&0
\end{pmatrix},$$

$(M')^2 =$

$$
\begin{pmatrix}
12 & 4 & 4 & 4 & 4 & 4 & 4 & 4 & 4 & 4 & 4 & 4 & 4 & 4 & 4 & 4 & 6 & 6 & 6 & 6 & 6 & 6 & 6 & 6 & 6 & 6 & 6 & 6 \\
4 & 12 & 4 & 4 & 4 & 6 & 6 & 6 & 6 & 4 & 4 & 4 & 6 & 6 & 6 & 6 & 6 & 6 & 6 & 6 & 4 & 4 & 4 & 4 & 4 & 4 & 4 & 4 \\
4 & 4 & 12 & 4 & 6 & 4 & 6 & 4 & 6 & 6 & 6 & 6 & 6 & 4 & 6 & 4 & 4 & 4 & 4 & 4 & 6 & 4 & 6 & 4 & 6 & 4 & 6 & 4 \\
4 & 4 & 4 & 12 & 6 & 6 & 4 & 6 & 4 & 6 & 6 & 6 & 4 & 6 & 4 & 6 & 4 & 4 & 4 & 4 & 4 & 6 & 4 & 6 & 4 & 6 & 4 & 6 \\
4 & 4 & 6 & 6 & 12 & 4 & 4 & 4 & 4 & 4 & 6 & 6 & 6 & 6 & 6 & 6 & 4 & 4 & 4 & 4 & 4 & 4 & 4 & 4 & 6 & 6 & 6 & 6 \\
4 & 6 & 4 & 6 & 4 & 12 & 4 & 6 & 6 & 6 & 6 & 4 & 4 & 4 & 6 & 6 & 4 & 6 & 6 & 4 & 4 & 6 & 4 & 6 & 4 & 4 & 4 & 4 \\
4 & 6 & 6 & 4 & 4 & 4 & 12 & 6 & 6 & 6 & 4 & 6 & 6 & 6 & 4 & 4 & 6 & 4 & 4 & 6 & 6 & 4 & 6 & 4 & 4 & 4 & 4 & 4 \\
4 & 6 & 4 & 6 & 4 & 6 & 6 & 12 & 4 & 6 & 6 & 4 & 6 & 6 & 4 & 4 & 6 & 4 & 4 & 6 & 4 & 6 & 4 & 6 & 4 & 4 & 4 & 4 \\
4 & 6 & 6 & 4 & 4 & 6 & 6 & 4 & 12 & 6 & 4 & 6 & 4 & 4 & 6 & 6 & 4 & 6 & 6 & 4 & 6 & 4 & 6 & 4 & 4 & 4 & 4 & 4 \\
4 & 4 & 6 & 6 & 4 & 6 & 6 & 6 & 6 & 12 & 6 & 6 & 4 & 4 & 4 & 4 & 4 & 4 & 4 & 4 & 6 & 6 & 6 & 6 & 4 & 4 & 4 & 4 \\
4 & 4 & 6 & 6 & 6 & 6 & 4 & 6 & 4 & 6 & 12 & 4 & 6 & 4 & 6 & 4 & 4 & 4 & 4 & 4 & 4 & 6 & 4 & 6 & 6 & 4 & 6 & 4 \\
4 & 4 & 6 & 6 & 6 & 4 & 6 & 4 & 6 & 6 & 4 & 12 & 4 & 6 & 4 & 6 & 4 & 4 & 4 & 4 & 6 & 4 & 6 & 4 & 4 & 6 & 4 & 6 \\
4 & 6 & 6 & 4 & 6 & 4 & 6 & 6 & 4 & 4 & 6 & 4 & 12 & 6 & 6 & 4 & 6 & 4 & 4 & 6 & 4 & 4 & 4 & 4 & 6 & 4 & 6 & 4 \\
4 & 6 & 4 & 6 & 6 & 4 & 6 & 6 & 4 & 4 & 4 & 6 & 6 & 12 & 4 & 6 & 6 & 4 & 4 & 6 & 4 & 4 & 4 & 4 & 4 & 6 & 4 & 6 \\
4 & 6 & 6 & 4 & 6 & 6 & 4 & 4 & 6 & 4 & 6 & 4 & 6 & 4 & 12 & 6 & 4 & 6 & 6 & 4 & 4 & 4 & 4 & 4 & 6 & 4 & 6 & 4 \\
4 & 6 & 4 & 6 & 6 & 6 & 4 & 4 & 6 & 4 & 4 & 6 & 4 & 6 & 6 & 12 & 4 & 6 & 6 & 4 & 4 & 4 & 4 & 4 & 4 & 6 & 4 & 6 \\
6 & 6 & 4 & 4 & 4 & 4 & 6 & 6 & 4 & 4 & 4 & 4 & 6 & 6 & 4 & 4 & 12 & 4 & 6 & 6 & 6 & 6 & 4 & 4 & 6 & 6 & 4 & 4 \\
6 & 6 & 4 & 4 & 4 & 6 & 4 & 4 & 6 & 4 & 4 & 4 & 4 & 4 & 6 & 6 & 4 & 12 & 6 & 6 & 4 & 4 & 6 & 6 & 4 & 4 & 6 & 6 \\
6 & 6 & 4 & 4 & 4 & 6 & 4 & 4 & 6 & 4 & 4 & 4 & 4 & 4 & 6 & 6 & 6 & 6 & 12 & 4 & 6 & 6 & 4 & 4 & 6 & 6 & 4 & 4 \\
6 & 6 & 4 & 4 & 4 & 4 & 6 & 6 & 4 & 4 & 4 & 4 & 6 & 6 & 4 & 4 & 6 & 6 & 4 & 12 & 4 & 4 & 6 & 6 & 4 & 4 & 6 & 6 \\
6 & 4 & 6 & 4 & 4 & 4 & 6 & 4 & 6 & 6 & 4 & 6 & 4 & 4 & 4 & 4 & 6 & 4 & 6 & 4 & 12 & 6 & 6 & 4 & 6 & 6 & 4 & 4 \\
6 & 4 & 4 & 6 & 4 & 6 & 4 & 6 & 4 & 6 & 6 & 4 & 4 & 4 & 4 & 4 & 6 & 4 & 6 & 4 & 6 & 12 & 4 & 6 & 6 & 6 & 4 & 4 \\
6 & 4 & 6 & 4 & 4 & 4 & 6 & 4 & 6 & 6 & 4 & 6 & 4 & 4 & 4 & 4 & 4 & 6 & 4 & 6 & 6 & 4 & 12 & 6 & 4 & 4 & 6 & 6 \\
6 & 4 & 4 & 6 & 4 & 6 & 4 & 6 & 4 & 6 & 6 & 4 & 4 & 4 & 4 & 4 & 4 & 6 & 4 & 6 & 4 & 6 & 6 & 12 & 4 & 4 & 6 & 6 \\
6 & 4 & 6 & 4 & 6 & 4 & 4 & 4 & 4 & 4 & 6 & 4 & 6 & 4 & 6 & 4 & 6 & 4 & 6 & 4 & 6 & 6 & 4 & 4 & 12 & 6 & 6 & 4 \\
6 & 4 & 4 & 6 & 6 & 4 & 4 & 4 & 4 & 4 & 4 & 6 & 4 & 6 & 4 & 6 & 6 & 4 & 6 & 4 & 6 & 6 & 4 & 4 & 6 & 12 & 4 & 6 \\
6 & 4 & 6 & 4 & 6 & 4 & 4 & 4 & 4 & 4 & 6 & 4 & 6 & 4 & 6 & 4 & 4 & 6 & 4 & 6 & 4 & 4 & 6 & 6 & 6 & 4 & 12 & 6 \\
6 & 4 & 4 & 6 & 6 & 4 & 4 & 4 & 4 & 4 & 4 & 6 & 4 & 6 & 4 & 6 & 4 & 6 & 4 & 6 & 4 & 4 & 6 & 6 & 4 & 6 & 6 & 12
\end{pmatrix}.
$$

5. Conclusion

In this paper we give a new look to the connection between association schemes, strongly regular graphs and linear codes as we use the projective dual transform to obtain structures from one another. The proposed approach is still tested only with known two weight codes with small parameters. The example also gives a known strongly regular graph. Continuation of the research in this area could lead to obtaining new structures of bigger orders, as it is possible to improve the algorithms by including more useful properties as well as parallel implementations.

Acknowledgements

This paper is partially supported by the Bulgarian National Science Fund under contract no. FSD-31-243-23/21.03.23.

References

[1] R.A. Bailey, *Association schemes: Designed experiments, algebra and combinatorics*, Cambridge Univ. Press, 2004.

[2] R.C. Bose, Strongly regular graphs, partial geometries and partially balanced designs, *Pacific J. Math.* **13** (1963), 389–419.

[3] R.C. Bose and D.M. Mesner, On linear associative algebras corresponding to association schemes of partially balanced designs, *Ann. Math. Statist.* **30** (1959), 21–38.

[4] R.C. Bose and T. Shimamoto, Classification and analysis of partially balanced incomplete block designs with two associate classes, *J. Amer. Statist. Assoc.* **47** (1952), 151–184.

[5] I. Bouyukliev, V. Fack, W. Willems and J. Winne, Projective two-weight codes with small parameters and their corresponding graphs, *Des. Codes Criptogr.* **41** (2006), 59–78.

[6] I. Bouyukliev and S. Bouyuklieva, Dual transform and projective self-dual codes, *Adv. Math. Commun.* **18** (2024), 328–341.

[7] A.E. Brouwer, P.J. Cameron, W.H. Haemers and D.A. Preece, Self-dual, not self-polar, *Discrete Math.* **305** (2006), 3051–3053.

[8] A.E. Brouwer and M. Van Eupen, The correspondence between Projective Codes and 2-weight Codes, *Des. Codes Criptogr.* **11** (1997), 261–266.

[9] A.E. Brouwer and H. Van Maldeghem, *Strongly regular graphs*, Cambridge Univ. Press, 2022.

[10] P. Delsarte, An algebraic approach to the association schemes of coding theory, *Philips Res. Rep. Suppl.* **10** (1973), 97pp.

[11] S. Dodunekov and J. Simonis, Codes and Projective Multisets, *Electron. J. Combin.* **5** (1998), Research paper 37, 23pp.

[12] F.J. MacWilliams and N.J.A. Sloane, *The theory of error-correcting codes*, North-Holland Publishing Co., Amsterdam-New York-Oxford, 1977.

[13] J.J. Seidel, Introduction to association schemes, *Sém. Lothar. Combin.* **26** (1991), B26g, 17pp.

Received February 16, 2024
Revised March 29, 2024

© 2025 World Scientific Publishing Company
Modern Approaches to Differential Geometry
and its Related Fields 123 – 136

KÄHLER GRAPHS WHOSE PRINCIPAL GRAPHS ARE OF TENSOR PRODUCT TYPE

Kohei SHIOTANI

Department of Computer Science, Nagoya Institute of Technology,
Nagoya 466-8555, Japan
E-mail: shiotani.kohei.nit@ezweb.ne.jp

Toshiaki ADACHI*

Department of Mathematics, Nagoya Institute of Technology,
Nagoya 466-8555, Japan
E-mail: adachi@nitech.ac.jp

A Kähler graph consists of two kinds of graphs having the same set of vertices. As candidates of model Kähler graphs which correspond to homogeneous spaces admitting magnetic fields, vertex transitive normal Kähler graphs are proposed in [5]. To give many examples of such Kähler graphs, we define some product operations. In this note, we restrict ourselves to the product Kähler graphs whose first kind graphs are of tensor product type, and study whether they are connected, are bipartite, are vertex-transitive and are normal.

Keywords: Kähler graphs; adjacency operators; commutative; connected; bipartite; eigenvalues.

1. Introduction

This note is a sequel to the paper [6]. A Kähler graph consists of three kinds of sets, which are sets of vertices, of principal edges and of auxiliary edges. Therefore, a Kähler graph is a compound of two graphs, the principal graph which is formed by principal edges and the auxiliary graph which is formed by auxiliary edges. It was introduced by the second author as a discrete model of Riemannian manifolds admitting magnetic fields ([3]). We consider that passes on the principal graph correspond to geodesics, and use the auxiliary graphs to express how these passes are bended by the influence of magnetic fields. For magnetic fields on Riemannian manifolds, see [1, 10], for example. To develop our study on Kähler graphs, the authors

*The second author is partially supported by Grant-in-Aid for Scientific Research (C) (No. 20K03581), Japan Society for the Promotion of Science.

think it is important to provide model Kähler graphs which correspond to homogeneous spaces. Being different from ordinary graphs, considering zeta functions for Kähler graphs ([5]), we find that the vertex-transitivity is not enough, and come to the conclusion that the normality is important. Here, a Kähler graph is said to be normal if its adjacency operators of the principal and auxiliary graphs are commutative.

The first step of our study on normal Kähler graphs is to provide many examples. In [7], Chen and the second author gave some examples by putting vertices on a circle and by giving cyclic-invariant edges. Another way to provide examples is to make use of product operations. For ordinary graphs we have some kinds of product operations, Cartesian product, strong product, tensor product, lexicographic product and so on. Since Kähler graphs consist of two kinds of graphs, combining these operations we can consider many kinds of product operations. In [6], we studied product Kähler graphs whose principal graphs are obtained by Cartesian product. Therefore, we study some other product Kähler graphs. Considering duality of Kähler graphs, in this paper we take five kinds of Kähler graphs whose principal graphs are obtained by tensor product. We study properties of these graphs such as connectivity, bipartiteness and normality by posing some conditions on factor graphs.

2. Kähler graphs

A (non-directed) graph G consists of a set V of vertices and a set E of edges, and is expressed as $G = (V, E)$. It is considered as a 1-dimensional CW-complex. Throughout of this paper, we suppose that every graph is simple. That is, it does not have loops and multiple edges. Here, a loop is an edge joining a vertex and itself, and multiple edges are edges joining the same pair of vertices. We shall start by recalling some notions on graphs and by giving some notations. We say that two vertices $v, v' \in V$ are *adjacent* to each other if there is an edge joining them. In this case, we denote as $v \sim v'$. A *path* on G is a chain of edges. We express a path as $(v_0, v_1, \dots, v_m)$ with $v_{i-1} \sim v_i$ $(i = 1, \dots, m)$, and say that it is of m-step. We say that G is connected if for arbitrary vertices v, v' there is a path joining them. For a vertex $v \in V$, the cardinality of the set $\{v' \in V \mid v' \sim v\}$ is denoted by $d_G(v)$ and is called the *degree* at v. When $d_G(v)$ is finite at each vertex v, we say that G is locally finite. A locally finite graph is said to be *regular* if the degree does not depend on the choice of vertices. For a locally finite graph, we set $C_c(V)$ as the set of all functions of V whose supports are finite. When G is finite, that is, when V is a finite set, hence, so is E, the set $C_c(V)$ coincides with the set $C(V)$ of all functions of V. We define the adjacency operator $\mathcal{A}_G$ and the transition operator $\mathcal{Q}_G$ on $C_c(V)$ by $(\mathcal{A}_G f)(v) = \sum_{v'\sim v} f(v')$ and $(\mathcal{Q}_G f)(v) = \big(1/d_G(v)\big)\sum_{v'\sim v} f(v')$. If we define inner products on $C_c(V)$ by $\langle f, g\rangle = \sum_{v\in V} f(v)g(v)$ and $(f,g) = \sum_{v\in V}\big(1/d_G(v)\big)f(v)g(v)$, these operators are selfadjoint with respect to these inner products, respectively. We say that a graph G is *bipartite* if the set V of vertices is decomposed into two disjoint subsets $V^+ \cup V^-$ so that there are no edges joining vertices belonging to V^+ and no edges joining vertices belonging to V^-. We shall call the decomposition $V = V^+ \cup V^-$ of the set of vertices the vertex-decomposition of a bipartite graph. When a bipartite graph is connected, between two vertices belonging to V^+ and between two vertices belonging to V^-, there are only even-step paths joining them, and between a vertex in V^+ and that in V^- there are only odd-step paths joining them. A bipartite graph is characterized by this property. A homomorphism φ between two graphs $G = (V, E)$, $H = (W, F)$ is a map of V to W satisfying $\varphi(v) \sim \varphi(v')$ if $v \sim v'$. A bijection of V to W is said to be an isomorphism if both φ and φ^{-1} are homomorphisms. We call a graph G *vertex-transitive* if for arbitrary vertices v, v' there is an isomorphism φ satisfying $\varphi(v) = v'$.

We say that a graph $G = (V, E)$ is *Kähler* if the set E of edges is divided into two disjoint subsets $E^{(p)}$ and $E^{(a)}$ so that at each vertex $v \in V$ there are at least two edges in $E^{(p)}$ and at least two edges in $E^{(a)}$ all of which are emanating from v. We call edges in $E^{(p)}$ *principal* and call edges in $E^{(a)}$ *auxiliary*. For a Kähler graph G, by using its principal and auxiliary edges, we define its principal graph $G^{(p)}$ by $(V, E^{(p)})$ and its auxiliary graph $G^{(a)}$ by $(V, E^{(a)})$. In order to distinguish Kähler graphs and non-Kähler graphs clearly, we sometimes call a non-Kähler graph an ordinary graph. When a Kähler graph G is locally finite, we define its $(1,1)$-probabilistic adjacency operator $\mathcal{A}_G^{(1,1)}$ as the composition $\mathcal{A}_{G^{(p)}}\mathcal{Q}_{G^{(a)}}$. We say a locally finite Kähler graph is *regular* if both of its principal and auxiliary graphs are regular. In this case, its $(1,1)$-probabilistic adjacency operator is given as $\big(1/d_{G^{(a)}}\big)\mathcal{A}_{G^{(p)}}\mathcal{A}_{G^{(a)}}$. Generally, the $(1,1)$-probabilistic adjacency operator of a Kähler graph is not selfadjoint. We say that a regular Kähler graph G is *normal* if its adjacency operators are commutative, i.e. $\mathcal{A}_{G^{(p)}}\mathcal{A}_{G^{(a)}} = \mathcal{A}_{G^{(a)}}\mathcal{A}_{G^{(p)}}$. Clearly, the $(1,1)$-probabilistic adjacency operator of a normal Kähler graph is selfadjoint. We call a Kähler graph vertex-transitive if for arbitrary vertices v, v' there is a map φ of the set of vertices onto itself such that it is an isomorphism of both of the principal and auxiliary graphs and satisfies $\varphi(v) = v'$. Magnetic spherical means, which are considered to correspond to adjacency operators, for Kähler magnetic fields on complex space forms have the same selfadjoint property (see [2]). Hence, the authors consider that vertex-transitive normal Kähler graphs are good candidates of graphs which correspond to homogeneous Hermitian spaces. For a Kähler graph $G = (V, E^{(p)} \cup E^{(a)})$, by setting $F^{(p)} = E^{(a)}$, $F^{(a)} = E^{(p)}$, we call $G^* = (V, F^{(p)} \cup F^{(a)})$ the *dual* Kähler graph of G. Trivially, a Kähler graph is normal if and only if its dual is normal. We therefore study Kähler graphs whose principal graphs are obtained by tensor product and whose auxiliary graphs are graphs which did not appear in the operations we studied in the previous paper [6].

3. Product operations

In order to make a product Kähler graph, we take two ordinary graphs and consider two kinds of product operations, one is for the principal graph and the other is for the auxiliary graph. Let G and H be two ordinary graphs. We express them as $G = (V, E)$ and $H = (W, F)$, and consider the product $V \times W$ as the set of vertices of a new graph. The tensor product $G \otimes H$ of G and H is defined by the rule that two vertices $(v, w), (v', w') \in V \times W$ are adjacent to each other if and only if $v \sim v'$ and $w \sim w'$. In this paper, we use this product for making principal graphs of Kähler graphs. Tensor products of two graphs have the following properties (see [8, 9], for example, and see also [6]).

Proposition 3.1. *Let G and H be two connected ordinary graphs.*

(1) *If one of these graphs is not bipartite, then their tensor product $G \otimes H$ is connected.*
(2) *If both of these graphs are bipartite, then $G \otimes H$ has two connected components.*
(3) *The product $G \otimes H$ is bipartite if and only if one of these graphs is bipartite.*

In order to define a new type of product Kähler graphs, we consider compatible product operations for making auxiliary graphs. For such operations, the use of complementary graphs may come to mind. The complement graph G^c of G is given as (V, E^c) with $E^c = \{(v, v') \in V \times V \mid v \not\sim v',\ v \neq v'\}$. By using complement graphs, we have the products $G \otimes H^c$ and $G ⓥ H$. Here, the set of vertices of the graph $G ⓥ H$ is given as the product of the sets of vertices of G and H, and the set of edges is the union of the sets of edges of $G^c \otimes H$ and $G \otimes H^c$. We shall call these operations $G \otimes H^c$ and $G \,\textcircled{\scriptsize v}\, H$ the *complement* product and the *vacant* product, respectively. But as is easily guessed, it is not easy to study these graphs only by using conditions on G and H. Here we give images of these product operations by taking two line graphs of 5 vertices. We show the adjacency of the vertex which stands in the center (see Figs. 2,3).

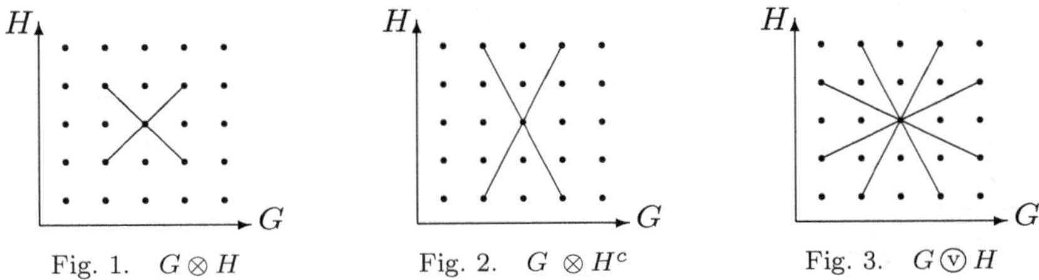

Fig. 1. $G \otimes H$ Fig. 2. $G \otimes H^c$ Fig. 3. $G \,\textcircled{\scriptsize v}\, H$

In order to control properties of product graphs by those of factor graphs, by combining the Cartesian product operation and the above operations, here we give five sets of edges, and define five graphs of product type, which are

- semi-Cartesian-complement product $G \,\boxed{\scriptstyle S}\, H$,
- partial Cartesian-complement product $G \,\boxed{\scriptstyle P}\, H$,
- Cartesian-complement product $G \,\boxed{\scriptstyle C}\, H$,
- semi-Cartesian-vacant product $G \,\boxed{\scriptstyle V}\, H$,
- Cartesian-vacant product $G \,\boxed{\scriptstyle M}\, H$,

according to the following rules on adjacency of two vertices (v, w), $(v', w') \in V \times W$:

(a) For a semi-Cartesian-complement product $G \,\boxed{\scriptstyle S}\, H$,
i) $v = v'$ and $w \sim w'$, or ii) $v \sim v'$ and $w \not\sim w'$, $w \neq w'$;
(b) For a partial-Cartesian-complement product $G \,\boxed{\scriptstyle P}\, H$,
$v \sim v'$ and $w \not\sim w'$;
(c) For a Cartesian-complement product $G \,\boxed{\scriptstyle C}\, H$,
i) $v = v'$ and $w \sim w'$, or ii) $v \sim v'$ and $w \not\sim w'$;
(d) For a semi-Cartesian-vacant product $G \,\boxed{\scriptstyle V}\, H$,
i) $v \sim v'$ and $w \not\sim w'$, $w \neq w'$, or ii) $w \sim w'$ and $v \not\sim v'$;
(e) For a Cartesian-vacant product $G \,\boxed{\scriptstyle M}\, H$,
i) $v \sim v'$ and $w \not\sim w'$, or ii) $w \sim w'$ and $v \not\sim v'$.

In order to give images of these product operations, by taking two line graphs of 5 vertices we show adjacency of each product type at the vertex which stands in the center (see Figs. 4–8). In this paper, in order to obtain normal Kähler graphs, we give product operations of graphs comprehensively. The second author considers that some of them correspond to discrete models of magnetic fields on manifolds admitting fibrations. Still, we did not study our product operations from this point of view.

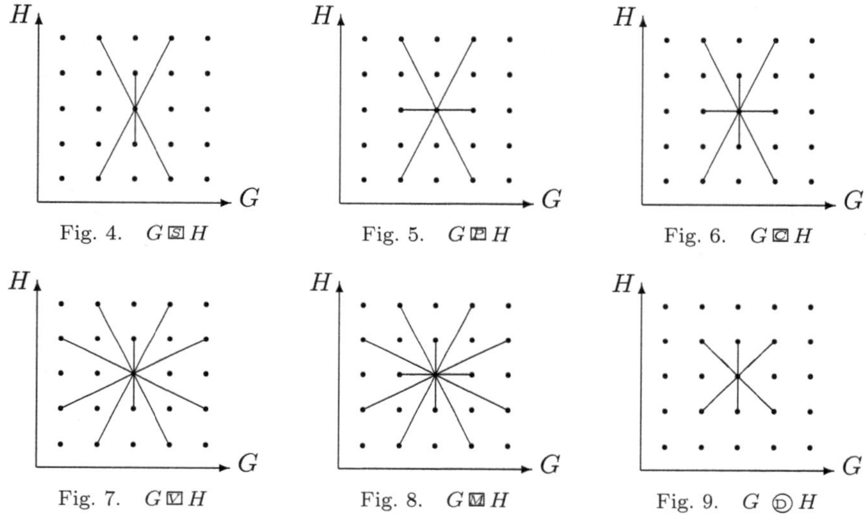

Fig. 4. $G \boxed{\scriptstyle S} H$ Fig. 5. $G \boxed{\scriptstyle P} H$ Fig. 6. $G \boxed{\scriptstyle C} H$

Fig. 7. $G \boxed{\scriptstyle V} H$ Fig. 8. $G \boxed{\scriptstyle M} H$ Fig. 9. $G \textcircled{\scriptsize D} H$

We study some basic properties of these graphs. When H is finite, degrees at a vertex $(v, w) \in V \times W$ are

$$\begin{aligned} d_{G \boxed{\scriptstyle S} H}(v, w) &= d_G(v)\big(n_H - d_H(w) - 1\big) + d_H(w), \\ d_{G \boxed{\scriptstyle P} H}(v, w) &= d_G(v)\big(n_H - d_H(w) - 1\big) + d_G(v), \\ d_{G \boxed{\scriptstyle C} H}(v, w) &= d_G(v)\big(n_H - d_H(w)\big) + d_H(w), \end{aligned}$$

and when both G and H are finite, degrees at a vertex (v, w) are

$$\begin{aligned} d_{G \boxed{\scriptstyle V} H}(v, w) &= \big(n_G - d_G(v)\big)d_H(w) + d_G(v)\big(n_H - d_H(w) - 1\big), \\ d_{G \boxed{\scriptstyle M} H}(v, w) &= \big(n_G - d_G(v)\big)d_H(w) + d_G(v)\big(n_H - d_H(w)\big). \end{aligned}$$

Here, n_G and n_H denote the cardinalities of the sets V and W, respectively.

Theorem 3.1. *Let $G = (V, E)$ be a connected locally finite graph and $H = (W, F)$ be a finite graph satisfying $\max_{w \in W} d_H(w) \leq n_H - 2$. We suppose that either G is not bipartite or H is connected. Then we have the following.*

(1) *Their semi-Cartesian-complement product $G \boxed{\scriptstyle S} H$ is connected.*

(2) *If G is not bipartite and $n_H \geq 3$, then $G \boxdot H$ is not bipartite.*
(3) *If H is not bipartite, then $G \boxdot H$ is not bipartite.*

Proof. (1) We take two distinct vertices (v, w), $(v', w') \in V \times W$ and construct a path joining them.

First we study the case $v = v'$. When w and w' are contained in the same connected component of H, we take a path $(w_0, w_1, \ldots, w_n)$ in H joining w and w'. Then $\big((v, w_0), (v, w_1), \ldots, (v, w_n)\big)$ is a path in $G \boxdot H$. When w and w' are not contained in the same connected component of H, then H is not connected, hence G is not bipartite. We take an odd-step path $(v_0, v_1, \ldots, v_{2m+1})$ in G whose origin and terminus are v. Then $\big((v_0, w), (v_1, w'), \ldots, (v_{2m}, w), (v_{2m+1}, w')\big)$ is a path in $G \boxdot H$. These paths join (v, w) and (v', w').

Next we study the case $v \sim v'$ in G. If $w \not\sim w'$ and $w \neq w'$, we have $(v, w) \sim (v', w')$ in $G \boxdot H$. If either $w \sim w'$ or $w = w'$, because $d_H(w') \leq n_H - 2$, we have $w'' \in W \setminus \{w, w'\}$ with $w' \not\sim w''$. Since $(v', w') \sim (v, w'')$ in $G \boxdot H$ and there is a path in $G \boxdot H$ joining (v, w) and (v, w'') by the above argument, we get a path joining (v, w) and (v', w') in $G \boxdot H$.

Thirdly, we study the case $v \not\sim v'$ and $v \neq v'$. Since G is connected, there is a path $(v_0, v_1, \ldots, v_m)$ from v to v'. We take $w'' \in W$ with $w'' \not\sim w$ and $w'' \neq w$. When m is even, $\big((v_0, w), (v_1, w''), (v_2, w), (v_3, w''), \ldots, (v_{m-1}, w''), (v_m, w)\big)$ is a path in $G \boxdot H$ from (v, w) to (v', w). Since we have a path from (v', w) to (v', w') by the above argument, we can join (v, w) and (v', w') in this case. When m is odd, $\big((v_0, w), (v_1, w''), (v_2, w), (v_3, w''), \ldots, (v_{m-1}, w), (v_m, w'')\big)$ is a path in $G \boxdot H$ from (v, w) to (v', w''). If $w'' \neq w'$, we have a path from (v', w'') to (v', w') by the above argument, hence we have a path joining (v, w) and (v', w') also in this case. Thus, we find that $G \boxdot H$ connected.

(2) We take a pair $(v, w), (v, w') \in V \times W$ of distinct vertices and show that there are an odd-step path and an even-step path joining them.

First we study the case $d_H \not\equiv 0$. Therefore, here we suppose $w \sim w'$. Then $(v, w) \sim (v, w')$ by definition. This means that we have a one-step path. On the other hand, if there is $w'' \in W$ satisfying $w'' \not\sim w$ and $w'' \not\sim w'$, by taking an even-step closed path $(v_0, \ldots, v_m)$ whose origin is v, we obtain an even-step path $\big((v_0, w), (v_1, w''), \ldots, (v_{m-2}, w), (v_{m-1}, w''), (v_m, w')\big)$ of (v, w) to (v, w'). If $w'' \in W$ satisfies $w'' \not\sim w$ and $w'' \sim w'$, by taking an odd-step closed path $(v_0, \ldots, v_m)$ whose origin is v, we obtain an even-step path $\big((v_0, w), (v_1, w''), \ldots, (v_{m-1}, w), (v_m, w''), (v, w')\big)$ of (v, w) to (v, w').

Next we study the case $d_H \equiv 0$, that is, all vertices of H are isolated. Therefore, we have $w \not\sim w'$ and $w \neq w'$. By taking an odd-step closed path $(v_0, v_1, \ldots, v_m)$ whose origin is v, we have an odd-step path $\big((v_0, w), (v_1, w'), \ldots, (v_{m-1} w), (v_m, w')\big)$ from (v, w) to (v, w'). On the other hand, since $d_H \equiv 0$ and $n_H \geq 3$, we have $w'' \in W \setminus \{w, w'\}$ satisfying $w'' \not\sim w$ and $w'' \not\sim w'$. By taking an even-step closed path $(v_0, \ldots, v_m)$ whose origin is v, we obtain an even-step path $\big((v_0, w), (v_1, w''), \ldots, (v_{m-2}, w), (v_{m-1}, w''), (v_m, w')\big)$ of (v, w) to (v, w'). Thus, we find that $G \boxdot H$ is not bipartite.

(3) We take a set $\{v_0\} \times W$ and define $(v_0, w) \sim (v_0, w')$ if and only if $w \sim w'$. Then we find that it is isomorphic to H and is a subgraph of $G \boxdot H$. Thus, the assertion is clear. □

Example 3.1. When both G and H are 4-circuits, their semi-Cartesian-complement product $G \boxdot H$ is connected and bipartite (see Fig. 10).

Example 3.2. When G is a 3-circuit and H is 4-circuits, as G is not bipartite, the product $G \boxdot H$ is not bipartite either (see Fig. 11).

Example 3.3. When G is a 4-circuit and H is 6-circuits, even though they are bipartite, $G \boxdot H$ is not bipartite.

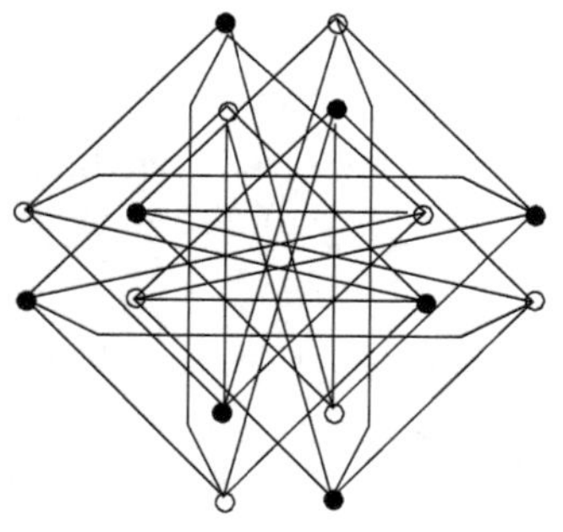

Fig. 10.

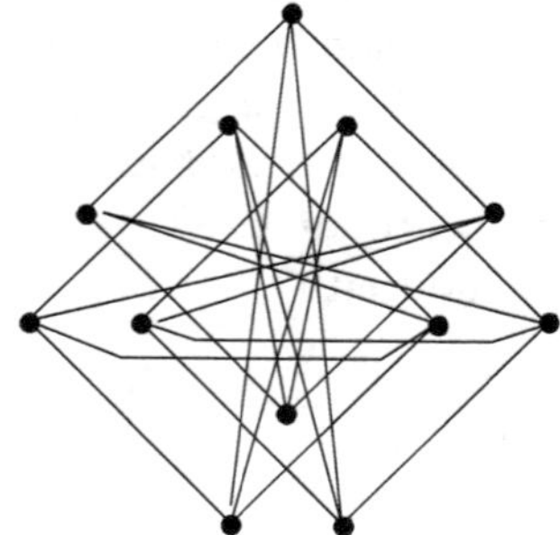

Fig. 11.

By the definition of other product operations, we have the following.

Corollary 3.1. *Let G be a connected locally finite graph and H be a finite graph satisfying $\max_{w \in W} d_H(w) \leq n_H - 2$. We suppose that either G is not bipartite or H is connected.*

(1) *Their Cartesian-complement product* $G \mathbin{\boxed{\scriptstyle C}} H$, *their semi-Cartesian-vacant product* $G \mathbin{\boxed{\scriptstyle V}} H$, *and their Cartesian-vacant product* $G \mathbin{\boxed{\scriptstyle M}} H$ *are connected.*

(2) *Under the same assumption as in* (2) *or in* (3) *of Theorem* 3.1, *these three graphs are not bipartite.*

Though the definitions of semi-Cartesian-complement product and partial Cartesian-complement product resemble each other, to show some properties of graphs of partial Cartesian-complement product type, we need a condition on complement graphs. Given two graphs $G = (V, E)$ and $H = (W, F)$, we consider the set $V \times W$ and define a rule of adjacency of two vertices (v, w) and (v', w') by

either i) $v = v'$ and $w \sim w'$ or ii) $v \sim v'$ and $w \sim w'$.

We denote by $G \mathbin{\textcircled{\tiny D}} H$ this graph of product type and call this operation tensor-semi-Cartesian product. Figure 9 shows the adjacency of this product when G and H are line graphs of 5 vertices.

Proposition 3.2. *Let G and H be connected locally finite graphs.*

(1) *Their tensor-semi-Cartesian product* $G \mathbin{\textcircled{\tiny D}} H$ *is connected.*

(2) $G \mathbin{\textcircled{\tiny D}} H$ *is bipartite if and only if* H *is bipartite.*

Proof. We express G and H as (V, E) and (W, F), respectively.
(1) Since $G \otimes H$ is a subgraph of $G \mathbin{\textcircled{\tiny D}} H$, by Proposition 3.1, it is enough to study only the case that both G and H are bipartite. Let $V = V^+ \cup V^-$ and $W = W^+ \cup W^-$ be the vertex-decompositions of G and H as bipartite graphs. In this case, the tensor product $G \otimes H$ has two connected components which are formed by $(V^+ \times W^+) \cup (V^- \times W^-)$ and $(V^+ \times W^-) \cup (V^- \times W^+)$. We take $v, v' \in V$ so that $v \sim v'$, and take $w, w' \in W$ so that there is a path of the form (w, w'', w'). If $(v, w) \in (V^+ \times W^+) \cup (V^- \times W^-)$, then $(v', w') \in (V^+ \times W^-) \cup (V^- \times W^+)$, and vice versa. In the graph $G \mathbin{\textcircled{\tiny D}} H$, we see $(v, w) \sim (v, w'') \sim (v', w')$. Thus, we find that $G \mathbin{\textcircled{\tiny D}} H$ is connected.
(2) When H is not bipartite, as H is a subgraph of $G \mathbin{\textcircled{\tiny D}} H$, it is clear that the product is not bipartite. On the other hand, when $\big((v_0, w_0), (v_1, w_1), \ldots, (v_n, w_n)\big)$ is a path in $G \mathbin{\textcircled{\tiny D}} H$, then we find that $(w_0, w_1, \ldots, w_n)$ is a path in H. This means that if H is bipartite, we find that $(V \times W^+) \cup (V \times W^-)$ is a vertex-decomposition of the set $V \times W$ of $G \mathbin{\textcircled{\tiny D}} H$ as a bipartite graph. □

Corollary 3.2. *Let G be a connected locally finite graph and H be a graph* *whose complementary graph H^c is connected and locally finite.*

(1) *Their partial Cartesian-complement product $G \mathbin{\boxed{\scriptstyle P}} H$ is connected.*
(2) *$G \mathbin{\boxed{\scriptstyle P}} H$ is bipartite if and only if H^c is bipartite.*

We note that if a finite graph $H = (W, F)$ satisfies $\max_{w \in W} d_H(w) < n_H/2$ then its complement graph H^c is connected.

Given two ordinary graphs G and H, we consider the following Kähler graphs of product type. Their principal graphs are the tensor product $G \otimes H$, and their auxiliary graphs are one of the products $G \mathbin{\boxed{\scriptstyle S}} H$, $G \mathbin{\boxed{\scriptstyle P}} H$, $G \mathbin{\boxed{\scriptstyle C}} H$, $G \mathbin{\boxed{\scriptstyle V}} H$ and $G \mathbin{\boxed{\scriptstyle M}} H$. When the auxiliary graph is $G \mathbin{\boxed{\scriptstyle S}} H$, we shall say that this Kähler graph is of tensor $*$ semi-Cartesian-complement product type. Similarly, we call other Kähler graphs of tensor $*$ partial-Cartesian-complement product type, of tensor $*$ Cartesian-complement product type, of tensor $*$ semi-Cartesian-vacant product type and of tensor $*$ Cartesian-vacant product type, respectively. By these definitions the following is clear.

Proposition 3.3. *For vertex-transitive ordinary graphs, their Kähler graphs of tensor$*$ semi-Cartesian-complement product type, tensor$*$ partial-Cartesian-complement product type, of tensor$*$ Cartesian-complement product type, of tensor$*$ semi-Cartesian-vacant product type and of tensor$*$ Cartesian-vacant product type are vertex-transitive.*

4. Adjacency operators

In this section, we check that constructed Kähler graphs are normal. We take two locally finite ordinary regular graphs $G = (V, E)$ and $H = (W, F)$. For $v \in V$ and $w \in W$, let $\delta_v \in C_c(V)$ and $\delta_w \in C_c(W)$ denote the characteristic functions defined by

$$\delta_v(x) = \begin{cases} 1, & \text{when } x = v, \\ 0, & \text{when } x \neq v, \end{cases} \qquad \delta_w(y) = \begin{cases} 1, & \text{when } y = w, \\ 0, & \text{when } y \neq w, \end{cases}$$

and define $\delta_{(v,w)} \in C_c(V \times W)$ by $\delta_{v,w}(x, y) = \delta_v(x)\delta_w(y)$. Since each element of $C_c(V \times W)$ is expressed by using characteristic functions, to show how adjacency operators act on $C_c(V \times W)$, it is enough to study the expression of their actions on characteristic functions. We find that the adjacency operator $\mathcal{A}_{G\otimes H}$ of the tensor product $G \otimes H$ satisfies

$$\big(\mathcal{A}_{G\otimes H}\delta_{(v,w)}\big)(x, y) = \sum_{x' \sim x;\, y' \sim y} \delta_{(v,w)}(x', y')$$

$$= \Big(\sum_{x'\sim x} \delta_v(x')\Big)\Big(\sum_{y'\sim y} \delta_w(y')\Big) = \big\{\big(\mathcal{A}_G\delta_v\big)(x)\big\}\big\{\big(\mathcal{A}_H\delta_w\big)(y)\big\}.$$

It is the common adjacency operator for principal graphs of the Kähler graphs which we consider in this paper. In order to check that our Kähler graphs of product type are normal, we study adjacency operators of product graphs defined in §3. These correspond to adjacency operators of auxiliary graphs of our Kähler graphs.

When H is finite, the adjacency operator $\mathcal{A}_{G \boxed{S} H}$ of the semi-Cartesian-complement product $G \boxed{S} H$ satisfies

$$\begin{aligned}
\big(\mathcal{A}_{G \boxed{S} H}\delta_{(v,w)}\big)(x,y) &= \sum_{x'\sim x;\, y'\not\sim y, y'\neq y} \delta_{(v,w)}(x',y') + \sum_{x'=x;\, y'\sim y} \delta_{(v,w)}(x',y') \\
&= \Big(\sum_{x'\sim x} \delta_v(x')\Big)\Big(\sum_{y'\not\sim y, y'\neq y} \delta_w(y')\Big) + \delta_v(x)\Big(\sum_{y'\sim y} \delta_w(y')\Big) \\
&= \big\{\big(\mathcal{A}_G\delta_v\big)(x)\big\}\big\{\big((\mathcal{M}_H - I_W - \mathcal{A}_H)\delta_w\big)(y)\big\} + \delta_v(x)\big\{\big(\mathcal{A}_H\delta_w\big)(y)\big\}.
\end{aligned}$$

Here, I_W is the identity of $C_c(W)$ and $\mathcal{M}_H$ shows the operator on $C(W)$ defined by $\mathcal{M}_H g(y) = \sum_{w\in W} g(w)$. Similarly, the adjacency operator $\mathcal{A}_{G \boxed{P} H}$ of the partial-Cartesian-complement product $G \boxed{P} H$ satisfies

$$\begin{aligned}
\big(\mathcal{A}_{G \boxed{P} H}\delta_{(v,w)}\big)(x,y) &= \sum_{x'\sim x;\, y'\not\sim y} \delta_{(v,w)}(x',y') = \Big(\sum_{x'\sim x} \delta_v(x')\Big)\Big(\sum_{y'\not\sim y} \delta_w(y')\Big) \\
&= \big\{\big(\mathcal{A}_G\delta_v\big)(x)\big\}\big\{\big((\mathcal{M}_H - \mathcal{A}_H)\delta_w\big)(y)\big\}.
\end{aligned}$$

The adjacency operator $\mathcal{A}_{G \boxed{C} H}$ of the Cartesian-complement product $G \boxed{C} H$ satisfies

$$\begin{aligned}
\big(\mathcal{A}_{G \boxed{C} H}\delta_{(v,w)}\big)(x,y) &= \sum_{x'\sim x;\, y'\not\sim y} \delta_{(v,w)}(x',y') + \sum_{x'=x;\, y'\sim y} \delta_{(v,w)}(x',y') \\
&= \Big(\sum_{x'\sim x} \delta_v(x')\Big)\Big(\sum_{y'\not\sim y} \delta_w(y')\Big) + \delta_v(x)\Big(\sum_{y'\sim y} \delta_w(y')\Big) \\
&= \big\{\big(\mathcal{A}_G\delta_v\big)(x)\big\}\big\{\big((\mathcal{M}_H - \mathcal{A}_H)\delta_w\big)(y)\big\} + \delta_v(x)\big\{\big(\mathcal{A}_H\delta_w\big)(y)\big\}.
\end{aligned}$$

Since H is regular, we have $\mathcal{M}_H\mathcal{A}_H = \mathcal{A}_H\mathcal{M}_H$. Therefore, we have the following.

Proposition 4.1. *Let G be a locally finite regular graph and H be a finite regular graph. Then their Kähler graphs of tensor$*$ semi-Cartesian-complement product type, of tensor$*$ partial-Cartesian-complement product type, and of tensor$*$ Cartesian-complement product type are normal.*

When both G and H are finite, the adjacency operator $\mathcal{A}_{G\boxslash H}$ of the semi-Cartesian-vacant product $G\boxslash H$ satisfies

$$\begin{aligned}
&\big(\mathcal{A}_{G\boxslash H}\delta_{(v,w)}\big)(x,y)\\
&\quad= \sum_{x'\sim x;\, y'\not\sim y, y'\neq y} \delta_{(v,w)}(x',y') + \sum_{x'\not\sim x;\, y'\sim y} \delta_{(v,w)}(x',y')\\
&\quad= \Big(\sum_{x'\sim x}\delta_v(x')\Big)\Big(\sum_{y'\not\sim y, y'\neq y}\delta_w(y')\Big) + \Big(\sum_{x'\not\sim x}\delta_v(x')\Big)\Big(\sum_{y'\sim y}\delta_w(y')\Big)\\
&\quad= \big\{\big(\mathcal{A}_G\delta_v\big)(x)\big\}\big\{\big((\mathcal{M}_H - I_W - \mathcal{A}_H)\delta_w\big)(y)\big\}\\
&\qquad+ \big\{\big((\mathcal{M}_G - \mathcal{A}_G)\delta_v\big)(x)\big\}\big\{\big(\mathcal{A}_H\delta_w\big)(y)\big\}.
\end{aligned}$$

The adjacency operator $\mathcal{A}_{G\boxtimes H}$ of the Cartesian-vacant product $G \boxtimes H$ satisfies

$$\begin{aligned}
&\big(\mathcal{A}_{G\boxtimes H}\delta_{(v,w)}\big)(x,y)\\
&\quad= \sum_{x'\sim x;\, y'\not\sim y} \delta_{(v,w)}(x',y') + \sum_{x'\not\sim x;\, y'\sim y} \delta_{(v,w)}(x',y')\\
&\quad= \Big(\sum_{x'\sim x}\delta_v(x')\Big)\Big(\sum_{y'\not\sim y}\delta_w(y')\Big) + \Big(\sum_{x'\not\sim x}\delta_v(x')\Big)\Big(\sum_{y'\sim y}\delta_w(y')\Big)\\
&\quad= \big\{\big(\mathcal{A}_G\delta_v\big)(x)\big\}\big\{\big((\mathcal{M}_H - \mathcal{A}_H)\delta_w\big)(y)\big\}\\
&\qquad+ \big\{\big((\mathcal{M}_G - \mathcal{A}_G)\delta_v\big)(x)\big\}\big\{\big(\mathcal{A}_H\delta_u\big)(y)\big\}.
\end{aligned}$$

Thus, we get the following.

Proposition 4.2. *Let G and H be finite regular graphs. Then their Kähler graphs of tensor∗ semi-Cartesian-vacant product type and of tensor∗ Cartesian-vacant product type are normal.*

Here, we make mention of eigenvalues of adjacency operators of graphs of our product types. Since the adjacency operators of the principal and auxiliary graphs of a normal Kähler graph are simultaneously diagonalizable, one can easily get eigenvalues of the $(1,1)$-probabilistic adjacency operator. Let $G = (V,E)$ and $H = (W,F)$ be finite connected ordinary graphs. We denote by $\lambda_1,\ldots,\lambda_{n_G}$ $(\lambda_1 \geq \cdots \geq \lambda_{n_G})$ and by $\mu_1,\ldots,\mu_{n_H}$ $(\mu_1 \geq \cdots \geq \mu_{n_H})$ the eigenvalues of $\mathcal{A}_G$ and $\mathcal{A}_H$, respectively. When G is regular we have $\lambda_1 = d_G$, and when H is regular we have $\mu_1 = d_H$. We denote by $f_1,\ldots,f_{n_G} \in C(V)$ and $g_1,\ldots,g_{n_H} \in C(W)$ eigenfunctions corresponding to these eigenvalues. By using expressions on the adjacency operators in this section, we have the following.

Proposition 4.3. *Let G and H be finite connected ordinary graphs. Suppose H is regular.*

(1) *The eigenvalues of $\mathcal{A}_{G \boxed{S} H}$ are*
$\lambda_i(n_H - 1 - d_H) + d_H$, $\mu_j - \lambda_i(1 + \mu_j)$ $(1 \leq i \leq n_G,\ 2 \leq j \leq n_H)$;
(2) *The eigenvalues of $\mathcal{A}_{G \boxed{P} H}$ are*
$\lambda_i(n_H - d_H)$, $-\lambda_i \mu_j$ $(1 \leq i \leq n_G,\ 2 \leq j \leq n_H)$;
(3) *The eigenvalues of $\mathcal{A}_{G \boxed{C} H}$ are*
$\lambda_i(n_H - d_H) + d_H$, $(1 - \lambda_i)\mu_j$ $(1 \leq i \leq n_G,\ 2 \leq j \leq n_H)$.

Proposition 4.4. *Let G and H be finite connected ordinary regular graphs.*

(1) *The eigenvalues of $\mathcal{A}_{G \boxed{V} H}$ are*
$d_G(n_H - 1 - d_H) + (n_G - d_G)d_H$, $-d_G(1 + 2\mu_j) + n_G \mu_j$,
$\lambda_i(n_H - 1 - 2d_H)$, $-\lambda_i(1 + 2\mu_j)$, $(2 \leq i \leq n_G,\ 2 \leq j \leq n_H)$;
(2) *The eigenvalues of $\mathcal{A}_{G \boxed{M} H}$ are*
$d_G(n_H - d_H) + (n_G - d_G)d_H$, $(n_G - 2d_G)\mu_j$,
$\lambda_i(n_H - 2d_H)$, $-2\lambda_i \mu_j$, $(2 \leq i \leq n_G,\ 2 \leq j \leq n_H)$.

Proposition 4.5. *For finite graphs G and H, the eigenvalues of $\mathcal{A}_{G \otimes H}$ are $\lambda_i \mu_j$ $(1 \leq i \leq n_G,\ 1 \leq j \leq n_H)$.*

Remark 4.1. In Propositions 4.3 and 4.4, corresponding to these eigenvalues, the functions $(f_i, g_j) \in C(V \times W)$ $(1 \leq i \leq n_G,\ 1 \leq j \leq n_H)$ defined by $(f_i, g_j)(x, y) = f_i(x) f_j(y)$ are eigenfunctions.

Example 4.1. When G is a 4-circuit and H is a 3-circuit, their adjacency operators have eigenvalues $2, 0, 0, -2$ and $2, -1, -1$, respectively. Therefore, the eigenvalues of $G \otimes H$ and $G \boxed{S} H$ are as in Tables 1 and 2. Hence, the eigenvalues of the $(1, 1)$-probabilistic adjacency operator are as in Table 3.

	2	0	0	−2
2	4	0	0	−4
−1	−2	0	0	2
−1	−2	0	0	2

Table 1

	2	0	0	−2
2	2	2	2	2
−1	−1	−1	−1	−1
−1	−1	−1	−1	−1

Table 2

	2	0	0	−2
2	8/3	0	0	−8/3
−1	2/3	0	0	−2/3
−1	2/3	0	0	−2/3

Table 3

For Kähler graphs we consider (p, q)-probabilistic adjacency operators as generating operators of random walks under influence of magnetic fields. When they are normal we can express these operators by using the adjacency operators of principal and auxiliary graphs and auxiliary degrees (see [4]). Therefore, Propositions 4.3, 4.4 and 4.5 show all eigenvalues of

(p, q)-probabilistic adjacency operators for these Kähler graphs of product type. In particular we can study their zeta functions by results in [5] and [11].

References

[1] T. Adachi, Kähler magnetic flows on a manifold of constant holomorphic sectional curvature, *Tokyo J. Math.* **18** (1995), 473–483.

[2] ———, Magnetic mean operators on a Kähler manifold, in *Topics in Almost Hermitian geometry and related fields*, Y. Matsushita et al. eds., World Scientific, Singapore, (2005), 30–40.

[3] ———, A discrete model for Kähler magnetic fields on a complex hyperbolic space, in *Trends in Differential Geometry, Complex Analysis and Mathematical Physics*, K. Sekigawa, V.S. Gerdijikov & S. Dimiev eds., World Scientific, Singapore, (2009), 1–9.

[4] ———, Eigenvalues of regular Kähler graphs having commutative adjacency operators, in *Recent Topics in Differential Geometry and its Related Fields*, T. Adachi & H. Hashimoto eds., World Scientific, Singapore, (2020), 83–106.

[5] ———, A note on zeta functions of Ihara type for normal Kähler graphs, *Discrete Math.* **345** (2022), paper no. 112688, 14 pp.

[6] ———, Kähler graphs whose principal graphs are of Cartesian product type, in *New Horizons in Differential Geometry and its Related Fields*, T. Adachi & H. Hashimoto eds., World Scientific, Singapore, (2022), 209–231

[7] T. Adachi and G. Chen, Regular and vertex-transitive Kähler graphs having commutative principal and auxiliary adjacency operators, *Graphs Combin.*, **36** (2020), 933–958.

[8] R. Hammack, W. Imrich and S. Klavžar, *Handbook of product graphs*, Discrete Mathematics and its Applications, 2011.

[9] W. Imrich and S. Klavžar, *Product graphs: Structure and recognition*, Wiley, 2000.

[10] T. Sunada, Magnetic flows on a Riemann surface, *Proc. KAIST Math. Workshop* **8** (1993), 93–108.

[11] T. Yaermaimaiti and T. Adachi, Zeta functions for a Kähler graph, *Kodai Math. J.*, **41** (2018), 227–239.

Received November 29, 2023
Revised February 26, 2024

Modern Approaches to Differential Geometry
and its Related Fields 137 – 158

BICONSERVATIVE HYPERSURFACES IN $\mathbb{E}_s^5$

Ram Shankar GUPTA

University School of Basic and Applied Sciences,
Guru Gobind Singh Indraprastha University,
Sector-16C, Dwarka, New Delhi–110078, India
E-mail: ramshankar.gupta@gmail.com

Andreas ARVANITOYEORGOS

Department of Mathematics, University of Patras,
GR-26500 Rion, Greece, and
Hellenic Open University
Aristotelous 18, GR-26335 Patras, Greece
E-mail: arvanito@math.upatras.gr

Savita RANI

International Centre for Theoretical Sciences
Tata Institute of Fundamental Research
Bengaluru–560 089, India
E-mail: savita.rani@icts.res.in

Marina STATHA

Department of Mathematics, University of Thessaly,
GR-35100 Lamia, Greece
E-mail: marina.statha@gmail.com

In this paper, we study biconservative hypersurfaces M_r^4 $(r = 0, 1, 2, 3, 4)$ in the pseudo-Euclidean space $\mathbb{E}_s^5$ $(s = 0, 1, 2, 3, 4, 5)$ with constant norm of second fundamental form. We found that every such hypersurface in $\mathbb{E}_s^5$ with diagonal shape operator has constant mean curvature and constant scalar curvature.

Keywords: Pseudo-Euclidean space; mean curvature vector; biconservative hypersurface; biharmonic submanifold.

1. Introduction

The classification of constant mean curvature hypersurfaces constitutes an important area of research not only in differential geometry but in relativity theory as well ([17, 18]). On the other hand, it is known that harmonic maps between Riemannian manifolds $\phi : (M, g) \to (N, h)$ are critical points of the energy functional $E(\phi) = \frac{1}{2}\int_M |d\phi|^2 v_g$, the biharmonic maps are critical points of the bienergy functional $E_2(\phi) = \frac{1}{2}\int_M |\tau(\phi)|^2 vg$, where $\tau = \text{trace}(\nabla d\phi)$ is the tension field of ϕ.

In 1924, Hilbert pointed out that the stress-energy tensor associated to a functional E is a conservative symmetric 2-covariant tensor S at the critical points of E, i.e. $\operatorname{div} S = 0$ ([14]). For the bienergy functional E_2, Jiang defined the stress-bienergy tensor S_2 and proved that it satisfies $\operatorname{div} S_2 = -\langle \tau_2(\phi), d\phi \rangle$, where τ_2 is the bitension field ([15]). Thus, if ϕ is biharmonic, then $\operatorname{div} S_2 = 0$. For biharmonic submanifolds, from the above relation, we see that $\operatorname{div} S_2 = 0$ if and only if the tangential part of the bitension field vanishes. In particular, an isometric immersion $\phi : (M, g) \to (N, h)$ is called *biconservative* if $\operatorname{div} S_2 = 0$. The terminology was introduced by R. Caddeo et al. in [2].

Biconservative submanifolds were studied and classified in $\mathbb{E}^4$ by Hasanis and Vlachos in [13], where the biconservative hypersurfaces were called H-hypersurfaces. In [19], Turgay obtained a complete classification of H-hypersurfaces with three distinct curvatures in Euclidean space of arbitrary dimension, and gave some explicit examples. Several other authors have studied biconservative submanifolds in Euclidean spaces, having obtained important results. In [9], the first author proved that every biconservative hypersurface M in $\mathbb{E}^5$ with constant norm of second fundamental form has constant mean curvature. In [11], the first and the second author examined biconservative hypersurfaces in space forms with four distinct principal curvatures whose second fundamental form has constant norm.

In the present article, we study biconservative hypersurfaces M^4_r in a pseudo-Euclidean space $\mathbb{E}^5_s$. So we review some previous works in this direction. In [5], Fu classified biconservative surfaces in $\mathbb{E}^3_1$. It turns out that these are surfaces of constant mean curvature, or a surface of revolution, or a null scroll. In [6], the same author gave classification of biconservative surfaces in a de Sitter 3-space $\mathbb{S}^3_1(1)$ and in an anti-de Sitter 3 space $\mathbb{H}^3_1(-1)$. In [20], Upadhyay and Turgay classified biconservative hypersurfaces of index 2 in $\mathbb{E}^5_2$ with diagonal shape operator having three distinct principal curvatures, and constructed an example of a biconservative hypersurface with four distinct principal curvatures.

In [7], Fu and Turgay gave a complete classification of biconservative hypersurfaces with diagonalizable shape operator in the Minkowski 4-space $\mathbb{E}^4_1$. In [4], Deepika proved that every biconservative Lorentz hypersurface in $\mathbb{E}^{n+1}_1$ with complex eigenvalues having at most five distinct principal curvatures has constant mean curvature. It was also proved that a biconservative Lorentz hypersurface with constant length of second fundamental form and whose shape operator has complex eigenvalues with six distinct principal curvatures has constant mean curvature.

In [12], the first author and Sharfuddin proved that every biconservative Lorentz hypersurface M^n_1 in $\mathbb{E}^{n+1}_1$ with complex eigenvalues has constant mean curvature. If it is biharmonic, then it is minimal. They also gave examples of such hypersurfaces. Finally, in [16], Kayhan and Turgay obtained classification of biconservative hypersurfaces in $\mathbb{E}^4_1$ with non diagonalizable shape operator, completing the study of [7]. In [10], the first and the second authors investigated hypersurfaces M^4_r $(r = 0, 1, 2, 3, 4)$ satisfying $\Delta\vec{H} = \lambda\vec{H}$, with a constant λ, in the pseudo-Euclidean space $\mathbb{E}^5_s$ $(s = 0, 1, 2, 3, 4, 5)$.

In view of the above developments, we study biconservative hypersurfaces M^4_r in $\mathbb{E}^5_s$ with constant norm of second fundamental form. Although, the shape operator of a Riemannian hypersurface is always diagonalizable, this is not true in the pseudo-Riemannian case. In all results of the paper, we assume that the shape operator of hypersurfaces is locally diagonalizable, i.e., has diagonal form with respect to a smooth local tangent orthonormal frame $\{e_1, ..., e_4\}$ and has smooth principal curvatures. The main result is the following:

Theorem 1.1. *Every biconservative hypersurface* M^4_r $(r = 0, 1, \ldots, 4)$ *in the pseudo-Euclidean space* $\mathbb{E}^5_s$ $(s = 0, 1, \ldots, 5)$ *with diagonal shape operator and constant norm of second fundamental form, has constant mean curvature.*

A submanifold satisfying $\Delta\vec{H} = 0$ is called *biharmonic submanifold.* A well known conjecture posed by Chen in 1991 states that *the only biharmonic submanifolds of Euclidean spaces are the minimal ones* ([3]). Several authors have contributed towards the proof of this conjecture with positive results in various cases. We refer to the recent article of Fu, Hong and Zhan [8], in which the authors settled the conjecture for hypersurfaces in $\mathbb{R}^6$, and the references therein for previous works. However, the conjecture is not always true for submanifolds in pseudo-Euclidean spaces, but it is conjectured by the second author, Defever and Kaimakamis ([1]) that it is true for hypersurfaces of pseudo-Euclidean spaces.

The structure of the paper is as follows. In Section 2, we give some preliminaries. In Section 3, we investigate biconservative hypersurfaces M^4_r with diagonal shape operator in $\mathbb{E}^5_s$. We also assume that the mean curvature is not constant and $\operatorname{grad} H \neq 0$. We obtain expressions of covariant derivatives of an orthonormal frame in terms of connection forms (Lemma 3.1). In Section 4, we use the condition that the second fundamental form has constant norm, and obtain further simplifications of ω_{ij}^k (Lemmas 4.1 and 4.2). Section 5 is devoted to the proof of Theorem 1.1. In the process, a technical tool about resultant of polynomials is extensively used, when we eliminate variables of polynomials.

2. Preliminaries

Let (M_r^4, g), $r = 0, 1, \dots, 4$, be a 4-dimensional hypersurface isometrically immersed in a 5-dimensional semi-Euclidean space $(\mathbb{E}_s^5, \overline{g})$, $s = 0, 1, \dots, 4, 5$ and $g = \overline{g}\big|_{M_r^4}$. We denote by ξ the unit normal vector to M_r^4 with $\overline{g}(\xi, \xi) = \varepsilon$, where $\varepsilon = \pm 1$, according as M_r^4 is pseudo-Riemannian or Riemannian hypersurface.

The Gauss and Codazzi equations are given by

$$R(X, Y)Z = g(\mathcal{A}Y, Z)\mathcal{A}X - g(\mathcal{A}X, Z)\mathcal{A}Y, \tag{1}$$

$$(\nabla_X \mathcal{A})Y = (\nabla_Y \mathcal{A})X, \tag{2}$$

respectively, where R is the curvature tensor, $\mathcal{A}$ is the shape operator, ∇ denotes the linear connection on M_r^4 in $\mathbb{E}_s^5$ and

$$(\nabla_X \mathcal{A})Y = \nabla_X \mathcal{A}Y - \mathcal{A}(\nabla_X Y), \tag{3}$$

for all vector fields $X, Y, Z \in \Gamma(TM_r^4)$ on M_r^4. The mean curvature is given by

$$\varepsilon H = \frac{1}{4} \operatorname{trace} \mathcal{A}. \tag{4}$$

The biharmonic equation can be decomposed into its normal and tangential parts. The necessary and sufficient conditions for M_r^4 to be biharmonic in E_s^5 are

$$\Delta H + \varepsilon H \operatorname{trace} \mathcal{A}^2 = 0, \tag{5}$$

$$\mathcal{A}(\operatorname{grad} H) + 2\varepsilon H \operatorname{grad} H = 0. \tag{6}$$

Definition 2.1. A hypersurface M_r^4 in $\mathbb{E}_s^5$ is called biconservative if equation (6) is satisfied.

3. Biconservative hypersurfaces in $\mathbb{E}_s^5$

In this section, we study biconservative hypersurfaces M_r^4 with diagonal shape operator in $\mathbb{E}_s^5$. We also assume that the mean curvature is not constant and $\operatorname{grad} H \neq 0$. This implies the existence of an open connected subset U of M with $\operatorname{grad}_p H \neq 0$ for all $p \in U$. From (6), it is easy to see that $\operatorname{grad} H$ is an eigenvector of the shape operator $\mathcal{A}$ with the corresponding principal curvature $-2\varepsilon H$. Without loss of generality, we choose e_1 in the direction of $\operatorname{grad} H$ and therefore the shape operator $\mathcal{A}$ of M_r^4 will take the following form with respect to a suitable frame $\{e_1, e_2, e_3, e_4\}$:

$$\mathcal{A} = \begin{pmatrix} -2\varepsilon H & & & \\ & \lambda_2 & & \\ & & \lambda_3 & \\ & & & \lambda_4 \end{pmatrix}. \tag{7}$$

Due to (6) the first principal curvature is $\lambda_1 = -2\varepsilon H$. Also, using (4) and (7), we obtain

$$\sum_{j=2}^{4} \lambda_j = 6\varepsilon H = -3\lambda_1. \tag{8}$$

The vector field $\operatorname{grad} H$ can be expressed as

$$\operatorname{grad} H = \sum_{i=1}^{4} e_i(H) e_i.$$

As we have taken e_1 parallel to $\operatorname{grad} H$, it follows that

$$e_1(H) \neq 0, \quad e_2(H) = 0, \quad e_3(H) = 0, \quad e_4(H) = 0. \tag{9}$$

We express

$$\nabla_{e_i} e_j = \sum_{k=1}^{4} \epsilon_k \omega_{ij}^k e_k, \quad i, j = 1, \ldots, 4, \tag{10}$$

where $\epsilon_k = \pm 1$.

Using (10) and the compatibility conditions $(\nabla_{e_k} g)(e_i, e_i) = 0$ and $(\nabla_{e_k} g)(e_i, e_j) = 0$, we obtain

$$\omega_{ki}^i = 0, \qquad \omega_{ki}^j + \omega_{kj}^i = 0, \tag{11}$$

for $i \neq j$, and $i, j, k = 1, \ldots, 4$. Taking $X = e_i, Y = e_j$ in (3) and using (7), (10), we get

$$(\nabla_{e_i} A) e_j = e_i(\lambda_j) e_j + \sum_{k=1}^{4} (\lambda_j - \lambda_k) \omega_{ij}^k \epsilon_k e_k.$$

Putting the value of $(\nabla_{e_i} A) e_j$ in (2), we find

$$e_i(\lambda_j) e_j + \sum_{k=1}^{n} (\lambda_j - \lambda_k) \omega_{ij}^k \epsilon_k e_k = e_j(\lambda_i) e_i + \sum_{k=1}^{n} (\lambda_i - \lambda_k) \omega_{ji}^k \epsilon_k e_k.$$

Then for $i \neq j = k$ and $i \neq j \neq k$ we obtain the equations

$$\epsilon_j e_i(\lambda_j) = (\lambda_i - \lambda_j)\omega_{ji}^j = (\lambda_j - \lambda_i)\omega_{jj}^i, \tag{12}$$

and

$$(\lambda_i - \lambda_j)\omega_{ki}^j = (\lambda_k - \lambda_j)\omega_{ik}^j, \tag{13}$$

respectively. Since $\lambda_1 = -2\varepsilon H$, from (9) we get

$$e_1(\lambda_1) \neq 0, \quad e_2(\lambda_1) = 0, \quad e_3(\lambda_1) = 0, \quad e_4(\lambda_1) = 0. \tag{14}$$

The following lemma gives some simplifications of the above derivatives.

Lemma 3.1. *Let M_r^4 be a biconservative hypersurface of non constant mean curvature with all distinct principal curvatures in $\mathbb{E}_s^5$, having the shape operator given by (7) with respect to suitable orthonormal frame $\{e_1, e_2, e_3, e_4\}$. Then, we have*

$$\nabla_{e_1} e_\ell = 0, \ \ \ell = 1,2,3,4, \qquad \nabla_{e_i} e_1 = -\omega_{ii}^1 \epsilon_i e_i, \ i = 2,3,4,$$

$$\nabla_{e_i} e_i = \sum_{\ell=1,\ell\neq i}^{4} \omega_{ii}^\ell \epsilon_\ell e_\ell, \ i = 2,3,4, \qquad \nabla_{e_i} e_j = \sum_{k=2,k\neq j}^{4} \omega_{ij}^k \epsilon_k e_k, \ i,j = 2,3,4,$$

where ω_{ij}^i satisfy (11) and (12).

Proof. Using (9), (10) and the fact that

$$0 = [e_i, e_j](H) = \nabla_{e_i} e_j(H) - \nabla_{e_j} e_i(H) = \epsilon_1 \omega_{ij}^1 e_1(H) - \epsilon_1 \omega_{ji}^1 e_1(H),$$

we find

$$\omega_{ij}^1 = \omega_{ji}^1, \tag{15}$$

for $i \neq j$ and $i, j = 2,3,4$.

Now, putting $i \neq 1, j = 1$ in (12), and using (14) and (11), we find

$$\omega_{1i}^1 = 0, \qquad i = 1,2,3,4. \tag{16}$$

Also, putting $k = 1, j \neq i$, and $i, j = 2,3,4$ in (13), and using (15), we get

$$\omega_{ij}^1 = \omega_{ji}^1 = \omega_{1i}^j = \omega_{i1}^j = 0, \quad j \neq i \text{ and } i,j = 2,3,4. \tag{17}$$

Finally, using (10), (16) and (17) we conclude the proof of the lemma. □

For later use we claim that $\lambda_1 \neq \lambda_j, j = 2,3,4$. Indeed, if $\lambda_j = \lambda_1$ for some $j \neq 1$, then from (12), we find that

$$\epsilon_j e_1(\lambda_j) = (\lambda_1 - \lambda_j)\omega_{j1}^j = 0,$$

which contradicts the first expression of (14).

Next, using Lemma 3.1 and the fact that $[e_i, e_1] = \nabla_{e_i} e_1 - \nabla_{e_1} e_i$, we obtain

$$e_i e_1 - e_1 e_i = -\epsilon_i \omega_{ii}^1 e_i, \quad i = 2, 3, 4. \tag{18}$$

Using (9), and acting (18) at λ_1, we find

$$e_i e_1(\lambda_1) = 0, \quad i = 2, 3, 4. \tag{19}$$

Finally, evaluating

$$g(R(e_1, e_i)e_1, e_i), \quad g(R(e_1, e_i)e_i, e_j), \quad g(R(e_i, e_j)e_i, e_1)$$

by using Lemma 3.1, (1) and (7) with respect to a suitable orthonormal frame $\{e_1, e_2, e_3, e_4\}$, we find that

$$e_1(\epsilon_i \omega_{ii}^1) - (\epsilon_i \omega_{ii}^1)^2 = \epsilon_1 \lambda_1 \lambda_i, \qquad i = 2, 3, 4, \tag{20}$$

$$e_1(\omega_{ii}^j) - \omega_{ii}^j \omega_{ii}^1 \epsilon_i = 0, \qquad i \neq j, \ i, j = 2, 3, 4, \tag{21}$$

$$e_j(\omega_{ii}^1) = \omega_{ii}^j (\epsilon_i \omega_{ii}^1 - \epsilon_j \omega_{jj}^1), \qquad i \neq j, \ i, j = 2, 3, 4. \tag{22}$$

4. Biconservative hypersurfaces in $\mathbb{E}^5_s$ with constant norm of second fundamental form

Let M be a biconservative hypersurface in $\mathbb{E}^5_s$ with constant norm of second fundamental form. We denote by β the squared norm of its second fundamental form. Then, from (7), we find that

$$\beta = \lambda_1^2 + \sum_{j=2}^{4} \lambda_j^2. \tag{23}$$

For simplicity, we put

$$b_2 := \epsilon_2 \omega_{22}^1, \quad b_3 := \epsilon_3 \omega_{33}^1, \quad b_4 := \epsilon_4 \omega_{44}^1.$$

The following lemma gives further simplification for ω_{ij}^k:

Lemma 4.1. *Let M_r^4 be a biconservative hypersurface of non constant mean curvature with all distinct principal curvatures in E^5_s, having the shape operator given by (7) with respect to a suitable orthonormal frame $\{e_1, e_2, e_3, e_4\}$ and let*

$$b_1 = (\lambda_4 - \lambda_2) b_3 + (\lambda_2 - \lambda_3) b_4 + (\lambda_3 - \lambda_4) b_2.$$

If the second fundamental form is of constant norm, then we have

$$\omega_{44}^2 = 0 = \omega_{33}^2, \quad \omega_{22}^3 = 0 = \omega_{44}^3, \quad \omega_{22}^4 = 0 = \omega_{33}^4.$$

Proof. Differentiating (8) and (23) with respect to e_2 and using (9), we get

$$e_2(\lambda_2) + e_2(\lambda_3) + e_2(\lambda_4) = 0, \tag{24}$$

$$\lambda_2 e_2(\lambda_2) + \lambda_3 e_2(\lambda_3) + \lambda_4 e_2(\lambda_4) = 0, \tag{25}$$

respectively. Eliminating $e_2(\lambda_2)$ from (24) and (25), we find

$$(\lambda_3 - \lambda_2)e_2(\lambda_3) + (\lambda_4 - \lambda_2)e_2(\lambda_4) = 0. \tag{26}$$

Putting the value of $e_2(\lambda_3)$ and $e_2(\lambda_4)$ from (12) in (26), we obtain

$$(\lambda_4 - \lambda_2)^2 \epsilon_4 \omega_{44}^2 + (\lambda_3 - \lambda_2)^2 \epsilon_3 \omega_{33}^2 = 0. \tag{27}$$

Differentiating (27) with respect to e_1 and using (12) and (21), we have

$$\begin{aligned} &\left[2(\lambda_2 - \lambda_1)b_2 + (2\lambda_1 + \lambda_2 - 3\lambda_4)\, b_4\right] (\lambda_4 - \lambda_2)\, \epsilon_4 \omega_{44}^2 \\ &\quad + \left[2(\lambda_2 - \lambda_1)b_2 + (2\lambda_1 + \lambda_2 - 3\lambda_3)\, b_3\right] (\lambda_3 - \lambda_2)\, \epsilon_3 \omega_{33}^2 = 0. \end{aligned} \tag{28}$$

Claim 1. $\omega_{33}^2 = 0 = \omega_{44}^2$.

Indeed, if $(\omega_{33}^2, \omega_{44}^2) \neq (0,0)$, then the value of the determinant formed by the coefficients of ω_{33}^2 and ω_{44}^2 in (27) and (28) will be zero. Therefore, we find

$$\begin{aligned} 2(\lambda_1 - \lambda_2)(\lambda_3 - \lambda_4)\, b_2 + (2\lambda_1 + \lambda_2 - 3\lambda_4)(\lambda_2 - \lambda_3)\, b_4 \\ - (2\lambda_1 + \lambda_2 - 3\lambda_3)(\lambda_2 - \lambda_4)\, b_3 = 0. \end{aligned} \tag{29}$$

Eliminating b_2 from (29) and b_1, we get

$$2(\lambda_1 - \lambda_2)\, b_1 = 3(\lambda_2 - \lambda_3)(\lambda_2 - \lambda_4)(b_3 - b_4). \tag{30}$$

Now, we consider two cases.

Case 1. $b_1 = 0$.

From (30), we obtain

$$b_4 = b_3. \tag{31}$$

Differentiating (31) with respect to e_1 and using (20) and (31), we find $\lambda_4 = \lambda_3$, a contradiction to all distinct principal curvatures.

Case 2. $b_1 \neq 0$.

Differentiating (30) with respect to e_2 and using (9) and (24), we get

$$\begin{aligned} &2(\lambda_1 - \lambda_2)e_2(b_1) + 2\big(e_2(\lambda_3) + e_2(\lambda_4)\big)b_1 \\ &\quad = 3\big[\big(-2e_2(\lambda_3) - e_2(\lambda_4)\big)(\lambda_2 - \lambda_4)(b_3 - b_4) \\ &\qquad + (\lambda_2 - \lambda_3)(\lambda_2 - \lambda_4)\big(e_2(b_3) - e_2(b_4)\big) \\ &\qquad + (\lambda_2 - \lambda_3)\big(-e_2(\lambda_3) - 2e_2(\lambda_4)\big)(b_3 - b_4)\big]. \end{aligned} \tag{32}$$

Now, using (12) and (22) in (32), we obtain that

$$f_1\epsilon_3\omega_{33}^2 + f_2\epsilon_4\omega_{44}^2 = 2\,(\lambda_1 - \lambda_2)\,e_2(b_1), \tag{33}$$

where

$$\begin{aligned} f_1 &= (\lambda_2 - \lambda_3)\big[3(\lambda_4 - \lambda_2)b_2 + 3(4\lambda_2 - \lambda_3 - 3\lambda_4)b_3 \\ &\qquad + 3(\lambda_3 - 3\lambda_2 + 2\lambda_4)b_4 + 2b_1\big], \\ f_2 &= (\lambda_2 - \lambda_4)\big[3(\lambda_2 - \lambda_3)b_2 + 3(3\lambda_2 - 2\lambda_3 - \lambda_4)b_3 \\ &\qquad + 3(3\lambda_3 + \lambda_4 - 4\lambda_2)b_4 + 2b_1\big]. \end{aligned}$$

Differentiating b_1 with respect to e_2 and using (12), (22) and (24), we find

$$f_3\epsilon_3\omega_{33}^2 + f_4\epsilon_4\omega_{44}^2 + (\lambda_3 - \lambda_4)\,e_2\,(b_2) = e_2(b_1), \tag{34}$$

where

$$\begin{aligned} f_3 &= (\lambda_3 - \lambda_4)\,b_2 + (-2\lambda_2 + \lambda_3 + \lambda_4)\,b_3 + 2\,(\lambda_2 - \lambda_3)\,b_4, \\ f_4 &= (\lambda_3 - \lambda_4)\,b_2 + 2\,(\lambda_4 - \lambda_2)\,b_3 + (2\lambda_2 - \lambda_3 - \lambda_4)\,b_4. \end{aligned}$$

Differentiating (8) with respect to e_1 and using (12), we find

$$(\lambda_2 - \lambda_1)\,b_2 + (\lambda_3 - \lambda_1)\,b_3 + (\lambda_4 - \lambda_1)\,b_4 = 6\epsilon e_1(H). \tag{35}$$

Differentiating (35) with respect to e_2 and using (9), (12), (22), (19), and (24), we find

$$(\lambda_2 - \lambda_1)\,e_2\,(b_2) = f_5\epsilon_3\omega_{33}^2 + f_6\epsilon_4\omega_{44}^2, \tag{36}$$

where

$$\begin{aligned} f_5 &= (\lambda_1 + \lambda_2 - 2\lambda_3)\,(b_3 - b_2), \\ f_6 &= (\lambda_1 + \lambda_2 - 2\lambda_4)\,(b_4 - b_2). \end{aligned}$$

Eliminating $e_2(b_2)$ from (34) and (36), we get

$$\begin{aligned} &\big[(\lambda_2 - \lambda_1)f_3 + (\lambda_3 - \lambda_4)f_5\big]\epsilon_3\omega_{33}^2 \\ &\quad + \big[(\lambda_2 - \lambda_1)f_4 + (\lambda_3 - \lambda_4)f_6\big]\epsilon_4\omega_{44}^2 = (\lambda_2 - \lambda_1)e_2(b_1). \end{aligned} \tag{37}$$

Eliminating $e_2(b_1)$ from (33) and (37), we find

$$f_7\epsilon_3\omega_{33}^2 + f_8\epsilon_4\omega_{44}^2 = 0, \tag{38}$$

where

$$\begin{aligned} f_7 &= f_1 + 2\,(\lambda_2 - \lambda_1)\,f_3 + 2\,(\lambda_3 - \lambda_4)\,f_5, \\ f_8 &= f_2 + 2\,(\lambda_2 - \lambda_1)\,f_4 + 2\,(\lambda_3 - \lambda_4)\,f_6. \end{aligned}$$

Now, we simplify f_7 and f_8. Eliminating b_2 from f_7 using b_1, we get

$$\begin{aligned}(\lambda_4-\lambda_3)\,f_7 \\ &= (b_3-b_4)(3\lambda_2^2-8\lambda_2\lambda_3-\lambda_3^2+2\lambda_2\lambda_4+10\lambda_3\lambda_4-6\lambda_4^2)(\lambda_2-\lambda_3) \\ &\quad + b_1(3\lambda_2^2+4\lambda_1\lambda_3-5\lambda_2\lambda_3-2\lambda_3^2-4\lambda_1\lambda_4-\lambda_2\lambda_4+5\lambda_3\lambda_4).\end{aligned} \tag{39}$$

Eliminating b_1 from (39) using (30), we obtain

$$2\,(\lambda_1-\lambda_2)\,(\lambda_4-\lambda_3)\,f_7=(\lambda_2-\lambda_3)^2\,(b_3-b_4)g_1, \tag{40}$$

where

$$g_1=3\lambda_2^2+4\lambda_2\,(\lambda_3-4\lambda_4)+2\lambda_1\,(3\lambda_2+\lambda_3-4\lambda_4)+3\lambda_4\,(-2\lambda_3+5\lambda_4).$$

Similarly, eliminating b_2 from f_8 using b_1, we get

$$\begin{aligned}(\lambda_3-\lambda_4)f_8 \\ &= (\lambda_2-\lambda_4)(b_3-b_4)(3\lambda_2^2-8\lambda_2\lambda_4-\lambda_4^2+2\lambda_2\lambda_3+10\lambda_3\lambda_4-6\lambda_3^2) \\ &\quad + b_1(3\lambda_2^2+4\lambda_1\lambda_4-5\lambda_2\lambda_4-2\lambda_4^2-4\lambda_1\lambda_3-\lambda_2\lambda_3+5\lambda_3\lambda_4).\end{aligned} \tag{41}$$

Eliminating b_1 from (41) using (30), we obtain

$$2\,(\lambda_1-\lambda_2)\,(\lambda_3-\lambda_4)\,f_8=(\lambda_2-\lambda_4)^2\,(b_3-b_4)\,g_2, \tag{42}$$

where

$$g_2=3\lambda_2^2+4\lambda_2\,(\lambda_4-4\lambda_3)+2\lambda_1\,(3\lambda_2+\lambda_4-4\lambda_3)+3\lambda_3\,(-2\lambda_4+5\lambda_3).$$

Eliminating f_7 and f_8 from (38) using (40) and (42), we obtain

$$(\lambda_2-\lambda_3)^2\,g_1\epsilon_3\omega_{33}^2-(\lambda_2-\lambda_4)^2\,g_2\epsilon_4\omega_{44}^2=0. \tag{43}$$

Eliminating $\epsilon_3\omega_{33}^2$ and $\epsilon_4\omega_{44}^2$ from (27) and (43), we have

$$2\lambda_2^2+5\lambda_3^2-4\lambda_3\lambda_4+5\lambda_4^2-4\lambda_2(\lambda_3+\lambda_4)+\lambda_1\big(4\lambda_2-2(\lambda_3+\lambda_4)\big)=0. \tag{44}$$

Now, eliminating λ_3 and λ_4 from (44) using (8) and (23), we find

$$7\beta=19\lambda_1^2-6\lambda_1\lambda_2+3\lambda_2^2. \tag{45}$$

Differentiating (45) with respect to e_2, we get $e_2(\lambda_2)=0$. Therefore, from (24) and (12), we obtain

$$(\lambda_3-\lambda_2)\,\epsilon_3\omega_{33}^2+(\lambda_4-\lambda_2)\,\epsilon_4\omega_{44}^2=0. \tag{46}$$

Eliminating $\epsilon_3\omega_{33}^2$ and $\epsilon_4\omega_{44}^2$ from (27) and (46), we get

$$(\lambda_3-\lambda_2)\,(\lambda_4-\lambda_2)\,(\lambda_3-\lambda_4)=0, \tag{47}$$

which gives a contradiction to distinct principal curvatures. Hence, we obtain that $\omega_{33}^2=0=\omega_{44}^2$, which proves the claim.

Claim 2. $\omega_{22}^3 = 0 = \omega_{44}^3$.

We proceed in a similar way as in the proof of Claim 1 by replacing the derivative with respect to e_2 with the derivative with respect to e_3. We then obtain the following three equations corresponding to (27), (28) and (29), respectively:

$$(\lambda_4 - \lambda_3)^2 \epsilon_4 \omega_{44}^3 + (\lambda_2 - \lambda_3)^2 \epsilon_2 \omega_{22}^3 = 0, \tag{48}$$

$$\begin{aligned}&\left[2(\lambda_3 - \lambda_1) b_3 + (2\lambda_1 + \lambda_3 - 3\lambda_4) b_4\right] (\lambda_4 - \lambda_3) \epsilon_4 \omega_{44}^3 \\ &\quad + \left[2(\lambda_3 - \lambda_1) b_3 + (2\lambda_1 + \lambda_3 - 3\lambda_2) b_2\right] (\lambda_2 - \lambda_3) \epsilon_2 \omega_{22}^3 = 0,\end{aligned}$$

$$\begin{aligned}2(\lambda_1 - \lambda_3)(\lambda_2 - \lambda_4) b_3 + (2\lambda_1 + \lambda_3 - 3\lambda_4)(\lambda_3 - \lambda_2) b_4 \\ - (2\lambda_1 + \lambda_3 - 3\lambda_2)(\lambda_3 - \lambda_4) b_2 = 0.\end{aligned} \tag{49}$$

Eliminating b_3 from (49) and b_1, we get

$$2(\lambda_1 - \lambda_3) b_1 = 3(\lambda_3 - \lambda_2)(b_2 - b_4)(\lambda_3 - \lambda_4).$$

Now, we consider two cases.

Case 1. $b_1 = 0$.

From (48), we obtain

$$b_4 = b_2. \tag{50}$$

Differentiating (50) with respect to e_1 and using (20) and (50), we find $\lambda_4 = \lambda_2$, a contradiction to all distinct principal curvatures.

Case 2. $b_1 \neq 0$.

In this case, proceeding in a similar way as in the proof of Claim 1 by replacing the derivative with respect to e_2 with the derivative with respect to e_3, and replacing elimination of b_2 with that of b_3, we obtain the following four equations corresponding to (43), (44), (45), and (46), respectively:

$$\begin{aligned}&(\lambda_3 - \lambda_2)^2 g_3 \epsilon_2 \omega_{22}^3 - (\lambda_3 - \lambda_4)^2 g_4 \epsilon_4 \omega_{44}^3 = 0, \\ &2\lambda_3^2 + 5\lambda_2^2 - 4\lambda_2\lambda_4 + 5\lambda_4^2 - 4\lambda_3(\lambda_2 + \lambda_4) + \lambda_1\left(4\lambda_3 - 2(\lambda_2 + \lambda_4)\right) = 0, \\ &7\beta = 19\lambda_1^2 - 6\lambda_1\lambda_3 + 3\lambda_3^2, \\ &(\lambda_2 - \lambda_3) \epsilon_2 \omega_{22}^3 + (\lambda_4 - \lambda_3) \epsilon_4 \omega_{44}^3 = 0,\end{aligned} \tag{51}$$

where

$$\begin{aligned}g_3 &= 3\lambda_3^2 + 4\lambda_3(\lambda_2 - 4\lambda_4) + 2\lambda_1(3\lambda_3 + \lambda_2 - 4\lambda_4) + 3\lambda_4(-2\lambda_2 + 5\lambda_4), \\ g_4 &= 3\lambda_3^2 + 4\lambda_3(\lambda_4 - 4\lambda_2) + 2\lambda_1(3\lambda_3 + \lambda_4 - 4\lambda_2) + 3\lambda_2(-2\lambda_4 + 5\lambda_2).\end{aligned}$$

Now, eliminating $\epsilon_2\omega_{22}^3$ and $\epsilon_4\omega_{44}^3$ from (48) and (51), we get (47), which gives a contradiction to distinct principal curvatures. Hence, we obtain that $\omega_{22}^3 = 0 = \omega_{44}^3$.

Claim 3. $\omega_{22}^4 = 0 = \omega_{33}^4$.

We proceed in a similar way as above in the proof of Claim 1 by replacing the derivative with respect to e_2 with the derivative with respect to e_4. We then obtain the following three equations corresponding to (27), (28) and (29), respectively:

$$(\lambda_2 - \lambda_4)^2 \epsilon_2\omega_{22}^4 + (\lambda_3 - \lambda_4)^2 \epsilon_3\omega_{33}^4 = 0, \tag{52}$$

$$\begin{aligned}&\left[2(\lambda_4-\lambda_1)b_4 + (2\lambda_1+\lambda_4 - 3\lambda_2)\, b_2\right] (\lambda_2-\lambda_4)\, \epsilon_2\omega_{22}^4\\ &\quad + \left[2(\lambda_4-\lambda_1)b_4 + (2\lambda_1+\lambda_4-3\lambda_3)\, b_3\right] (\lambda_3-\lambda_4)\, \epsilon_3\omega_{33}^4 = 0,\end{aligned}$$

$$\begin{aligned}2\,(\lambda_1 - \lambda_4)\,(\lambda_3 - \lambda_2)\, b_4 + (2\lambda_1 + \lambda_4 - 3\lambda_2)\,(\lambda_4 - \lambda_3)\, b_2&\\ - (2\lambda_1 + \lambda_4 - 3\lambda_3)\,(\lambda_4 - \lambda_2)\, b_3 = 0.&\end{aligned} \tag{53}$$

Eliminating b_4 from (53) and b_1, we get

$$2\,(\lambda_1 - \lambda_4)\, b_1 = 3\,(\lambda_4 - \lambda_3)\,(\lambda_4 - \lambda_2)\,(b_3 - b_2). \tag{54}$$

Now, we consider two cases.

Case 1. $b_1 = 0$.

From (54), we obtain

$$b_2 = b_3. \tag{55}$$

Differentiating (55) with respect to e_1 and using (20) and (55), we find $\lambda_2 = \lambda_3$, a contradiction to all distinct principal curvatures.

Case 2. $b_1 \neq 0$.

In this case, proceeding in a similar way as in the proof of Claim 1 by replacing the derivative with respect to e_2 with the derivative with respect to e_4 and replacing elimination of b_2 with that of b_4, we obtain the following four equations corresponding to (43), (44), (45), and (46), respectively:

$$\begin{aligned}&(\lambda_4 - \lambda_3)^2 g_5\epsilon_3\omega_{33}^4 - (\lambda_4 - \lambda_2)^2 g_6\epsilon_2\omega_{22}^4 = 0,\\ &2\lambda_4^2 + 5\lambda_3^2 - 4\lambda_3\lambda_2 + 5\lambda_2^2 - 4\lambda_4\,(\lambda_3+\lambda_2) + \lambda_1\big(4\lambda_4-2(\lambda_3+\lambda_2)\big) = 0,\\ &7\beta = 19\lambda_1^2 - 6\lambda_1\lambda_4 + 3\lambda_4^2,\\ &(\lambda_3 - \lambda_4)\,\epsilon_3\omega_{33}^4 + (\lambda_2 - \lambda_4)\,\epsilon_2\omega_{22}^4 = 0,\end{aligned} \tag{56}$$

where

$$g_5 = 3\lambda_4^2 + 4\lambda_4(\lambda_3 - 4\lambda_2) + 2\lambda_1(3\lambda_4 + \lambda_3 - 4\lambda_2) + 3\lambda_2(-2\lambda_3 + 5\lambda_2),$$

$$g_6 = 3\lambda_4^2 + 4\lambda_4(\lambda_2 - 4\lambda_3) + 2\lambda_1(3\lambda_4 + \lambda_2 - 4\lambda_3) + 3\lambda_3(-2\lambda_2 + 5\lambda_3).$$

Eliminating $\epsilon_3\omega_{33}^4$ and $\epsilon_2\omega_{22}^4$ from (52) and (56), we get (47), which gives a contradiction to distinct principal curvatures. Hence, we obtain that $\omega_{33}^4 = 0 = \omega_{22}^4$, whereby proof of Lemma 4.1 is complete. □

Next, we have:

Lemma 4.2. *Let M_r^4 be a biconservative hypersurface of non constant mean curvature with all distinct principal curvatures in E_s^5, having the shape operator given by (7) with respect to a suitable orthonormal frame $\{e_1, e_2, e_3, e_4\}$. Set*

$$b_1 = (\lambda_3 - \lambda_4)b_2 + (\lambda_4 - \lambda_2)b_3 + (\lambda_2 - \lambda_3)b_4.$$

If the second fundamental form is of constant norm, then we have the following:

(***a***) *If $b_1 \neq 0$, it is*

$$\omega_{23}^4 = \omega_{32}^4 = \omega_{42}^3 = \omega_{24}^3 = \omega_{34}^2 = \omega_{43}^2 = 0; \tag{57}$$

(***b***) *If $b_1 = 0$, we have that*

$$\begin{aligned} &b_i = \alpha\lambda_i + \phi \qquad (i = 2, 3, 4),\\ &e_1(\alpha) = \alpha\phi + \lambda_1(\epsilon_1 + \alpha^2), \quad e_1(\phi) = \phi^2 + \alpha\lambda_1\phi, \end{aligned} \tag{58}$$

for some smooth functions α and ϕ.

Proof. (a) Evaluating $g(R(e_1, e_3)e_2, e_4)$ and $g(R(e_1, e_2)e_3, e_4)$ by using (1), (7), Lemma 3.1 and Lemma 4.1, we find

$$e_1(\omega_{32}^4) - \omega_{32}^4 b_3 = 0, \quad e_1(\omega_{23}^4) - \omega_{23}^4 b_2 = 0, \tag{59}$$

respectively. Putting $j = 4, k = 2, i = 3$ in (13), we get

$$(\lambda_2 - \lambda_4)\omega_{32}^4 = (\lambda_3 - \lambda_4)\omega_{23}^4. \tag{60}$$

Differentiating (60) with respect to e_1, and using (59), we find

$$\begin{aligned} &\big(e_1(\lambda_2) - e_1(\lambda_4)\big)\omega_{32}^4 + (\lambda_2 - \lambda_4)\omega_{32}^4\, b_3\\ &\quad = \big(e_1(\lambda_3) - e_1(\lambda_4)\big)\omega_{23}^4 + (\lambda_3 - \lambda_4)\omega_{23}^4\, b_2. \end{aligned} \tag{61}$$

Using (12) in (61), we obtain

$$\omega_{32}^4(b_2 - b_4) = \omega_{23}^4(b_3 - b_4). \tag{62}$$

Eliminating ω_{23}^4 using (60) and (62), we find

$$\omega_{32}^4\, b_1 = 0. \tag{63}$$

From (63), we obtain $\omega_{32}^4 = 0$ as $b_1 \neq 0$. Using $\omega_{32}^4 = 0$ in (60), gives $\omega_{23}^4 = 0$. From (11), we get $\omega_{34}^2 = -\omega_{32}^4$ and $\omega_{23}^4 = -\omega_{24}^3$. Therefore, we obtain $\omega_{34}^2 = 0$ and $\omega_{24}^3 = 0$, which by use of (13) gives $\omega_{43}^2 = 0$ and $\omega_{42}^3 = 0$.
(b) Since $b_1 = 0$, it follows that

$$\frac{b_2 - b_3}{\lambda_2 - \lambda_3} = \frac{b_4 - b_3}{\lambda_4 - \lambda_3} = \frac{b_2 - b_4}{\lambda_2 - \lambda_4} = \alpha, \tag{64}$$

for some smooth function α. From (64), we get

$$b_i = \alpha\lambda_i + \phi, \quad i = 2, 3, 4, \tag{65}$$

for some smooth function ϕ. Differentiating (65) with respect to e_1 and using (12), (20) and (65), we find

$$e_1(\alpha) = \alpha\phi + \lambda_1(\epsilon_1 + \alpha^2), \quad e_1(\phi) = \phi^2 + \alpha\lambda_1\phi,$$

whereby completing the proof of the lemma. □

5. Proof of Theorem 1.1

Suppose that M_r^4 is a biconservative hypersurface with non constant mean curvature. Depending upon principal curvatures, we consider the following cases.

Case I. Four distinct principal curvatures.

Evaluating $g(R(e_2, e_3)e_2, e_3)$, $g(R(e_2, e_4)e_2, e_4)$, and $g(R(e_3, e_4)e_3, e_4)$ by using (1), (7), Lemma 3.1 and Lemma 4.1, we find that

$$-\epsilon_1 b_2 b_3 + (\omega_{32}^4\omega_{23}^4 - \omega_{34}^2\omega_{43}^2 - \omega_{42}^3\omega_{24}^3)\epsilon_2\epsilon_3\epsilon_4 = \lambda_2\lambda_3, \tag{66}$$

$$-\epsilon_1 b_2 b_4 + (\omega_{42}^3\omega_{24}^3 - \omega_{34}^2\omega_{43}^2 - \omega_{32}^4\omega_{23}^4)\epsilon_2\epsilon_3\epsilon_4 = \lambda_2\lambda_4, \tag{67}$$

$$-\epsilon_1 b_3 b_4 + (\omega_{34}^2\omega_{43}^2 - \omega_{24}^3\omega_{42}^3 - \omega_{23}^4\omega_{32}^4)\epsilon_2\epsilon_3\epsilon_4 = \lambda_3\lambda_4, \tag{68}$$

respectively.

Depending upon b_1, we consider the following cases.

Case A. $b_1 \neq 0$.

Using (57) of Lemma 4.2 in (66), (67) and (68), we obtain

$$\epsilon_1 b_2 b_3 = -\lambda_2\lambda_3, \quad \epsilon_1 b_2 b_4 = -\lambda_2\lambda_4, \quad \epsilon_1 b_3 b_4 = -\lambda_3\lambda_4. \tag{69}$$

Subcase A1. At least one of the principal curvatures λ_i, $i = 2, 3, 4$ is zero, say $\lambda_2 = 0$. Then from (8), (23) and (69), we find that

$$3\lambda_1 + \lambda_3 + \lambda_4 = 0, \tag{70}$$

$$\beta - \lambda_1^2 - \lambda_3^2 - \lambda_4^2 = 0, \tag{71}$$

$$\epsilon_1 b_3 b_4 = -\lambda_3\lambda_4. \tag{72}$$

Eliminating λ_4 from (70) and (71), we obtain

$$10\lambda_1^2 + 6\lambda_3\lambda_1 + 2\lambda_3^2 = \beta. \tag{73}$$

Differentiating (70) and (71) with respect to e_1 and using (12), we find

$$(\lambda_3 - \lambda_1)b_3 + (\lambda_4 - \lambda_1)b_4 = -3e_1(\lambda_1), \tag{74}$$

$$\lambda_3(\lambda_3 - \lambda_1)b_3 + \lambda_4(\lambda_4 - \lambda_1)b_4 = -\lambda_1 e_1(\lambda_1), \tag{75}$$

respectively. Eliminating $e_1(\lambda_1)$ from (74) using (75), we obtain

$$b_4\left(-\lambda_1^2 + 4\lambda_4\lambda_1 - 3\lambda_4^2\right) = b_3\left(\lambda_1^2 - 4\lambda_3\lambda_1 + 3\lambda_3^2\right). \tag{76}$$

Differentiating (76) with respect to e_1 and using (12) and (20), we find

$$\begin{aligned} 0 &= e_1(\lambda_1)\big(2b_3(\lambda_1 - 2\lambda_3) + 2b_4(\lambda_1 - 2\lambda_4)\big) \\ &\quad + b_3^2\left(5\lambda_1^2 - 14\lambda_3\lambda_1 + 9\lambda_3^2\right) + b_4^2\left(5\lambda_1^2 - 14\lambda_4\lambda_1 + 9\lambda_4^2\right) \\ &\quad + \lambda_1\big((\lambda_3+\lambda_4)\lambda_1^2 - 4\left(\lambda_3^2+\lambda_4^2\right)\lambda_1 + 3\left(\lambda_3^3+\lambda_4^3\right)\big)\epsilon_1. \end{aligned} \tag{77}$$

Eliminating $e_1(\lambda_1)$ from (77) using (74), we obtain

$$\begin{aligned} &3\lambda_1\big((\lambda_3+\lambda_4)\lambda_1^2 - 4\left(\lambda_3^2+\lambda_4^2\right)\lambda_1 + 3\left(\lambda_3^3+\lambda_4^3\right)\big)\epsilon_1 \\ &\quad = b_3^2(-17\lambda_1^2 + 48\lambda_3\lambda_1 - 31\lambda_3^2) \\ &\qquad + 2b_4b_3(-2\lambda_1^2 + 3(\lambda_3+\lambda_4)\lambda_1 - 4\lambda_3\lambda_4) \\ &\qquad + b_4^2\left(-17\lambda_1^2 + 48\lambda_4\lambda_1 - 31\lambda_4^2\right). \end{aligned} \tag{78}$$

Eliminating b_4 from (78) using (72), we obtain

$$\begin{aligned} &b_3^2\Big(3(\lambda_3+\lambda_4)\lambda_1^3 - 4\left(3\lambda_3^2 + \lambda_4\lambda_3 + 3\lambda_4^2\right)\lambda_1^2 \\ &\qquad + (9\lambda_3^3 + 6\lambda_4\lambda_3^2 + 6\lambda_4^2\lambda_3 + 9\lambda_4^3)\lambda_1 - 8\lambda_3^2\lambda_4^2\Big)\epsilon_1 \\ &\qquad + \lambda_3^2\lambda_4^2\left(17\lambda_1^2 - 48\lambda_4\lambda_1 + 31\lambda_4^2\right) \\ &= b_3^4\left(-17\lambda_1^2 + 48\lambda_3\lambda_1 - 31\lambda_3^2\right). \end{aligned} \tag{79}$$

Eliminating b_4 from (76) using (72), we obtain

$$\lambda_3\lambda_4\left(\lambda_1^2 - 4\lambda_4\lambda_1 + 3\lambda_4^2\right)\epsilon_1 = b_3^2\left(\lambda_1^2 - 4\lambda_3\lambda_1 + 3\lambda_3^2\right). \tag{80}$$

Eliminating b_3 from (79) using (80), we obtain

$$\begin{aligned}\lambda_3(\lambda_1-\lambda_4)\lambda_4\Big(&(\lambda_3+\lambda_4)\lambda_1^5+(-7\lambda_3^2+4\lambda_4\lambda_3-7\lambda_4^2)\lambda_1^4\\&+(15\lambda_3^3-23\lambda_4\lambda_3-23\lambda_4^2\lambda_3+15\lambda_4^3)\lambda_1^3\\&+(-9\lambda_3^4+34\lambda_4\lambda_3^3+118\lambda_4^2\lambda_3^2+34\lambda_4^3\lambda_3-9\lambda_4^4)\lambda_1^2\\&-3\lambda_3\lambda_4(8\lambda_3^3+43\lambda_4\lambda_3^2+43\lambda_4^2\lambda_3+8\lambda_4^3)\lambda_1\\&+3\lambda_3^2\lambda_4^2(31\lambda_3^2-8\lambda_4\lambda_3+31\lambda_4^2)\Big)=0.\end{aligned} \tag{81}$$

If we suppose that $\lambda_1=\lambda_4$ holds, by using (70) and (71), we find that λ_1 is a real constant, which is a contradiction. Hence we have $\lambda_1\neq\lambda_4$. We drop the factor $\lambda_3(\lambda_1-\lambda_4)\lambda_4$ from (81), and eliminate λ_4 from (81) by using (70). We then find

$$\begin{aligned}21\lambda_3^6+189\lambda_1\lambda_3^5+671\lambda_1^2\lambda_3^4+1191\lambda_1^3\lambda_3^3&\\+786\lambda_1^4\lambda_3^2-450\lambda_1^5\lambda_3&=120\lambda_1^6.\end{aligned} \tag{82}$$

Eliminating λ_3 from (82) using (73), we find

$$21\beta^3-422\beta^2\lambda_1^2+1540\beta\lambda_1^4+4840\lambda_1^6=0,$$

which is a polynomial equation in λ_1 with real coefficients. Therefore, λ_1 is a real constant, which is a contradiction.

Subcase A2. $\lambda_i\neq 0$, $i=2,3,4$. Then from (69), we get

$$b_2^2+\epsilon_1\lambda_2^2=0. \tag{83}$$

If $\epsilon_1=1$, then from (83), we get $b_2=0=\lambda_2$, a contradiction to our assumption $\lambda_i\neq 0$, $i=2,3,4$. Hence, $\epsilon_1\neq 1$.

Differentiating (8) and (23) with respect to e_1 and using (12), we find

$$(\lambda_2-\lambda_1)b_2+(\lambda_3-\lambda_1)b_3+(\lambda_4-\lambda_1)b_4=-3e_1(\lambda_1), \tag{84}$$

$$\lambda_2(\lambda_2-\lambda_1)b_2+\lambda_3(\lambda_3-\lambda_1)b_3+\lambda_4(\lambda_4-\lambda_1)b_4=-\lambda_1e_1(\lambda_1), \tag{85}$$

respectively. Eliminating $e_1(\lambda_1)$ from (85) using (84), we obtain

$$\begin{aligned}(3\lambda_2-\lambda_1)(\lambda_2-\lambda_1)b_2+(3\lambda_3-\lambda_1)(\lambda_3-\lambda_1)b_3&\\+(3\lambda_4-\lambda_1)(\lambda_4-\lambda_1)b_4&=0.\end{aligned} \tag{86}$$

Multiplying (86) with b_2 and using (69), we find

$$\begin{aligned}&(3\lambda_2-\lambda_1)(\lambda_2-\lambda_1)b_2^2\\&\quad=(3\lambda_3-\lambda_1)(\lambda_3-\lambda_1)\epsilon_1\lambda_2\lambda_3+(3\lambda_4-\lambda_1)(\lambda_4-\lambda_1)\epsilon_1\lambda_2\lambda_4.\end{aligned} \tag{87}$$

Eliminating b_2^2 from (87) using (83), we obtain

$$(\lambda_2+\lambda_3+\lambda_4)\lambda_1^2+3(\lambda_2^3+\lambda_3^3+\lambda_4^3)=4\lambda_1(\lambda_2^2+\lambda_3^2+\lambda_4^2). \tag{88}$$

Eliminating λ_4 from (88) using (8), we obtain

$$(35\lambda_1 + 9\lambda_3)\,\lambda_2^2 + \left(105\lambda_1^2 + 62\lambda_3\lambda_1 + 9\lambda_3^2\right)\lambda_2 = \lambda_1\left(-120\lambda_1^2 - 105\lambda_3\lambda_1 - 35\lambda_3^2\right). \tag{89}$$

Eliminating λ_4 from (23) using (8), we obtain

$$10\lambda_1^2 + 6\lambda_2\lambda_1 + 6\lambda_3\lambda_1 + 2\lambda_2^2 + 2\lambda_3^2 + 2\lambda_2\lambda_3 = \beta. \tag{90}$$

Eliminating λ_3 from (89) using (90), we find

$$110\lambda_1^3 + 90\lambda_2\lambda_1^2 + 54\lambda_2^2\lambda_1 + 18\lambda_2^3 - \beta\,(35\lambda_1 + 9\lambda_2) = 0, \tag{91}$$

which is a polynomial equation

$$G(\lambda_1, \lambda_2) = 0, \tag{92}$$

in terms of λ_1, λ_2. Differentiating (92) with respect to e_1, we get

$$G_1 e_1(\lambda_1) + G_2 e_1(\lambda_2) = 0, \tag{93}$$

where $G_1 = \frac{\partial G}{\partial \lambda_1}$, $G_2 = \frac{\partial G}{\partial \lambda_2}$.

Eliminating $e_1(\lambda_1)$ from (93) using (84) and using (12), we find

$$(3G_2 - G_1)(\lambda_2 - \lambda_1)b_2 = \big((\lambda_3 - \lambda_1)b_3 + (\lambda_4 - \lambda_1)b_4\big)G_1. \tag{94}$$

Multiplying (94) with b_2 and using (69) and (83), we obtain

$$L\lambda_2 = (\lambda_3 - \lambda_1)\lambda_3 + (\lambda_4 - \lambda_1)\lambda_4,$$

where $L = (3G_2 - G_1)(\lambda_2 - \lambda_1)/G_1$. It can be written as

$$4\lambda_2\,(\lambda_2 - \lambda_1)\left(2\beta - 15\lambda_1^2 + 36\lambda_2\lambda_1 + 27\lambda_2^2\right) + \left(\lambda_3^2 + \lambda_4^2 - \lambda_1(\lambda_3 + \lambda_4)\right)\left(-35\beta + 330\lambda_1^2 + 180\lambda_2\lambda_1 + 54\lambda_2^2\right) = 0. \tag{95}$$

Eliminating λ_4 from (95) using (8), we obtain

$$6\Big(660\lambda_1^4 + 5\,(151\lambda_2 + 66\lambda_3)\,\lambda_1^3 + \left(339\lambda_2^2 + 290\lambda_3\lambda_2 + 110\lambda_3^2\right)\lambda_1^2 + 3\lambda_2\left(33\lambda_2^2 + 38\lambda_3\lambda_2 + 20\lambda_3^2\right)\lambda_1 + 9\lambda_2^2\left(3\lambda_2^2 + 2\lambda_3\lambda_2 + 2\lambda_3^2\right)\Big) = \beta\Big(420\lambda_1^2 + (253\lambda_2 + 210\lambda_3)\,\lambda_1 + 27\lambda_2^2 + 70\lambda_3^2 + 70\lambda_2\lambda_3\Big). \tag{96}$$

Eliminating λ_3 from (96) by using (90), we get

$$35\beta^2 + \lambda_1^2\left(246\lambda_2^2 - 260\beta\right) + \lambda_1\left(90\lambda_2^3 - 137\beta\lambda_2\right) = 97\beta\lambda_2^2 + 660\lambda_1^4 + 750\lambda_2\lambda_1^3 + 54\lambda_2^4. \tag{97}$$

Eliminating λ_2 from (97) by using (91), we have

$$\begin{aligned} &413343\beta^6 + 187348300\beta^4\lambda_1^4 + 15830560800\beta^2\lambda_1^8 + 22372416000\lambda_1^{12} \\ &\quad = 61008638\beta^5\lambda_1^2 + 3900182024\beta^3\lambda_1^6 + 27773539200\beta\lambda_1^{10}, \end{aligned}$$

which is a polynomial equation in λ_1 with real coefficients. Hence λ_1 is a real constant, which is a contradiction.

Case B. $b_1 = 0$. From (11) and (13), we obtain

$$(\lambda_2 - \lambda_3)\omega_{42}^3 = (\lambda_4 - \lambda_3)\omega_{24}^3 = (\lambda_2 - \lambda_4)\omega_{32}^4. \tag{98}$$

From (11) and (98), we find

$$\omega_{34}^2\omega_{43}^2 + \omega_{42}^3\omega_{24}^3 + \omega_{32}^4\omega_{23}^4 = 0. \tag{99}$$

Adding (66), (67), (68) and using (99), we obtain

$$b_2b_3 + b_2b_4 + b_3b_4 = -\epsilon_1(\lambda_2\lambda_3 + \lambda_2\lambda_4 + \lambda_3\lambda_4). \tag{100}$$

Using (8), (23), and (58) of Lemma 4.2 in (100), we find

$$6\phi^2 - 12\phi\alpha\lambda_1 + (\alpha^2 + \epsilon_1)(10\lambda_1^2 - \beta) = 0. \tag{101}$$

On the other hand, using (8), (23), and (58) in (35), we obtain

$$6\lambda_1\phi - \alpha(\beta + 2\lambda_1^2) = 3\epsilon e_1(\lambda_1). \tag{102}$$

Differentiating (101) with respect to e_1 and using (58) and (102), we get

$$\begin{aligned} &\epsilon_1\phi^3 - 18\alpha\epsilon_1\lambda_1\phi^2 - 36\epsilon\epsilon_1\alpha\phi^2\lambda_1 + 60\lambda_1^2\epsilon\phi - 6\epsilon_1\phi\alpha^2\lambda_1^2 - 18\lambda_1^2\phi \\ &\quad + 72\epsilon\epsilon_1\phi\alpha^2\lambda_1^2 - 20\alpha\lambda_1^3\epsilon + 30\epsilon_1\alpha^3\lambda_1^3 - 20\epsilon\epsilon_1\alpha^3\lambda_1^3 + 30\alpha\lambda_1^3 \\ &= \beta(-6\alpha^2\epsilon\epsilon_1\phi + 3\alpha^2\epsilon_1\phi + 10\epsilon\alpha\lambda_1 + 3\epsilon_1\lambda_1\alpha^3 + 10\epsilon\epsilon_1\lambda_1\alpha^3 + 3\alpha\lambda_1). \end{aligned} \tag{103}$$

Differentiating (103) with respect to e_1 and using (58) and (102), we get

$$f(\alpha, \phi, \lambda_1) = 0. \tag{104}$$

Eliminating α from (103) and (101) and from (101) and (104), we obtain polynomial equations which we denote by

$$h_1(\phi, \lambda_1) = 0, \quad \text{and} \quad h_2(\phi, \lambda_1) = 0,$$

respectively. Finally, eliminating ϕ from $h_1(\phi, \lambda_1) = 0$ and $h_2(\phi, \lambda_1) = 0$, we get a polynomial equation $F(\lambda_1) = 0$ in λ_1 with constant coefficients. Thus, the real function λ_1 satisfies a polynomial equation $F(\lambda_1) = 0$ with constant coefficients, and therefore, it must be a constant. This is a contradiction.

Case II. Three distinct principal curvatures.

Suppose that M^4_r is a biconservative hypersurface with three distinct principal curvatures $\lambda_1 = -2\varepsilon H$, $\lambda_2 = \lambda = \lambda_3$, λ_4, having constant norm of second fundamental forms. Without loss of generality, we choose e_1 in the direction of grad H and therefore the shape operator $\mathcal{A}$ of the hypersurface will take the following form with respect to a suitable frame $\{e_1, e_2, e_3, e_4\}$:

$$\mathcal{A}_H e_1 = -2\epsilon H e_1, \quad \mathcal{A}_H e_2 = \lambda e_2, \quad \mathcal{A}_H e_3 = \lambda e_3, \quad \mathcal{A}_H e_4 = \lambda_4 e_4. \tag{105}$$

Using (105) in (8) and (23), we get

$$2\lambda + \lambda_4 = -3\lambda_1, \tag{106}$$

$$2\lambda^2 + \lambda_4^2 = \beta - \lambda_1^2. \tag{107}$$

Eliminating λ_4 from (107) using (106), we obtain

$$\beta = 10\lambda_1^2 + 12\lambda_1\lambda + 6\lambda^2. \tag{108}$$

Similarly, eliminating λ from (107) by using (106), we get

$$2\beta = 11\lambda_1^2 + 6\lambda_1\lambda_4 + 3\lambda_4^2. \tag{109}$$

Differentiating (108) with respect to e_4 and using (12), we find

$$e_4(\lambda) = 0, \quad \omega_{33}^4 = \omega_{22}^4 = 0. \tag{110}$$

Differentiating (109) with respect to e_2 and e_3 and using (12), we have

$$e_2(\lambda_4) = e_3(\lambda_4) = 0, \quad \omega_{44}^2 = \omega_{44}^3 = 0. \tag{111}$$

Also, putting $k = 4$, $j = 2$, $i = 3$ in (13) and using (11) and (13), we get

$$\omega_{34}^2 = \omega_{32}^4 = \omega_{23}^4 = \omega_{24}^3 = 0. \tag{112}$$

Evaluating $g(R(e_3, e_4)e_3, e_4)$ by using (1), (105), (110)~(112), we obtain

$$b_3 b_4 = -\epsilon_1 \lambda\lambda_4. \tag{113}$$

Differentiating (106) and (107) with respect to e_1 and using (13), we find

$$3e_1(\lambda_1) + 2(\lambda - \lambda_1)b_3 + (\lambda_4 - \lambda_1)b_4 = 0, \tag{114}$$

$$\lambda_1 e_1(\lambda_1) + 2\lambda(\lambda - \lambda_1)b_3 + \lambda_4(\lambda_4 - \lambda_1)b_4 = 0. \tag{115}$$

Eliminating $e_1(\lambda_1)$ from (114) and (115), we find

$$b_4\left(-\lambda_1^2 + 4\lambda_4\lambda_1 - 3\lambda_4^2\right) = b_3\left(6\lambda^2 - 8\lambda_1\lambda + 2\lambda_1^2\right). \tag{116}$$

Eliminating b_4 from (113) and (116), we get

$$2b_3^2\left(3\lambda^2 - 4\lambda_1\lambda + \lambda_1^2\right) = \lambda\lambda_4\left(\lambda_1^2 - 4\lambda_4\lambda_1 + 3\lambda_4^2\right)\epsilon_1. \tag{117}$$

Differentiating (117) with respect to e_1 and using (13) and (20), we find

$$\begin{aligned} & e_1(\lambda_1)\big(b_3^2\,(4\lambda_1-8\lambda)+2\lambda\lambda_4(2\lambda_4-\lambda_1)\epsilon_1\big) \\ & \quad +4b_3(3\lambda^2-4\lambda_1\lambda+\lambda_1^2)(b_3^2+\lambda\lambda_1\epsilon_1) \\ & \quad +b_4\lambda\,(\lambda_1-\lambda_4)\left(\lambda_1^2-8\lambda_4\lambda_1+9\lambda_4^2\right)\epsilon_1 \\ & \quad +b_3\,(\lambda-\lambda_1)\left(4b_3^2\,(3\lambda-2\lambda_1)-\lambda_4\left(\lambda_1^2-4\lambda_4\lambda_1+3\lambda_4^2\right)\epsilon_1\right)=0. \end{aligned} \tag{118}$$

Eliminating $e_1(\lambda_1)$ from (114) and (118), we get

$$\begin{aligned} & b_3\big(88b_3^2\lambda^2-132b_3^2\lambda_1\lambda+44b_3^2\lambda_1^2+36\lambda_1\lambda^3\epsilon_1-48\lambda_1^2\lambda^2\epsilon_1 \\ & \quad -8\lambda_4^2\lambda^2\epsilon_1+4\lambda_1\lambda_4\lambda^2\epsilon_1+12\lambda_1^3\lambda\epsilon_1-9\lambda_4^3\lambda\epsilon_1+20\lambda_1\lambda_4^2\lambda\epsilon_1 \\ & \quad -7\lambda_1^2\lambda_4\lambda\epsilon_1+9\lambda_1\lambda_4^3\epsilon_1-12\lambda_1^2\lambda_4^2\epsilon_1+3\lambda_1^3\lambda_4\epsilon_1\big) \\ & =b_4\big(-4b_3^2\lambda_1^2+8b_3^2\lambda\lambda_1+4b_3^2\lambda_4\lambda_1-8b_3^2\lambda\lambda_4-3\lambda\lambda_1^3\epsilon_1 \\ & \quad +29\lambda\lambda_4\lambda_1^2\epsilon_1-57\lambda\lambda_4^2\lambda_1\epsilon_1+31\lambda\lambda_4^3\epsilon_1\big). \end{aligned} \tag{119}$$

Eliminating b_4 from (113) and (119), we obtain

$$\begin{aligned} & 44b_3^4\left(2\lambda^2-3\lambda_1\lambda+\lambda_1^2\right) \\ & \quad +b_3^2\big(-\left(48\lambda^2+11\lambda_4\lambda+12\lambda_4^2\right)\lambda_1^2 \\ & \qquad +3(12\lambda^3+4\lambda_4\lambda^2+8\lambda_4^2\lambda+3\lambda_4^3)\lambda_1 \\ & \qquad +3\,(4\lambda+\lambda_4)\,\lambda_1^3-\lambda\lambda_4^2\,(16\lambda+9\lambda_4)\big)\epsilon_1 \\ & \quad +\lambda^2\lambda_4\left(-3\lambda_1^3+29\lambda_4\lambda_1^2-57\lambda_4^2\lambda_1+31\lambda_4^3\right)=0. \end{aligned} \tag{120}$$

Eliminating b_3 from (117) and (120), we obtain

$$\begin{aligned} \lambda(\lambda_1-\lambda_4)\lambda_4\Big(& -(2\lambda+\lambda_4)\,\lambda_1^5+\left(14\lambda^2-6\lambda_4\lambda+7\lambda_4^2\right)\lambda_1^4 \\ & +\left(-30\lambda^3+37\lambda_4\lambda^2+32\lambda_4^2\lambda-15\lambda_4^3\right)\lambda_1^3 \\ & +(18\lambda^4-68\lambda_4\lambda^3-173\lambda_4^2\lambda^2-38\lambda_4^3\lambda+9\lambda_4^4)\lambda_1^2 \\ & +3\lambda\lambda_4(16\lambda^3+86\lambda_4\lambda^2+41\lambda_4^2\lambda+10\lambda_4^3)\lambda_1 \\ & -3\lambda^2\lambda_4^2(62\lambda^2-16\lambda_4\lambda+35\lambda_4^2)\Big)=0. \end{aligned} \tag{121}$$

Since we suppose that we have three distinct principal curvatures, we have $\lambda\,(\lambda_1-\lambda_4)\,\lambda_4\neq 0$. We drop the factor $\lambda\,(\lambda_1-\lambda_4)\,\lambda_4$ from (121), and eliminate λ_4 from (121) and (106). We then have

$$117\lambda^6+567\lambda_1\lambda^5+1068\lambda_1^2\lambda^4+888\lambda_1^3\lambda^3+85\lambda_1^4\lambda^2-275\lambda_1^5\lambda=50\lambda_1^6. \tag{122}$$

Eliminating λ from (122) and (108), we have

$$\begin{aligned} & 169\beta^6+56360\beta^4\lambda_1^4+1548800\beta\lambda_1^{10}+3097600\lambda_1^{12} \\ & \quad =2\beta^2\lambda_1^2\left(2807\beta^3+69776\beta\lambda_1^4+275360\lambda_1^6\right), \end{aligned}$$

which is a polynomial equation in λ_1 with constant coefficients. Hence λ_1 is a constant, which is a contradiction.

Case III. Two distinct principal curvatures.

Suppose that M_r^4 is a biconservative hypersurface with two distinct principal curvatures with constant norm of second fundamental form. Without losing generality, we choose e_1 in the direction of grad H and, therefore, the shape operator $\mathcal{A}$ of hypersurfaces will take the following form with respect to a suitable frame $\{e_1, e_2, e_3, e_4\}$:

$$\mathcal{A}_H e_1 = -2\epsilon H e_1, \quad \mathcal{A}_H e_i = \lambda e_i, \quad i = 2, 3, 4. \tag{123}$$

Using (123) in (8) and (23), we get

$$\lambda = -\lambda_1, \tag{124}$$

$$3\lambda^2 = \beta - \lambda_1^2, \tag{125}$$

respectively. From (124) and (125), we get λ_1 is a constant, which gives H constant, a contradiction.

Combining Cases I, II, and III, the proof of Theorem 1.1 is complete.

Acknowledgments

The first author is partially supported by FRGS grant for the year 2023-24, F. No. GGSIPU/RDC/FRGS/2023/1448/10. The fourth author was partially supported by a grant from the Empirikeion Foundation of Athens, Greece.

References

[1] A. Arvanitoyeorgos, F. Defever and G. Kaimakamis, Hypersurfaces of E_s^4 with proper mean curvature vector, *J. Math. Soc. Japan* **59**(3) (2007), 797–809.

[2] R. Caddeo, S. Montaldo, C. Oniciuc and P. Piu, Surfaces in three-dimensional space forms with divergence-free stress-bienergy tensor, *Ann. Mat. Pura Appl.* **193**(2) (2014), 529–550.

[3] B.Y. Chen, Some open problems and conjectures on submanifolds of finite type, *Soochow J. Math.* **17**(2) (1991), 169–188.

[4] Deepika, On biconservative Lorentz hypersurface with non-diagonalizable shape operator, *Mediterr. J. Math.* **14**(3) (2017), Article 127, 18 pp.

[5] Y. Fu, On bi-conservative surfaces in Minkowski 3-space, *J. Geom. Phys.* **66** (2013), 71–79.

[6] Y. Fu, Explicit classification of biconservative surfaces in Lorentz 3-space forms, *Ann. Mat. Pura Appl.* **194**(3) (2015), 805–822.

[7] Y. Fu and N.C. Turgay, Complete classification of biconservative hypersurfaces with diagonalizable shape operator in the Minkowski 4-space, *Intern. J. Math.* **27**(5) (2016), Article 1650041.

[8] Y. Fu, M-C. Hong and Z. Zhan, Biharmonic conjectures on hypersurfaces in a space form, *Trans. Amer. Math. Soc.* **376**(12) (2023), 8411–8445.

[9] R.S. Gupta, Biconservative hypersurfaces in Euclidean 5-space, *Bull. Iran. Math. Soc.* **45** (2019), 1117–1133.

[10] R.S. Gupta and A. Arvanitoyeorgos, Hypersurfaces satisfying $\Delta\vec{H} = \lambda\vec{H}$ in $\mathbb{E}^5_s$, *J. Math. Anal. Appl.* **525**(2) (2023), Article 127182.

[11] R.S. Gupta and A. Arvanitoyeorgos, Biconservative hypersurfaces in space forms $\bar{M}^{n+1}(c)$, *Mediterr. J. Math.* **19** (2022), Article 256, 20pp.

[12] R.S. Gupta and A. Sharfuddin, Biconservative Lorentz hypersurfaces in $\mathbb{E}^{n+1}_1$ with complex eigenvalues, *Rev. Un. Mat. Argentina* **60**(2) (2019), 595–610.

[13] Th. Hasanis and Th. Vlachos, Hypersurfaces in E^4 with harmonic mean curvature vector field, *Math. Nachr.* **172** (1995), 145–169.

[14] D. Hilbert, Die Grundlagen der Physik, *Math. Ann.* **92**(1-2) (1924), 1–32.

[15] G. Y. Jiang, The conservation law for 2-harmonic maps between Riemannian manifolds (Chinese), *Acta Math. Sinica* **30** (1987), 220–225.

[16] A. Kayhan and N.C. Turgay, Biconservative hypersurfaces in $\mathbb{E}^4_1$ with non-diagonalizable shape operator, *Mediterr. J. Math.* **20**(3) (2023), Article 105, 17pp.

[17] J.E. Marsden and F.J. Tipler, Maximal hypersurfaces and foliations of constant mean curvature in general relativity, *Phys. Rep.* **66**(3) (1980), 109–139.

[18] S. Stumbles, Hypersurfaces of constant mean extrinsic curvature, *Ann. Physics* **133**(1) (1981) 28–56.

[19] N.C. Turgay, H-hypersurfaces with three distinct principal curvatures in the Euclidean spaces, *Ann. Mat. Pura Appl.* **194**(6) (2015), 1795–1807.

[20] A. Upadhyay and N.C. Turgay, A classification of biconservative hypersurfaces in a pseudo-Euclidean space, *J. Math. Anal. Appl.* **444**(2) (2016), 1703–1720.

Received January 2, 2024
Revised March 25, 2024

Modern Approaches to Differential Geometry
and its Related Fields 159 – 181

CANONICAL FORM THEORY IN GEOMETRY — FOUNDATIONS AND APPLICATIONS

Osamu IKAWA

Department of Mathematics and Physical Sciences,
Faculty of Arts and Sciences, Kyoto Institute of Technology,
Matsugasaki, Sakyo-ku, Kyoto 606-8585, Japan
E-mail: ikawa@kit.ac.jp

In this paper, we deal with the geometry of orbits of hyperpolar actions, which include Hermann actions, on compact Riemannian symmetric spaces.

Keywords: σ-action; hyperpolar action; symmetric triad.

1. Introduction

This paper gives a survey of Lie group actions on manifolds which is focused on works around the author's research.

There are various canonical form theories in linear algebra. The canonical form of the matrix makes linear transformations easier to handle. One of the most impressive facts is that any real symmetric matrix can be diagonalized by an orthogonal matrix. From a slightly different point of view, it can be said that the special orthogonal group of degree n naturally acts on the vector space of symmetric matrices of degree n, and each orbit of the action intersects on the space of diagonal matrices (see Example 2.2 below). From this viewpoint, there should be a canonical form theory in geometry that orbit intersects a well-shaped submanifold when a Lie group acts on a manifold. The well-shaped submanifold that the author envisions is a flat totally geodesic submanifold, which is called a section. The element of the section is the "canonical form". Canonical form theory in geometry is expected to make Lie group actions on manifolds easier to handle.

First, assume that a Lie group G acts on a manifold X. We denote by $G\backslash X$ the orbit space, that is, the space of all G-orbits. We are interested in the following problems (i), (ii) and (iii).

(i) Give a good representation of the orbit space $G\backslash X$.
(ii) Show properties of each orbit in X.

(iii) Study applications of the canonical form theory of $G\backslash X$.

Our study on (i) and (ii) gives foundations of canonical form theory. Concerning the problem (i), it is desirable that the orbit space should be represented as clearly as possible. For example, it is easy to understand if the orbit space is identified with a relatively simple subset like a simplex of a Euclidean space, and the regular orbits correspond to the interior points of the subset. Throughout this paper, a *regular orbit* means maximal dimensional orbit, and a *singular orbit* means the orbit which is not a regular orbit. Concerning the problem (ii), one of the properties of orbits is the dimension and codimension of orbits. When X is a Riemannian manifold and G is an isometry group of X, then we are interested in geometric properties of each orbit. For example, totally geodesic orbits, weakly reflective orbits (see [10] for definition), austere orbits (see [6] for definition), and minimal orbits.

The simplest case is the case when G acts on X transitively. In this case, $G\backslash X$ consists of only one element. In general, the smaller G is, the more difficult the problems (i), (ii) and (iii) become to solve. In what follows we give examples such that $G\backslash X$ is not a single point.

The organization of this paper is as follows. In Section 2, we will give simple but important examples regarding (i) and (ii) above. In Section 3, we will give simple but important examples regarding (iii) above. The examples given in Sections 2 and 3 serve as the starting point for our research. Some examples are revisited in generalized form in Section 4. In Section 4, we will define hyperpolar actions and Hermann actions on compact Riemannian symmetric spaces. We explain that the Hermann action is a hyperpolar action, and deal with (i), (ii) and (iii) above in the case of Hermann actions. In Subsection 4.2, we deal with a concrete example (see Propositions 4.1, 4.2 and 4.3 below). In Section 5, we will discuss future plans.

The author is grateful to the anonymous referee for his advice, and he also thanks the editors, Professor Toshiaki Adachi and Professor Hideya Hashimoto, for help and encouragement during preparations of the manuscript.

2. Examples

In this section, we give examples of our study on the problems (i) and (ii) described in Section 1 and the "canonical forms".

Example 2.1. Denote by $S^2 = \{x \in \mathbb{R}^3 \mid |x| = 1\}$ the two dimensional sphere. The group $SO(2)$ of rotations with the line segment through the north pole $N(0,0,1)$ and the south pole $S(0,0,-1)$ as the axis of rotation induces an isometric action on S^2:

$$SO(2) = \left\{ R(\theta) := \begin{pmatrix} \cos\theta & -\sin\theta & 0 \\ \sin\theta & \cos\theta & 0 \\ 0 & 0 & 1 \end{pmatrix} \,\middle|\, \theta \in \mathbb{R} \right\}.$$

Since the mapping of the closed interval $[0,\pi]$ to the orbit space $SO(2)\backslash S^2$ defined by

$$[0,\pi] \to SO(2)\backslash S^2;\ \alpha \mapsto O_\alpha := SO(2)M_\alpha, \quad M_\alpha := \begin{pmatrix} \sin\alpha \\ 0 \\ \cos\alpha \end{pmatrix}$$

is bijective, we can identify $SO(2)\backslash S^2$ with $[0,\pi]$. Hence, we get:

(i) $[0,\pi]$ is a good representation of $SO(2)\backslash S^2$,
(ii) $O_0 = \{N\}$, $O_\pi = \{S\}$, $O_{\pi/2}$ = the great circle (equator). For $0 < \alpha < \pi, \alpha \neq \pi/2$, the orbit O_α is a small circle.

In this case, M_α is considered as a *canonical form* of the $SO(2)$-action.

We give a representation of the orbit space $SO(2)\backslash S^2$ which is not good. The mapping defined by

$$[0,2\pi] \to SO(2)\backslash S^2;\ \alpha \mapsto O_\alpha := SO(2)M_\alpha, \quad M_\alpha := \begin{pmatrix} \sin\alpha \\ 0 \\ \cos\alpha \end{pmatrix}$$

is surjective. And we have $O_\alpha = O_\beta$ if and only if $\alpha + \beta = 2\pi$. Define a mapping f of $[0,\pi]$ to $[0,2\pi]$ by

$$f(\alpha) = \begin{cases} \alpha & (\alpha \in \mathbb{Q}), \\ 2\pi - \alpha & (\alpha \notin \mathbb{Q}). \end{cases}$$

where $\mathbb{Q}$ is the set of rational numbers. Then the mapping $f([0,2\pi]) \to SO(2)\backslash S^2; \alpha \mapsto O_\alpha$ is bijective. Thus $f([0,2\pi])$ is a representation of $SO(2)\backslash S^2$ which is not good.

Example 2.1 is extended to a sphere of general dimension in Example 4.1.

Example 2.2. The special orthogonal group $SO(n)$ of degree n acts on the space $\mathrm{Sym}(n) := \{X \in M_n(\mathbb{R}) \mid {}^tX = X\}$ of symmetric matrices as follows:

$$\rho(g)X = gXg^{-1} = gX\,{}^t g \qquad (g \in SO(n), X \in \mathrm{Sym}(n)).$$

Here, for a matrix X, we denote by ${}^tX$ its transposed matrix. It is well-known in linear algebra that, for each $X \in \mathrm{Sym}(n)$, there exists $g \in SO(n)$ such that $\rho(g)X$ is a diagonal matrix. Hence, when we study $SO(n)$-orbits, we may assume that the starting point of the orbit is a diagonal matrix. Since we can exchange the entries in the diagonal elements of the diagonal matrix by $SO(n)$-action, we may assume that the starting point of the orbit of $SO(n)$-action is in the element of the set defined by

$$A := \left\{ X(x_1, \ldots, x_n) := \begin{pmatrix} x_1 & & \\ & \ddots & \\ & & x_n \end{pmatrix} \middle| \, x_1 \leq \cdots \leq x_n \right\}. \tag{1}$$

Since $X(x_1, \ldots, x_n), X(y_1, \ldots, y_n) \in A$ are transformed with each other by the $SO(n)$-action if and only if $X(x_1, \ldots, x_n) = X(y_1, \ldots, y_n)$, we get:

(i) The set A is a good representation of $SO(n)\backslash\mathrm{Sym}(n)$.

In this case, an element of A is considered as a *canonical form* of the $SO(n)$-action.

Through the identification of A and $SO(n)\backslash\mathrm{Sym}(n)$, the boundary points correspond to the singular orbits. Furthermore, we can stratify A. It is easy to verify this in the case when $n = 2,\ 3$.

When $n = 2$, the boundary ∂A and the interior $\mathrm{int}(A)$ of the orbit space A is given by

$$\partial A = \left\{ X(x,x) = \begin{pmatrix} x & \\ & x \end{pmatrix} = xE_2 \,\middle|\, x \in \mathbb{R} \right\}, \quad \mathrm{int}(A) = \{X(x,y) \mid x < y\}.$$

Then we have the stratification of A as $A = \partial A \cup \mathrm{int}(A)$. The orbit through xE_2 consists of only one element. Hence, the orbits passing through the boundary point are singular. When $n = 3$, define subsets $A_0, \ldots, A_3$ of A by

$$A_0 = \{X(x,x,x) = xE_3 \mid x \in \mathbb{R}\}, \quad A_1 = \{X(x,x,y) \mid x < y\},$$
$$A_2 = \{X(x,y,y) \mid x < y\}, \qquad A_3 = \mathrm{int}(A) = \{X(x,y,z) \mid x < y < z\}.$$

Then we have the stratification of A as $A = A_0 \cup A_1 \cup A_2 \cup A_3$ and $\partial(A) = A_0 \cup A_1 \cup A_2$. The elements $X \in A$ corresponds to a singular orbit if and only if X is in ∂A.

This example is treated again in Example 4.2 from a different perspective.

Example 2.3. Denote by $M_{m,n}(F)$ ($F = \mathbb{R}, \mathbb{C}$) the vector space over F consisting of (m,n) matrices whose entries are in F. The group $GL(m,F) \times GL(n,F)$ acts on $M_{m,n}(F)$ as follows:

$$(g,h) \cdot X = gXh^{-1} \qquad (g \in GL(m,F), h \in GL(n,F), X \in M_{m,n}(F)).$$

Let X and Y be elements of $M_{m,n}(F)$. It is well-known in linear algebra that X and Y are transformed with each other by $GL(m,F) \times GL(n,F)$-action if and only if $\mathrm{rank}(X) = \mathrm{rank}(Y)$. Since for any $X \in M_{m,n}(F)$ there exist $g \in GL(m,F)$ and $h \in GL(n,F)$ such that

$$gXh^{-1} = \begin{pmatrix} E_r & O \\ O & O \end{pmatrix},$$

we can identify the orbit space with

$$\left\{ \begin{pmatrix} E_r & O \\ O & O \end{pmatrix} \;\middle|\; r = 0, 1, \ldots, \min\{m,n\} \right\} \cong \{0, 1, \ldots, \min\{m,n\}\}.$$

Hence, we get:

(i) $\{0, 1, \ldots, \min\{m,n\}\}$ is a good representation of the orbit space $GL(m,F) \times GL(n,F) \backslash M_{m,n}(F)$.

In this case, the matrix $\begin{pmatrix} E_r & O \\ O & O \end{pmatrix}$ is considered as a *canonical form* of the $GL(m,F) \times GL(n,F)$-action.

We denote by K the isotropy subgroup of $GL(m,F) \times GL(n,F)$ at $\begin{pmatrix} E_r & O \\ O & O \end{pmatrix}$. Then we have

$$K = \left\{ \left(\begin{pmatrix} g_{11} & g_{12} \\ O_{m-r,r} & g_{22} \end{pmatrix}, \begin{pmatrix} g_{11} & O_{r,n-r} \\ h_{21} & h_{22} \end{pmatrix} \right) \;\middle|\; \begin{array}{l} g_{11} \in GL(r,F), \\ g_{22} \in GL(m-r,F), \\ h_{22} \in GL(n-r,F), \\ g_{12} \in M_{r,m-r}(F), \\ h_{21} \in M_{n-r,r}(F) \end{array} \right\}.$$

Since $\dim_F K = r^2 + m(m-r) + n(n-r)$, the dimension of the orbit passing through the point $\begin{pmatrix} E_r & O \\ O & O \end{pmatrix}$ is equal to $r(m+n-r)$. Hence we see that X and Y in $M_{m,n}(F)$ are transformed with each other by $GL(m,F) \times GL(n,F)$-action if and only if those dimensions of orbits coincide.

As we have seen in the three examples above, the canonical form of an orbit is a good starting point which represents the orbit.

3. Applications of canonical forms

Concerning the problem (iii), we give two elementary applications of group actions.

Example 3.1. For $a, b, c, d \in \mathbb{R}^3$, the identity

$$\begin{vmatrix} \langle a, c\rangle & \langle a, d\rangle \\ \langle b, c\rangle & \langle b, d\rangle \end{vmatrix} = \langle a \times b, c \times d\rangle \tag{2}$$

holds, where $\langle\ ,\ \rangle$ is the canonical inner product on $\mathbb{R}^3$, and $a \times b$ is the exterior product of a and b.

We show the identity (2) in two ways for comparison. One proof uses a group action, while the other does not.

When $a = e_1 = (1, 0, 0)$ and $b = e_2 = (0, 0, 1)$, we have

$$\begin{aligned}\begin{vmatrix} \langle e_1, c\rangle & \langle e_1, d\rangle \\ \langle e_2, c\rangle & \langle e_2, d\rangle \end{vmatrix} &= \langle e_1, c\rangle\langle e_2, d\rangle - \langle e_2, c\rangle\langle e_1, d\rangle = \langle e_3, c \times d\rangle \\ &= \langle e_1 \times e_2, c \times d\rangle. \end{aligned} \tag{3}$$

The natural action of $SO(3)$ on $\mathbb{R}^3$ induces a transitive action of $SO(3)$ on S^2. The isotropy subgroup of $SO(3)$ at e_1 is isomorphic to $SO(2)$. Since we have

$$\langle ga, gb\rangle = \langle a, b\rangle, \qquad g(a \times b) = (ga) \times (gb) \quad (g \in SO(3), a, b \in \mathbb{R}^3),$$

both sides of (2) is $SO(3)$-invariant. Since both sides of (2) is quadruple linear with respect to the variables a, b, c and d, and is alternative with respect to a and b, and is also alternative with respect to c and d, in order to prove (2), we may assume that a, b, c and d are unit vectors and $a \perp b, c \perp d$. Then there exists $g \in SO(3)$ such that $ga = e_1$ and $gb = e_2$. Hence the assertion reduces to the case (3).

Next, we give a proof which does not use the group actions. Since, for $a, b, c, d, f \in \mathbb{R}^3$, we have

$$\langle f, c \times d\rangle = \langle c, d \times f\rangle, \quad d \times (a \times b) = \langle b, d\rangle a - \langle a, d\rangle b,$$

we get

$$\langle a \times b, c \times d\rangle = \langle c, d \times (a \times b)\rangle = \langle b, d\rangle\langle c, a\rangle - \langle a, d\rangle\langle c, b\rangle = \begin{vmatrix} \langle a, c\rangle & \langle a, d\rangle \\ \langle b, c\rangle & \langle b, d\rangle \end{vmatrix}.$$

Example 3.2. For $x = \begin{pmatrix} x_1 \\ \vdots \\ x_n \end{pmatrix} \in \mathbb{R}^n$ and $k \in \mathbb{R}$, the identity

$$\begin{vmatrix} x_1^2 + k & x_1 x_2 & \cdots & x_1 x_n \\ x_2 x_1 & x_2^2 + k & \cdots & x_2 x_n \\ \vdots & \vdots & \ddots & \vdots \\ x_n x_1 & x_n x_2 & \cdots & x_n^2 + k \end{vmatrix} = k^{n-1}(k + |x|^2) \tag{4}$$

holds.

We show the identity (4) in two ways for comparison.

The right side of (4) is equal to $|x\,{}^t x + kE_n|$. For $g \in SO(n)$, we have

$$|gx\,{}^t(gx) + kE_n| = |gx\,{}^t x\,{}^t g + kE_n| = |g(x\,{}^t x + kE_n){}^t g| = |x\,{}^t x + kE_n|.$$

Hence the both sides of (4) are $SO(n)$-invariant. Since $SO(n)$ acts on S^{n-1} transitively, there exists $g \in SO(n)$ such that $gx = |x|e_n$. Then we get

$$|x\,{}^t x + kE_n| = |gx\,{}^t(gx) + kE_n| = \begin{vmatrix} k & & & \\ & \ddots & & \\ & & k & \\ & & & k+|x|^2 \end{vmatrix} = k^{n-1}(k + |x|^2).$$

The following proof does not use the group actions. Put $y = \begin{pmatrix} y_1 \\ \vdots \\ y_n \end{pmatrix} \in \mathbb{R}^n$. We may assume that $x \neq 0$. Note that $(x{}^t x + kE_n)y = \langle x, y\rangle x + ky$. Define an $(n-1)$-dimensional subspace V of $\mathbb{R}^n$ by $V = (\mathbb{R}x)^\perp$. Then we have the orthogonal direct sum decomposition of $\mathbb{R}^n$: $\mathbb{R}^n = \mathbb{R}x \oplus V$. By the above equation, we get

$$(x\,{}^t x + kE_n)x = (|x|^2 + k)x, \quad (x\,{}^t x + kE_n)y = ky \quad (y \in V),$$

which implies the assertion.

4. Hyperpolar actions on compact symmetric spaces

In Subsection 4.1, we define hyperpolar actions and Hermann actions on compact Riemannian symmetric spaces. This is a main subject of this paper. We explain that the Hermann action is a hyperpolar action, and deal with the problems (i), (ii) and (iii) mentioned in Section 1 in the case of Hermann actions. In Subsection 4.2, we deal with a concrete example.

4.1. *Hyperpolar actions and Hermann actions*

An isometric action of a compact Lie group on a Riemannian manifold is called *hyperpolar*, if there exists a connected closed flat submanifold, called a *section*, that meets all orbits orthogonally. A section is automatically totally geodesic. The dimension of the section is called the *cohomogeneity* of the hyperpolar action.

For example, the isotropy action on a compact Riemannian symmetric space M is a hyperpolar action whose section is a maximal flat totally geodesic submanifold. The cohomogeneity is called the rank of M. If we take a biinvariant Riemannian metric on a compact connected Lie group G, the action of G on G itself defined by

$$\tau_g(x) = gxg^{-1} \tag{5}$$

is a hyperpolar action whose section is a maximal torus of G. Hermann action defined below is a generalization of these actions.

Definition 4.1. [Hermann action] Let (G, K_1, K_2) be a compact symmetric triad. That is, there exist two involutive automorphisms θ_1 and θ_2 on the compact connected Lie group G such that the closed subgroup K_i of G lies between $F(\theta_i, G)$ and the identity component $F(\theta_i, G)_0$ of $F(\theta_i, G)$ for $i = 1, 2$. Here, $F(\theta_i, G)$ denotes the fixed-point subgroup of θ_i in G. Take an $\mathrm{Aut}(G)$-invariant Riemannian metric $\langle\ ,\ \rangle$ on G. Then the coset manifold $M_i = G/K_i$ is a compact Riemannian symmetric space with respect to the G-invariant Riemannian metric induced from $\langle\ ,\ \rangle$ on G, which is also denoted by the same symbol $\langle\ ,\ \rangle$. The natural isometric action of K_2 on M_1 is called a *Hermann action.*

When $K_1 = K_2$, the Hermann action is nothing but the isotropy action on the compact Riemannian symmetric space. The study of the geometry of orbits of Hermann actions were initiated by O. Goertsches and G. Thorbergsson ([4]). Hermann action is a hyperpolar action. We shall verify this in the case of the isotropy action on the n-dimensional sphere.

Example 4.1. The special orthogonal group $G = SO(n+1)$ of degree $n+1$ isometrically acts on the n-dimensional sphere $S^n = \{x \in \mathbb{R}^{n+1} \mid |x| = 1\}$ in $\mathbb{R}^{n+1}$. This action is transitive. The isotropy subgroup K of G at $e_{n+1} = {}^t(0, \dots, 0, 1) \in S^n$ is given by

$$K = SO(n) \times \{1\} = \left\{ \begin{pmatrix} g & 0 \\ 0 & 1 \end{pmatrix} \middle| \, g \in SO(n) \right\}.$$

Hence S^n is expressed by $S^n = G/K$. Define an involutive automorphism θ on G by

$$\theta(g) = \begin{pmatrix} E_n & 0 \\ 0 & -1 \end{pmatrix} g \begin{pmatrix} E_n & 0 \\ 0 & -1 \end{pmatrix}.$$

Then the fixed-point subgroup $F(\theta)$ of θ on G equal to $S(O(n) \times O(1))$. Since K is equal to the identity component $F(\theta)_0$ of $F(\theta)$, we find that (G, K) is a compact symmetric pair. Hence $S^n = G/K$ is a compact Riemannian symmetric space. We show that the action of K on S^n is hyperpolar. Denote by $\mathfrak{g}$ and $\mathfrak{k}$ the Lie algebras of G and K, respectively. Denote by $\mathfrak{g} = \mathfrak{k} \oplus \mathfrak{m}$ the canonical decomposition of $\mathfrak{g}$. Then we have

$$\mathfrak{m} = \left\{ \begin{pmatrix} O_n & x \\ -{}^t x & 0 \end{pmatrix} \middle| \, x \in \mathbb{R}^n \right\}.$$

If we set

$$\mathfrak{a} = \left\{ \alpha \begin{pmatrix} O_n & e_1 \\ -{}^t e_1 & 0 \end{pmatrix} \middle| \, \alpha \in \mathbb{R} \right\}, \quad \text{where} \quad e_1 = \begin{pmatrix} 1 \\ 0 \\ \vdots \\ 0 \end{pmatrix} \in \mathbb{R}^n,$$

then $\mathfrak{a}$ is a maximal abelian subspace of $\mathfrak{m}$. If we set

$$A := \exp \mathfrak{a} = \left\{ \begin{pmatrix} \cos\alpha & 0 & \sin\alpha \\ 0 & E_{n-1} & 0 \\ -\sin\alpha & 0 & \cos\alpha \end{pmatrix} \middle| \, \alpha \in \mathbb{R} \right\},$$

then A is a toral subgroup of G. Denote by $\pi : G \to S^n$ the natural projection given by $g \mapsto ge_{n+1}$. Then we have

$$\pi(A) = \left\{ O_\alpha := \begin{pmatrix} \sin\alpha \\ 0_{n-1} \\ \cos\alpha \end{pmatrix} \middle| \, \alpha \in [0, 2\pi] \right\},$$

which is a great circle passing through e_{n+1} and $-e_{n+1}$. Since $SO(n)$ acts on S^{n-1} transitively, we see that

$$S^n = K\pi(A), \tag{6}$$

that is, every K-orbit meets $\pi(A)$. Furthermore every K-orbit meets $\pi(A)$ orthogonally, which implies that the action of K on S^n is a hyperpolar action whose section is $\pi(A)$. The orbits O_α and $O_{2\pi-\alpha}$ are transformed with each other by the K-action. Let α and β be in the closed interval $[0, \pi]$. The orbits O_α and O_β are transformed with each other by the K-action if and only if $\alpha = \beta$. Hence the mapping defined by $[0, \pi] \to K \backslash S^n; \alpha \mapsto O_\alpha$ is bijective. Thus, we get from (i) in Section 1 that the orbit space $K \backslash S^n$ can be identified with $[0, \pi]$. From (6), we see

$$G = KAK. \tag{7}$$

In the case when $n = 2$, this example reduces to Example 2.1. The expression (8) mentioned below is a generalization of (7).

Returning to the general case. We explain a section of a Hermann action. Denote by $\mathfrak{g}, \mathfrak{k}_1$ and $\mathfrak{k}_2$ the Lie algebras of G, K_1 and K_2, respectively. Two involutive automorphisms θ_1 and θ_2 induce involutive automorphisms of $\mathfrak{g}$, also denoted by θ_1 and θ_2, respectively. We have two canonical decompositions of $\mathfrak{g}$:

$$\mathfrak{g} = \mathfrak{k}_1 \oplus \mathfrak{m}_1 = \mathfrak{k}_2 \oplus \mathfrak{m}_2, \qquad \text{where} \quad \mathfrak{m}_i = \{X \in \mathfrak{g} \mid \theta_i(X) = -X\}.$$

Take a maximal abelian subspace $\mathfrak{a}$ of $\mathfrak{m}_1 \cap \mathfrak{m}_2$. It is known that $A := \exp \mathfrak{a}$ is closed in G ([5]). Hence A is a toral subgroup of G. Denote by $\pi_1 : G \to M = G/K_1$ the natural projection. Then $\pi_1(A) \subset M_1$ is a section of the Hermann action (see [5] for detail). And the cohomogeneity is equal to $\dim \mathfrak{a}$. In particular, we have

$$G = K_2 A K_1 = K_1 A K_2, \tag{8}$$

which implies that the map $\mathfrak{a} \to K_2 \backslash G/K_1$ defined by $H \mapsto K_2 \pi_1(\exp H)$ is surjective. Define an equivalence relation $\sim$ on $\mathfrak{a}$ by the rule that $H_1 \sim H_2$ if and only if $K_2\pi_1(\exp H_1) = K_2\pi_1(\exp H_2)$. In his paper [13], T. Matsuki defined a group $\tilde{J}$ acting on $\mathfrak{a}$ such that $H_1 \sim H_2$ if and only if H_1 and H_2 are transformed with each other by an element of $\tilde{J}$. Then the orbit space $K_2 \backslash G/K_1$ can be identified with $\mathfrak{a}/\!\sim\, = \mathfrak{a}/\tilde{J}$, that is,

$$K_2 \backslash G/K_1 \cong \mathfrak{a}/\!\sim\, = \mathfrak{a}/\tilde{J}. \tag{9}$$

When G is simple, and θ_1 and θ_2 commute with each other, the geometry of orbits of Hermann action was investigated in [7]. In this case, the author studied the problem (i) in Section 1 and proved that there exists a simplex P_0 in $\mathfrak{a}$ such that the mapping defined by

$$\overline{P_0} \to \mathfrak{a}/\!\sim\,; \quad H \mapsto [H]$$

is surjective, where $\overline{P_0}$ is the closure of P_0. In many cases the map above is bijective. Hence $\overline{P_0}$ is a substitute of the orbit space. And a point in the interior of $\overline{P_0}$, that is, a point in P_0, corresponds to a regular orbit. Hence $\overline{P_0}$ is a good representation of the orbit space $K_2 \backslash G/K_1$. Here, a simplex P_0 is constructed from a symmetric triad $(\widetilde{\Sigma}, \Sigma, W)$ obtained from (G, K_1, K_2). The symmetric triad is a generalization of the notion of irreducible root system. Concerning the problem (ii) in Section 1, the author determined which orbit is austere, and which orbit is totally geodesic. Furthermore the author proved that there exists a unique minimal orbit in each strata of $\overline{P_0}$. Concerning the problem (iii) in Section 1, for example we can prove the Wirtinger's inequality by using the compact symmetric triad $(G, K_1, K_2) = (SO(2n), SO(2a) \times SO(2n-2a), U(n))$ (see Section 5). In the next subsection, we explain the problems (i), (ii) and (iii) in a concrete example.

In the case when θ_1 and θ_2 do not commute with each other, S. Ohno studied the geometry of orbits of Hermann actions in [14].

Definition 4.2. [σ-action] Let σ be an automorphism of a compact connected Lie group G. The action of G on G itself defined by $g \cdot x = gx\sigma(g)^{-1}$ is called a *σ-action.*

When σ is the identity, then the σ-action is nothing but the action defined by (5).

Returning to the general case, take a biinvariant Riemannian metric $\langle \ , \ \rangle$ on G. Then σ-action is a hyperpolar action. We explain its section. Denote by $F(\sigma, G)$ the fixed-point subgroup of G. A maximal torus A of $F(\sigma, G)$ is a section of the σ-action. In particular, we have

$$G = \bigcup_{g \in G} gA\sigma(g)^{-1}.$$

We note that σ-action is a kind of Hermann actions (see [8], for instance).

When G is simple and σ is an involutive automorphism, the geometry of orbits of σ-actions were studied in [8]. The σ-actions when G is simple and the order of σ is greater than or equal to 3 are essentially limited to the case where $G = Spin(8)$ and σ is a triality automorphism. Recently the geometry of the σ-action in this case were studied in [9].

In his paper [11], Kollross showed that Hermann actions and cohomogeneity one actions exhaust all hyperpolar actions on the compact irreducible Riemannian symmetric spaces. Hence, the Hermann actions account for a large part of the hyperpolar actions on compact irreducible Riemann symmetric spaces. The study of a cohomogeneity one action on a compact irreducible symmetric space which is not a Hermann action is found in [16] and [3].

Hermann action is a hyperpolar action on a compact symmetric space, which is constructed by a compact symmetric triad, while we can construct a hyperpolar action on a Euclidean space by using a compact symmetric pair as follows. Let (G, K) be a compact symmetric pair. Denote by $\mathfrak{g}$ and $\mathfrak{k}$ the Lie algebras of G and K, respectively. There exists an involutive automorphism θ on G such that the fixed-point subgroup $F(\theta)$ and its identity component $F(\theta)_0$ satisfy $F(\theta)_0 \subset K \subset F(\theta)$. Denote by $\mathfrak{g} = \mathfrak{k} \oplus \mathfrak{m}$ the canonical decomposition of $\mathfrak{g}$ with respect to θ. Let $\mathfrak{a}$ be a maximal abelian subspace of $\mathfrak{m}$. Then the action of $\mathrm{Ad}_G(K)$ on $\mathfrak{m}$ is a hyperpolar action whose section is $\mathfrak{a}$. In particular, we have

$$\mathfrak{m} = \mathrm{Ad}_G(K)\mathfrak{a} \tag{10}$$

holds. Denote by $S = \{X \in \mathfrak{m} \mid |X| = 1\}$ the hypersphere in $\mathfrak{m}$. Then the action of $\mathrm{Ad}_G(K)$ on S is hyperpolar action whose section is $\mathfrak{a} \cap S$.

Consider Example 2.2 again from this viewpoint.

Example 4.2. Denote by G the unitary group $U(n)$ of degree n. We define an involutive automorphism θ of G by $\theta(g) = \bar{g}$. Then we have $F(\theta) = O(n)$ and $F(\theta)_0 = SO(n) =: K$. Thus, (G, K) is a compact symmetric pair. Denote by $\mathfrak{g} = \mathfrak{k} \oplus \mathfrak{m}$ the canonical decomposition of $\mathfrak{g}$ with respect to θ. Then we have

$$\mathfrak{m} = \left\{\sqrt{-1}X \;\middle|\; X \in \mathrm{sym}(n)\right\} \cong \mathrm{sym}(n).$$

The action of $\mathrm{Ad}_G(K)$ on $\mathfrak{m}$ is equivalent to the action of K on $\mathrm{sym}(n)$ which was treated in Example 2.2. Define a subspace $\mathfrak{a}$ of $\mathfrak{m}$ by

$$\mathfrak{a} = \left\{\sqrt{-1}D \;\middle|\; D \text{ is a real diagonal matrix of degree } n\right\}.$$

The isomorphism $\mathfrak{m} \cong \mathrm{sym}(n)$ induces an isomorphism

$$\mathfrak{a} \cong \left\{D \;\middle|\; D \text{ is a real diagonal matrix of degree } n\right\}.$$

Since $\mathfrak{a}$ is a maximal abelian subspace of $\mathfrak{m}$, we obtain that the equality (10) holds. This means that every real symmetric matrix can be diagonalized by a special orthogonal matrix.

4.2. $(SO(2p+2q), S(O(2p) \times O(2q)), U(p+q))$

In this subsection, we retain the notations in the previous sections.

We consider a compact symmetric triad $(G, K_1, K_2) = (SO(2p + 2q), S(O(2p) \times O(2q)), U(p+q))$. We may assume that $p \leq q$. Define an involutive automorphism θ_1 on G by

$$\theta_1(g) = I_{2p,2q}\, g\, I_{2p,2q}, \quad \text{where} \quad I_{2p,2q} = \begin{pmatrix} I_{2p} & \\ & -I_{2q} \end{pmatrix}.$$

Then we have $\mathfrak{k}_1 = \mathfrak{so}(2p) \oplus \mathfrak{so}(2q)$ and

$$\mathfrak{m}_1 = \left\{ \begin{pmatrix} & X \\ -{}^tX & \end{pmatrix} \,\middle|\, X \in M_{2p,2q}(\mathbb{R}) \right\}.$$

Define another involutive automorphism θ_2 on G by

$$\theta_2(g) = J_{p+q}\, g\, J_{p+q}^{-1} \quad \text{where} \quad J_{p+q} = \begin{pmatrix} & I_p & & \\ -I_p & & & \\ & & & I_q \\ & & -I_q & \end{pmatrix}.$$

Then we have $\theta_1\theta_2 = \theta_2\theta_1$ and

$$\mathfrak{k}_2 = \left\{ \begin{pmatrix} a & -b & c & -d \\ b & a & d & c \\ -{}^tc & -{}^td & e & -f \\ {}^td & -{}^tc & f & e \end{pmatrix} \,\middle|\, \begin{array}{l} a \in \mathfrak{so}(p), e \in \mathfrak{so}(q), \\ {}^tb = b \in M_{p,p}(\mathbb{R}), \\ {}^tf = f \in M_{q,q}(\mathbb{R}), \\ c, d \in M_{p,q}(\mathbb{R}) \end{array} \right\},$$

$$\mathfrak{m}_2 = \left\{ \begin{pmatrix} a & b & c & d \\ b & -a & d & -c \\ -{}^tc & -{}^td & e & f \\ -{}^td & {}^tc & f & -e \end{pmatrix} \,\middle|\, \begin{array}{l} a, b \in \mathfrak{so}(p), \\ e, f \in \mathfrak{so}(q), \\ c, d \in M_{p,q}(\mathbb{R}) \end{array} \right\}.$$

The Lie algebra $\mathfrak{k}_2$ is isomorphic to $\mathfrak{u}(p+q)$ as follows:

$$\mathfrak{k}_2 \cong \mathfrak{u}(p+q) = \left\{ \begin{pmatrix} a + \sqrt{-1}b & c + \sqrt{-1}d \\ -{}^tc + \sqrt{-1}{}^td & e + \sqrt{-1}f \end{pmatrix} \right\}.$$

From these we have

$$\mathfrak{m}_1 \cap \mathfrak{m}_2 = \left\{ \begin{pmatrix} & & c & d \\ & & d & -c \\ -{}^tc & -{}^td & & \\ -{}^td & {}^tc & & \end{pmatrix} \,\middle|\, c, d \in M_{p,q}(\mathbb{R}) \right\},$$

$$\mathfrak{k}_1 \cap \mathfrak{k}_2 = \left\{ \begin{pmatrix} a & -b & & \\ b & a & & \\ & & e & -f \\ & & f & e \end{pmatrix} \,\middle|\, \begin{array}{l} a \in \mathfrak{so}(p),\ e \in \mathfrak{so}(q), \\ {}^tb = b \in M_{p,p}(\mathbb{R}), \\ {}^tf = f \in M_{q,q}(\mathbb{R}) \end{array} \right\},$$

and

$$\mathfrak{k}_1 \cap \mathfrak{m}_2 = \left\{ \begin{pmatrix} a & b & & \\ b & -a & & \\ & & e & f \\ & & f & -e \end{pmatrix} \;\middle|\; \begin{array}{l} a, b \in \mathfrak{so}(p), \\ e, f \in \mathfrak{so}(q) \end{array} \right\},$$

$$\mathfrak{m}_1 \cap \mathfrak{k}_2 = \left\{ \begin{pmatrix} & & c & -d \\ & & d & c \\ -{}^t c & -{}^t d & & \\ {}^t d & -{}^t c & & \end{pmatrix} \;\middle|\; c, d \in M_{p,q}(\mathbb{R}) \right\}.$$

If we set

$$\mathfrak{a} = \sum_{i=1}^{p} \mathbb{R} \begin{pmatrix} & & E_{ii} & \\ & & & -E_{ii} \\ -E_{ii} & & & \\ & E_{ii} & & \end{pmatrix},$$

then $\mathfrak{a}$ is a maximal abelian subspace of $\mathfrak{m}_1 \cap \mathfrak{m}_2$. Hence the cohomogeneity of the Hermann action is equal to p. We define $e_i \in \mathfrak{a}$ by the equation $\langle e_i, H \rangle = x_i$ for every

$$H = \sum_{i=1}^{p} x_i \begin{pmatrix} & & E_{ii} & \\ & & & -E_{ii} \\ -E_{ii} & & & \\ & E_{ii} & & \end{pmatrix} \in \mathfrak{a}.$$

Define a closed subgroup G_{12} of G by $G_{12} := \{g \in G \mid \theta_1(g) = \theta_2(g)\}$, and a closed subgroup of K_{12} of G_{12} by $K_{12} := \{g \in G_{12} \mid \theta_1(g) = \theta_2(g) = g\}$. Then $\big((G_{12})_0, K_{12} \cap (G_{12})_0\big)$ is a compact symmetric pair, where $(G_{12})_0$ is the identity component of G_{12}, and θ_1 is an involution on $(G_{12})_0$ which defines $\big((G_{12})_0, K_{12} \cap (G_{12})_0\big)$. The canonical decomposition of the Lie algebra of $(G_{12})_0$ is given by

$$\mathfrak{g}_{12} = (\mathfrak{k}_1 \cap \mathfrak{k}_2) \oplus (\mathfrak{m}_1 \cap \mathfrak{m}_2).$$

Since $\mathfrak{a}$ is a maximal abelian subspace of $\mathfrak{m}_1 \cap \mathfrak{m}_2$, we can define a restricted root system Σ of $\big((G_{12})_0, K_{12} \cap (G_{12})_0\big)$ with respect to $\mathfrak{a}$.

Lemma 4.1. *When $p = q$, the restricted root system is*

$$\begin{aligned} \Sigma = & \big\{ \pm 2e_i \text{ (with multiplicity 1)} \;\big|\; 1 \le i \le p \big\} \\ & \bigcup \big\{ \pm e_i \pm e_j \text{ (with multiplicity 2)} \;\big|\; 1 \le i < j \le p \big\}. \end{aligned}$$

When $p < q$, it is

$$\Sigma = \big\{\pm 2e_i \text{ (with multiplicity 1)}, e_i \text{ (with multiplicity } 2(q-p)) \;\big|\; 1 \le i \le p\big\}$$
$$\bigcup \big\{\pm e_i \pm e_j \text{ (with multiplicity 2)} \;\big|\; 1 \le i < j \le p\big\}.$$

Proof. For $1 \le j \le p$, set

$$S_{2e_j} = \begin{pmatrix} & -E_{jj} & & \\ E_{jj} & & & \\ & & & -E_{jj} \\ & & E_{jj} & \end{pmatrix} \in \mathfrak{k}_1 \cap \mathfrak{k}_2,$$

$$T_{2e_j} = \begin{pmatrix} & & & -E_{jj} \\ & & -E_{jj} & \\ & E_{jj} & & \\ E_{jj} & & & \end{pmatrix} \in \mathfrak{m}_1 \cap \mathfrak{m}_2.$$

Then, for each $H \in \mathfrak{a}$, we have

$$[H, S_{2e_j}] = \langle 2e_j, H\rangle T_{2e_j}, \quad [H, T_{2e_j}] = -\langle 2e_j, H\rangle S_{2e_j}.$$

For $1 \le j \le p$, $1 \le i \le q-p$, set

$$S^{\,i}_{e_j,1} = \begin{pmatrix} 0 & & & \\ & 0 & & \\ & & E_{j,p+i} - E_{p+i,j} & \\ & & & E_{j,p+i} - E_{p+i,j} \end{pmatrix} \in \mathfrak{k}_1 \cap \mathfrak{k}_2,$$

$$T^{\,i}_{e_j,1} = \begin{pmatrix} & & -E_{j,p+i} & \\ & & & E_{j,p+i} \\ E_{p+i,j} & & & \\ & -E_{p+i,j} & & \end{pmatrix} \in \mathfrak{m}_1 \cap \mathfrak{m}_2.$$

Then, for each $H \in \mathfrak{a}$, we have

$$[H, S^{\,i}_{e_j,1}] = \langle e_j, H\rangle T^{\,i}_{e_j,1}, \quad [H, T^{\,i}_{e_j,1}] = -\langle e_j, H\rangle S^{\,i}_{e_j,1}.$$

For $1 \le j \le p$, $1 \le i \le q-p$, set

$$S^{\,i}_{e_j,2} = \begin{pmatrix} 0 & & & \\ & 0 & & \\ & & & -(E_{j,p+i} + E_{p+i,j}) \\ & & E_{j,p+i} + E_{p+i,j} & \end{pmatrix} \in \mathfrak{k}_1 \cap \mathfrak{k}_2,$$

$$T^{\,i}_{e_j,2} = \begin{pmatrix} & & & -E_{j,p+i} \\ & & -E_{j,p+i} & \\ & E_{p+i,j} & & \\ E_{p+i,j} & & & \end{pmatrix} \in \mathfrak{m}_1 \cap \mathfrak{m}_2.$$

Then, for each $H \in \mathfrak{a}$, we have

$$[H, S_{e_j,2}^{\,i}] = \langle e_j, H\rangle T_{e_j,2}^{\,i}, \quad [H, T_{e_j,2}^{\,i}] = -\langle e_j, H\rangle S_{e_j,2}^{\,i}.$$

For $1 \leq j < k \leq p$, set

$$S_{e_j+e_k}^{\,1} = \begin{pmatrix} & -(E_{jk}+E_{kj}) & & \\ E_{jk}+E_{kj} & & & \\ & & & -(E_{jk}+E_{kj}) \\ & & E_{jk}+E_{kj} & \end{pmatrix} \in \mathfrak{k}_1 \cap \mathfrak{k}_2,$$

$$T_{e_j+e_k}^{\,1} = \begin{pmatrix} & & & -(E_{jk}+E_{kj}) \\ & & -(E_{jk}+E_{kj}) & \\ & E_{jk}+E_{kj} & & \\ E_{jk}+E_{kj} & & & \end{pmatrix} \in \mathfrak{m}_1 \cap \mathfrak{m}_2,$$

$$S_{e_j+e_k}^{\,2} = \begin{pmatrix} E_{jk}-E_{kj} & & & \\ & E_{jk}-E_{kj} & & \\ & & E_{kj}-E_{jk} & \\ & & & E_{kj}-E_{jk} \end{pmatrix} \in \mathfrak{k}_1 \cap \mathfrak{k}_2,$$

$$T_{e_j+e_k}^{\,2} = \begin{pmatrix} & & E_{kj}-E_{jk} & \\ & & & E_{jk}-E_{kj} \\ E_{kj}-E_{jk} & & & \\ & E_{jk}-E_{kj} & & \end{pmatrix} \in \mathfrak{m}_1 \cap \mathfrak{m}_2.$$

Then, for each $H \in \mathfrak{a}$, $\ell = 1, 2$, we have

$$[H, S_{e_j+e_k}^{\,\ell}] = \langle e_j + e_k, H\rangle T_{e_j+e_k,2}^{\,\ell}, \quad [H, T_{e_j+e_k}^{\,\ell}] = -\langle e_j + e_k, H\rangle S_{e_j+e_k}^{\,\ell}.$$

For $1 \leq j < k \leq p$, set

$$S_{e_j-e_k}^{\,1} = \begin{pmatrix} & -(E_{jk}+E_{kj}) & & \\ E_{jk}+E_{kj} & & & \\ & & & E_{jk}+E_{kj} \\ & & -(E_{jk}+E_{kj}) & \end{pmatrix} \in \mathfrak{k}_1 \cap \mathfrak{k}_2,$$

$$T_{e_j-e_k}^{\,1} = \begin{pmatrix} & & & E_{jk}-E_{kj} \\ & & E_{jk}-E_{kj} & \\ & E_{jk}-E_{kj} & & \\ E_{jk}-E_{kj} & & & \end{pmatrix} \in \mathfrak{m}_1 \cap \mathfrak{m}_2,$$

$$S_{e_j-e_k}^{\,2} = \begin{pmatrix} E_{jk}-E_{kj} & & & \\ & E_{jk}-E_{kj} & & \\ & & E_{jk}-E_{kj} & \\ & & & E_{jk}-E_{kj} \end{pmatrix} \in \mathfrak{k}_1 \cap \mathfrak{k}_2,$$

$$T^{2}_{e_j-e_k}=\begin{pmatrix} & & E_{kj}+E_{jk} & \\ & & & -(E_{jk}+E_{kj}) \\ -(E_{kj}+E_{jk}) & & & \\ & E_{jk}+E_{kj} & & \end{pmatrix}\in\mathfrak{m}_1\cap\mathfrak{m}_2.$$

Then, for each $H\in\mathfrak{a}$, $\ell=1,2$, we have

$$[H,S^{\ell}_{e_j-e_k}]=\langle e_j-e_k,H\rangle T^{\ell}_{e_j-e_k,2},\quad [H,T^{\ell}_{e_j-e_k}]=-\langle e_j-e_k,H\rangle S^{\ell}_{e_j-e_k}.$$

Thus, we get the assertion. □

Since we have $[\mathfrak{a},(\mathfrak{k}_1\cap\mathfrak{m}_2)\oplus(\mathfrak{k}_2\cap\mathfrak{m}_1)]\subset(\mathfrak{k}_1\cap\mathfrak{m}_2)\oplus(\mathfrak{k}_2\cap\mathfrak{m}_1)$, we can define weights of $\mathfrak{a}$ acting on $(\mathfrak{k}_1\cap\mathfrak{m}_2)\oplus(\mathfrak{k}_2\cap\mathfrak{m}_1)$. Denote by W the set of non-zero weights. Define subspaces $V(\mathfrak{k}_1\cap\mathfrak{m}_2)$ and $V(\mathfrak{m}_1\cap\mathfrak{k}_2)$ of $\mathfrak{k}_1\cap\mathfrak{m}_2$ and $\mathfrak{m}_1\cap\mathfrak{k}_2$ respectively by

$$\begin{aligned} V(\mathfrak{k}_1\cap\mathfrak{m}_2)&=\{X\in\mathfrak{k}_1\cap\mathfrak{m}_2 \mid [\mathfrak{a},X]=\{0\}\},\\ V(\mathfrak{m}_1\cap\mathfrak{k}_2)&=\{X\in\mathfrak{m}_1\cap\mathfrak{k}_2 \mid [\mathfrak{a},X]=\{0\}\}.\end{aligned}$$

Then the weight space corresponding to weight 0 coincides with $V(\mathfrak{k}_1\cap\mathfrak{m}_2)\oplus V(\mathfrak{m}_1\cap\mathfrak{k}_2)$. The subspaces $V(\mathfrak{k}_1\cap\mathfrak{m}_2)$ and $V(\mathfrak{m}_1\cap\mathfrak{k}_2)$ are expressed as follows:

$$\begin{aligned} V(\mathfrak{k}_1\cap\mathfrak{m}_2)=&\sum_{1\le i<j\le q-p}\mathbb{R}\begin{pmatrix} 0 & & & \\ & 0 & & \\ & & E_{p+i,p+j}-E_{p+j,p+i} & \\ & & & E_{p+j,p+i}-E_{p+i,p+j} \end{pmatrix}\\ &+\sum_{1\le i<j\le q-p}\mathbb{R}\begin{pmatrix} 0 & & & \\ & 0 & & \\ & & & E_{p+i,p+j}-E_{p+j,p+i} \\ & & E_{p+i,p+j}-E_{p+j,p+i} & \end{pmatrix},\end{aligned}$$

$$V(\mathfrak{m}_1\cap\mathfrak{k}_2)=\sum_{i=1}^{p}\mathbb{R}\begin{pmatrix} & & E_{ii} & \\ & & & E_{ii} \\ -E_{ii} & & & \\ & -E_{ii} & & \end{pmatrix}+\sum_{i=1}^{p}\mathbb{R}\begin{pmatrix} & & & -E_{ii} \\ & & E_{ii} & \\ & -E_{ii} & & \\ E_{ii} & & & \end{pmatrix}.$$

Lemma 4.2. *The set of non-zero weights is given as*

$$\begin{aligned} W=&\{\pm e_i\pm e_j \text{ (with multiplicity 2)} \mid 1\le i<j\le p\}\\ &\bigcup\{\pm e_j \text{ (with multiplicity } 2(q-p)) \mid 1\le j\le p\}.\end{aligned}$$

Proof. For $1 \leq i \leq p$, set

$$X^1_{e_i \pm e_j} = \begin{pmatrix} E_{ij}-E_{ji} & & & \\ & E_{ji}-E_{ij} & & \\ & & \pm(E_{ji}-E_{ij}) & \\ & & & \pm(E_{ij}-E_{ji}) \end{pmatrix} \in \mathfrak{k}_1 \cap \mathfrak{m}_2,$$

$$Y^1_{e_i \pm e_j} = \begin{pmatrix} & & E_{ji} \mp E_{ij} & \\ & & & E_{ji} \mp E_{ij} \\ \pm E_{ji}-E_{ij} & & & \\ & \pm E_{ji}-E_{ij} & & \end{pmatrix} \in \mathfrak{k}_2 \cap \mathfrak{m}_1.$$

Then, for each $H \in \mathfrak{a}$, we have

$$[H, X^1_{e_i \pm e_j}] = \langle e_i \pm e_j, H\rangle Y^1_{e_i \pm e_j}, \quad [H, Y^1_{e_i \pm e_j}] = -\langle e_i \pm e_j, H\rangle X^1_{e_i \pm e_j}.$$

For $1 \leq i \leq p$, set

$$X^2_{e_i \pm e_j} = \begin{pmatrix} & E_{ij}-E_{ji} & & \\ E_{ij}-E_{ji} & & & \\ & & & \pm(E_{ij}-E_{ji}) \\ & & \pm(E_{ij}-E_{ji}) & \end{pmatrix} \in \mathfrak{k}_1 \cap \mathfrak{m}_2,$$

$$Y^2_{e_i \pm e_j} = \begin{pmatrix} & & & -E_{ji} \pm E_{ij} \\ & & E_{ji} \mp E_{ij} & \\ & -E_{ij} \pm E_{ji} & & \\ E_{ij} \mp E_{ji} & & & \end{pmatrix} \in \mathfrak{k}_2 \cap \mathfrak{m}_1.$$

Then, for each $H \in \mathfrak{a}$, we have

$$[H, X^2_{e_i \pm e_j}] = \langle e_i \pm e_j, H\rangle Y^2_{e_i \pm e_j}, \quad [H, Y^2_{e_i \pm e_j}] = -\langle e_i \pm e_j, H\rangle X^2_{e_i \pm e_j}.$$

For $1 \leq i \leq p$, $1 \leq j \leq q-p$, set

$$X^j_{i,1} = \begin{pmatrix} 0 & & & \\ & 0 & & \\ & & E_{i,p+j}-E_{p+j,i} & \\ & & & E_{p+j,i}-E_{i,p+j} \end{pmatrix} \in \mathfrak{k}_1 \cap \mathfrak{m}_2,$$

$$Y^j_{i,1} = \begin{pmatrix} & & E_{i,p+j} & \\ & & & E_{i,p+j} \\ -E_{p+j,i} & & & \\ & -E_{p+j,i} & & \end{pmatrix} \in \mathfrak{k}_2 \cap \mathfrak{m}_1,$$

$$X^j_{i,2} = \begin{pmatrix} 0 & & & \\ & 0 & & \\ & & & E_{i,p+j}-E_{p+j,i} \\ & & E_{i,p+j}-E_{p+j,i} & \end{pmatrix} \in \mathfrak{k}_1 \cap \mathfrak{m}_2,$$

$$Y_{i,2}^{j} = \begin{pmatrix} & & & E_{i,p+j} \\ & & -E_{i,p+j} & \\ & E_{p+j,i} & & \\ -E_{p+j,i} & & & \end{pmatrix} \in \mathfrak{k}_2 \cap \mathfrak{m}_1.$$

Then, for each $H \in \mathfrak{a}$, $\ell = 1, 2$, we have

$$[H, X_{i,\ell}^{j}] = \langle e_i, H \rangle Y_{i,\ell}^{|,j}, \quad [H, Y_{i,\ell}^{|,j}] = -\langle e_i, H \rangle X_{i,\ell}^{j}.$$

Thus, we get the assertion. □

If we set $\widetilde{\Sigma} := \Sigma \cup W$, then the triple $(\widetilde{\Sigma}, \Sigma, W)$ satisfies the axiom of symmetric triad. By Lemma 2.12 and Theorem 2.19 in [7], the symmetric triad $(\widetilde{\Sigma}, \Sigma, W)$ defines the simplex in $\mathfrak{a}$ given by

$$P_0 = \left\{ \frac{\pi}{|e_1|^2} \sum_{i=1}^{p} x_i e_i \;\middle|\; x_1 > \cdots > x_p > 0, x_1 + x_2 < \frac{1}{2} \right\}.$$

The set of vertices of P_0 is given by

$$\left\{0, H_1 = \frac{\pi}{2|e_1|^2} e_1 \right\} \bigcup \left\{ H_j = \frac{\pi}{4|e_1|^2} \sum_{i=1}^{j} e_i \;\middle|\; 2 \leq j \leq p \right\}.$$

A general theory of symmetric triad in [7] implies the following propositions. Concerning the problem (i) in Section 1, we have

Proposition 4.1. *The closure $\overline{P_0}$ of P_0 is identified with the orbit space $K_2 \backslash G / K_1$ by the following mapping:*

$$\overline{P_0} \to K_2 \backslash G / K_1; H \mapsto [(\exp H) K_1],$$

$[(\exp H) K_1]$ *is the K_2-orbit through* $(\exp H) K_1 \in G / K_1$.

Concerning the problem (ii) in Section 1, we have

Proposition 4.2. *Let H be in $\overline{P_0}$. When $p = q$, then we have the following.*

(1) *The orbit corresponding to H is totally geodesic if and only if H is one of $0, H_1$ and H_p.*
(2) *The orbit corresponding to H is austere which is not totally geodesic if and only if $H = H_j$ $(2 \leq j \leq p - 1)$.*

Proposition 4.3. *Let H be in $\overline{P_0}$. When $p < q$, then we have the following.*

(1) *The orbit corresponding to H is totally geodesic if and only if H is either 0 or H_1.*
(2) *The orbit corresponding to H is austere which is not totally geodesic if and only if H is either $\frac{1}{2} H_1$ or H_j $(1 \leq j \leq p)$.*

5. Future plans

Before discussing future plans, we note the following well-known facts.

The unitary group $U(n)$ of degree n naturally acts on the complex n-dimensional Euclidean space $\mathbb{C}^n$. This action induces an action on the Grassmann manifold $G_k^{\mathbb{R}}(\mathbb{C}^n)$ consisting of all real k-dimensional subspaces of $\mathbb{C}^n$. Since the map $G_k^{\mathbb{R}}(\mathbb{C}^n) \to G_{2n-k}^{\mathbb{R}}(\mathbb{C}^n)$ defined by $W \mapsto W^{\perp}$ is $U(n)$-equivariant, we may assume that $1 \leq k \leq n$ when we consider $U(n)$-action on $G_k^{\mathbb{R}}(\mathbb{C}^n)$. Denote by J the complex structure on $\mathbb{R}^{2n}$ induced from the identification on $\mathbb{R}^{2n}$ and $\mathbb{C}^n$. If we denote by $\langle\ ,\ \rangle$ the canonical inner product on $\mathbb{R}^{2n}$, then we have

$$\langle Jx, Jy\rangle = \langle x, y\rangle \qquad (x, y \in \mathbb{R}^{2n}).$$

The Kaehler form Ω on $(\mathbb{R}^{2n}, J)$ is a skew symmetric bilinear form on $\mathbb{R}^{2n}$ defined by $\Omega(x, y) = \langle x, Jy\rangle$. The absolute value $|\Omega(x, y)|$ of $\Omega(x, y)$ induces a function on $G_2^{\mathbb{R}}(\mathbb{C}^n)$. We denote it by $|\Omega|$. That is, $|\Omega(V)| = |\Omega(x, y)|$, where $\{x, y\}$ is an orthonormal basis of $V \in G_2^{\mathbb{R}}(\mathbb{C}^n)$. Thus, for a natural number m with $m \leq n$, the absolute value $|\Omega^m|$ of the $2m$-form Ω^{2m} can be regarded as a function on $G_{2m}^{\mathbb{R}}(\mathbb{C}^n)$. Then $|\Omega^m|$ is $U(n)$-invariant. Wirtinger's inequality says that $|\Omega^m(V)| \leq m!$ for each $V \in G_{2m}^{\mathbb{R}}(\mathbb{C}^n)$, and equality holds if and only if V is a complex subspace. The inequality follows immediately from the fact that for any $V \in G_{2m}^{\mathbb{R}}(\mathbb{C}^n)$, there exist $g \in U(n)$ and $\theta = (\theta_1, \ldots, \theta_m) \in [0, \pi/2]^m = [0, \pi/2] \times \cdots \times [0, \pi/2]$ satisfying $0 \leq \theta_1 \leq \cdots \leq \theta_m \leq \pi/2$ such that

$$gV = V(\theta) := \sum_{j=1}^{m} \mathrm{span}_{\mathbb{R}}\langle e_j, (\cos\theta_j)ie_j + (\sin\theta_j)e_{m+j}\rangle, \tag{11}$$

where $\{e_1, \ldots, e_n, e_{n+1} = ie_1, \ldots, e_{2n} = ie_n\}$ is the canonical orthonormal basis of $\mathbb{R}^{2n}$. Furthermore, $V(\theta)$ and $V(\varphi)$ are transformed with each other by the $U(n)$-action if and only if $\theta = \varphi$. Thus, the orbit space can be identified with

$$\begin{aligned}&\{V(\theta) \mid 0 \leq \theta_1 \leq \cdots \leq \theta_m \leq \pi/2\}\\ &\quad \cong \{\theta \in [0, \pi/2]^m \mid 0 \leq \theta_1 \leq \cdots \leq \theta_m \leq \pi/2\},\end{aligned}$$

and $V(\theta)$ is considered as a canonical form of the $U(n)$-action on $G_{2m}^{\mathbb{R}}(\mathbb{C}^n)$. Since the action of $U(n)$ on $G_{2m}^{\mathbb{R}}(\mathbb{C}^n)$ is a Hermann action, we can prove (11) by using (8) in Section 4. We can get Wirtinger's inequality immediately by using (11). Note that (11) is also obtained by applying the canonical form theory on skew symmetric bilinear form to the skew symmetric matrix

$$\begin{pmatrix} \Omega(u_1, u_1) & \cdots & \Omega(u_1, u_{2m}) \\ \vdots & & \vdots \\ \Omega(u_{2m}, u_1) & \cdots & \Omega(u_{2m}, u_{2m}) \end{pmatrix}$$

of degree $2m$, where $\{u_i\}_{1 \leq i \leq 2m}$ is an orthonormal basis of $V \in G^{\mathbb{R}}_{2m}(\mathbb{C}^n)$.

However, there is no known answer to the following Problem 1 in which $\mathbb{C}^n$ and $U(n)$ are replaced by $\mathbb{H}^n$ and $Sp(n) \times Sp(1)$, respectively.

Problem 1. Denote by $Sp(n)$ the compact symplectic group of degree n, and by $\mathbb{H}$ the field of quaternions. The group $Sp(n) \times Sp(1)$ naturally acts on $\mathbb{H}^n$. This action induces an action on the Grassmann manifold $G^{\mathbb{R}}_k(\mathbb{H}^n)$ consisting of all real k-dimensional subspaces of $\mathbb{H}^n$. In his paper [1], Berger found an $Sp(n) \times Sp(1)$-invariant 4-form Ω on $\mathbb{H}^n$ such that $|\Omega^m(V)| \leq m!$ for every $V \in G^{\mathbb{R}}_{4m}(\mathbb{H}^n)$, and the equality holds if and only if V is a quaternion subspace of $\mathbb{H}^n$. The proof requires $Sp(n) \times Sp(1)$-invariant integrals (see also [12] and [15]).

However it is unknown that the canonical forms of the action of $Sp(n) \times Sp(1)$ on $G^{\mathbb{R}}_k(\mathbb{H}^n)$ except the few cases mentioned below. Since the map $G^{\mathbb{R}}_k(\mathbb{H}^n) \to G^{\mathbb{R}}_{4n-k}(\mathbb{H}^n)$ defined by $W \mapsto W^{\perp}$ is $Sp(n) \times Sp(1)$-equivariant, we may assume that $1 \leq k \leq 2n$ when we consider $Sp(n) \times Sp(1)$-action on $G^{\mathbb{R}}_k(\mathbb{H}^n)$. When $k = 1$, then the action is transitive. When $k = 2$, then the action is of cohomogeneity one ([11]), and the orbit space can be identified with

$$\{V(\theta) := \mathrm{span}_{\mathbb{R}}\langle e_1, e_1 i \cos\theta + e_2 \sin\theta \rangle \mid 0 \leq \theta \leq \pi/2\} \subset G^{\mathbb{R}}_2(\mathbb{H}^n).$$

However, when $3 \leq k \leq 2n$, such a result is unknown. The compact connected Lie groups $U(2n)$, $O(4n)$, $Sp(n)$ and $Sp(n) \times Sp(1)$ transitively act on the hypersphere of $\mathbb{H}^n = \mathbb{C}^{2n} = \mathbb{R}^{4n}$ (see Section 7.13 in [2]). Comparing the dimensions of these Lie groups, $Sp(n) \times Sp(1)$ is much smaller than the other two. In fact, the dimensions of $Sp(n) \times Sp(1)$ ranges from 1/2 to 1/4 of those of other groups. This makes the problem difficult. Knowing canonical forms are considered the next step after Berger's discovery. 2-forms have canonical forms in the sense of linear algebra, but 4-forms have no known canonical forms in the same sense. This point makes the problem more difficult.

Problem 2. Symmetric triads with multiplicities were used effectively to study the orbit spaces of the Hermann actions and the properties of each orbit. In fact we explained in Section 4 that the symmetric triad with multiplicities defines a simplex P_0 such that the closure $\overline{P_0}$ of P_0 is a good representation of the orbit space of the Hermann action. Under this identification of $\overline{P_0}$ and the orbit space, we can see which orbit is totally geodesic, and which orbit is austere. Totally geodesic orbits correspond to vertices of the simplex, but the converse does not true in general. If we stratify the simplex into vertices, edges, faces, and so on, we can see that each stratum has a unique minimal orbit (see [7]). Based on the above fact, the element in $\bar{P}$ can be considered as a canonical form of the Hermann action. A minimal orbit which does not correspond to a vertex is unstable as a minimal submanifold.

However, there are some problems that cannot be determined simply by using symmetric triads with multiplicities. For example,

(1) studying the stability of the minimal orbits which correspond to the vertices of the orbit space,
(2) when does the austere orbit becomes weakly reflective?

Acknowledgments

This research was partially supported by Grant-in-Aid for Scientific Research(C), 22K03285.

References

[1] M. Berger, Du cote de chez Pu, *Ann. Sci. Ecole Norm. Sup.* **5** (1972), 1–41.
[2] A. L. Besse, *Einstein manifolds*, Springer, (1987).
[3] K. Enoyoshi, Principal curvatures of homogeneous hypersurfaces in a Grassmann manifold $\widetilde{Gr}_3(\mathrm{Im}(\mathbb{O}))$ by the G_2-action, *Tokyo J. Math.* **42** (2019) 571–584.
[4] O. Goertsches and G. Thorbergsson, On the geometry of orbits of Hermann actions, *Geom. Dedicata* **129** (2007), 101–118.
[5] E. Heintze, R. S. Palais, C. Terng and G. Thorbergsson, Hyperpolar actions on symmetric spaces, in *Geometry, topology and physics*, 214–245, Conf. Proc. Lecture Notes Geom. Topology, IV, Int. Press, Cambridge, MA, (1995).
[6] R. Harvey and H. B. Lawson, Jr., Caribrated geometries, *Acta Math.* **148** (1982), 47–157.
[7] O. Ikawa, The geometry of symmetric triads and orbit spaces of Hermann actions, *J. Math. Soc. Japan* **63** (2011), 79–136.
[8] O. Ikawa, σ-actions and symmetric triads, *Tohoku Math. J.* **70** (2018), 547–565.
[9] O. Ikawa and K. Mashimo, Geometric properties of orbits of σ-action induced from the triality automorphism, in preparation.

[10] O. Ikawa, T. Sakai and H. Tasaki, Weakly reflective submanifolds and austere submanifolds, *J. Math. Soc. Japan* **61** (2009) 437–481.

[11] A. Kollross, A classification of hyperpolar and cohomogeneity one actions, *Trans. Amer. Math. Soc.* **354** (2002), 571–612.

[12] V. Y. Kraines, Topology of quaternionic manifolds, *Trans. Amer. Math. Soc.* **122** (1966), 357–367.

[13] T. Matsuki, Double coset decompositions of reductive Lie groups arising from two involutions, *J. Algebra* **197** (1997), 49–91.

[14] S. Ohno, Geometric properties of orbits of Hermann actions, *Tokyo J. Math.* **46** (2023), 63–91, DOI: 10.3836/tjm/1502179367

[15] H. Tasaki, Certain minimal or homologically volume minimizing submanifolds in compact symmetric spaces, *Tsukuba J. Math.* **9** (1985), 117–131, DOI: 10.21099/tkbjm/1496160196

[16] L. Verhóczi, Special cohomogeneity one isometric actions on irreducible symmetric spaces of type I and II, *Beit. Alg. Geom.* **44** (2003), 57–74.

Received December 18, 2023
Revised April 9, 2024

Modern Approaches to Differential Geometry
and its Related Fields 183 – 194

AN OVERVIEW ON *-RICCI SOLITONS

George KAIMAKAMIS

Hellenic Army Academy,
Leoforos Varis–Koropiou, GR-16673 Athens, Greece
E-mail: gmiamis@gmail.com

Konstantina PANAGIOTIDOU

Ministry of Education, Religious Affairs and Sports, Greece
E-mail: konpanagiotidou@gmail.com

In [30], Tachibana introduced the notion of *-Ricci tensor on almost Hermitian manifolds. Motivated by the previous work, Hamada [18] defined *-Ricci tensors of real hypersurfaces in non-flat complex space forms. Using *-Ricci tensors, in [20], the authors introduced the notion of *-Ricci solitons of real hypersurfaces in non-flat complex space forms. This short note aims to present the progress in the study of *-Ricci soliton.

Keywords: *-Ricci tensor; *-Ricci soliton.

1. Introduction

The *Ricci tensor*, S, of a Riemannian manifold M is a tensor field of type (1,1) and is given by

$$g(SX, Y) = \text{trace}\{Z \mapsto R(Z, X)Y\}.$$

for all tangent vectors $X, Y \in T_pM$ at an arbitrary point $p \in M$. A Riemannian manifold (M, g) is called *Einstein*, if its Ricci tensor S satisfies the relation

$$S = \lambda g,$$

where λ is a constant.

Let M be a real hypersurface in a non-flat complex space form. A non-flat complex space is either a complex projective space $\mathbb{C}P^n$, or a complex hyperbolic space $\mathbb{C}H^n$ and is denoted by $M_n(c)$, with c being the constant holomorphic sectional curvature. The constant c is positive in the case of $\mathbb{C}P^n$, and is negative in the case of $\mathbb{C}H^n$. In [2], Cecil and Ryan proved that there is no Einstein real hypersurfaces in $\mathbb{C}P^n$. The non-existence of Einstein real hypersurfaces in $\mathbb{C}H^n$ was proved by Montiel in [24].

The non-flat complex space form $M_n(c)$ is equipped with a Kähler structure (J, G), with J being the complex structure and G the Kähler metric. The Kähler structure induces on the real hypersurface M an almost contact metric structure (ϕ, ξ, η, g). This structure consists of

- the vector field ξ called *structure vector field* or *Reeb vector field*,
- the tensor field of type (1,1) ϕ called *structure tensor*,
- the 1-form η,
- the metric g induced by G.

These members satisfy the following relations

$$\phi^2 X = -X + \eta(X)\xi, \quad \eta \circ \phi = 0, \quad \phi\xi = 0, \quad \eta(\xi) = 1,$$
$$g(\phi X, \phi Y) = g(X, Y) - \eta(X)\eta(Y),$$

with X, Y any vector fields on M. The tangent bundle TM of the real hypersurface M is decomposed as

$$TM = \mathbb{R}\xi \oplus \mathcal{D},$$

where $\mathcal{D} = \{X \in TM : \eta(X) = 0\}$. This $\mathcal{D}$ is called *holomorphic distribution.*

Definition 1.1. A real hypersurface M is called *Hopf hypersurface* if ξ is an eigenvector of the shape operator A of M.

In [18], Hamada defined the **-Ricci tensor* S^* of a real hypersurface in a non-flat complex space form by

$$g(S^* X, Y) = \frac{1}{2}\left(\text{trace}\{\varphi \circ R(X, \varphi Y)\}\right), \quad \text{for } X,\, Y \in TM.$$

The **-scalar curvature* is denoted by ρ^* and is defined to be the trace of S^*. If ρ^* is constant and a real hypersurface M in a non-flat complex space form satisfies the relation

$$g(S^* X, Y) = \frac{\rho^*}{2(n-1)} g(X, Y), \text{ for all } X,\, Y \text{ orthogonal to } \xi,$$

then M is called **-Einstein.* The study of *-Einstein real hypersurfaces in non-flat complex space forms was initiated by Hamada. In [18], he classified *-Einstein Hopf hypersurfaces in $M_n(c)$. Furthermore, he proved that ruled real hypersurfaces in non-flat complex space forms are *-Einstein. Finally, in [19], the classification of *-Einstein real hypersurfaces in non-flat complex space forms was completed.

In [20], the authors stated the following definition:

Definition 1.2. A Riemannian metric g on M is called **-Ricci soliton*, if

$$\frac{1}{2}\mathcal{L}_V g + Ric^* - \lambda g = 0, \tag{1}$$

with a constant λ. Here, the tensor Ric^* is defined by $Ric^*(X,Y) = g(S^*X,Y)$, and $\mathcal{L}_V$ denotes the Lie derivative along the vector field V. The vector field V is called *potential vector field* of the *-Ricci soliton.

In the same work [20], the following theorems were proved.

Theorem 1.1. *There do not exist real hypersurfaces in $\mathbb{C}P^n$, $n \geq 2$, admitting a *-Ricci soliton, with potential vector being the structure vector field ξ.*

Theorem 1.2. *Let M be a real hypersurface in $\mathbb{C}H^n$, $n \geq 2$, of constant holomorphic sectional curvature $c = -4$. Suppose that M admits a *-Ricci soliton whose potential vector is the structure vector field ξ. Then it is locally congruent to a geodesic hypersphere whose radius r satisfies $2n = \coth^2(r)$,*

2. Literature Review

In this section, we present results concerning the *-Ricci soliton of real hypersurfaces and some classes of almost contact manifolds such as Sasakian manifolds, Kenmotsu manifolds and so on.

2.1. *Real hypersurfaces*

Answering problems posed by the authors in [20], Chen studied in [3] real hypersurfaces in non-flat complex space forms admitting *-Ricci soliton with potential vector field V which is either belonging to the holomorphic distribution $\mathcal{D}$ or is a principal curvature vector field.

For the case that the potential vector field is principal, he studied real hypersurfaces in a 2-dimensional non-flat complex space form and showed the following.

Theorem 2.1. *Let M be a hypersurface of a non-flat complex space form $\widetilde{M}^2(c)$ with a *-Ricci soliton whose potential vector field is a principal* *curvature vector field associated with the function χ of non-null principal curvature. If χ is constant along ξ and $A\xi$, then M is a homogeneous Hopf hypersurface having constant principal curvatures. That is,*

1) *in the case of $\mathbb{C}P^2$, M is an open part of either a tube around the complex quadric or a geodesic hypersphere:*
2) *in the case of $\mathbb{C}H^2$, M is an open part of one of the following; a horosphere, a geodesic hypersphere, a tube around a totally geodesic $\mathbb{C}H^1$, and a tube around a totally geodesic real hyperbolic space $\mathbb{R}H^2$.*

Theorem 2.2. *Let M be a hypersurface of a complex projective space $\mathbb{C}P^2$. If it admits a *-Ricci soliton whose potential vector field V is formed by principal curvature vectors associated with null principal curvature at each point, then M is an open part of a tube around the complex quadric.*

For the case that the potential vector field belongs to $\mathcal{D}$, he showed

Theorem 2.3. *Let M be a hypersurface of a complex projective space $\mathbb{C}P^2$ with a *-Ricci soliton whose potential vector field $V \in \mathcal{D}$. If the principal curvatures of M are constant along ξ and $A\xi$, then it is locally congruent to a geodesic hypersphere. Moreover, if $g(A\xi, \xi) = 0$, then V is Killing.*

Corresponding to the decomposition $TM = \mathbb{R}\xi \oplus \mathcal{D}$, we denote as $A\xi = \alpha\xi + U$ with a smooth function α on M and a vector field U satisfying $U \in \mathcal{D}$. He studied a real hypersurface M in non-flat complex space form $M_n(c)$, $n \geq 2$, admitting *-Ricci soliton whose potential vector field V coincides with U. He proved that in this case V is the zero vector field and hence M is a *-Ricci flat Hopf hypersurface. Using this, he showed the following.

Theorem 2.4. *Let M be a real hypersurface in a non-flat complex space form $M_n(c), n \geq 2$.*

(1) *In the case of $\mathbb{C}P^n$, there are no real hypersurfaces admitting a *-Ricci soliton whose potential vector field coincides with U, the component of $A\xi$ tangent to the holomorphic distribution $\mathcal{D}$,*
(2) *In the case of $\mathbb{C}H^n$, if M admits a *-Ricci soliton whose potential vector field coincides with U, then it is locally congruent to a geodesic hypersphere.*

In [4], Chen also studied real hypersurfaces in complex two-plane Grassmannians $G_2(\mathbb{C}^{m+2}), m \geq 3$, in terms of *-Ricci tensor. He proved the following theorems.

Theorem 2.5. *Let M be a Hopf hypersurface in $G_2(\mathbb{C}^{m+2}), m \geq 3$ admitting a *-Ricci soliton. Then its *-Ricci tensor is commuting (i.e. $\phi \circ S^* = S^* \circ \phi$), and it is locally congruent to an open part of a tube around a totally geodesic quaternionic projective space $\mathbb{H}P^n$ in $G_2(\mathbb{C}^{m+2})$, where $m = 2n$.*

Theorem 2.6. *There do not exist real hypersurfaces of $G_2(\mathbb{C}^{m+2}), m \geq 3$, admitting a *-Ricci soliton whose potential vector field coincides with the Reeb vector field ξ.*

Finally, in [5], Chen defined the *-Ricci tensor on real hypersurfaces in a complex quadric Q^m, and studied real hypersurfaces in terms of this tensor. It was proved that there are no Hopf hypersurfaces in $Q^m, m \geq 3$, whose *-Ricci tensor either satisfies the commuting condition $\phi \circ S^* = S^* \circ \phi$, or is parallel. He also showed the following result corresponding to Theorems 2.4 and 2.6.

Theorem 2.7. *If a real hypersurface in a complex quadric $Q^m, m \geq 4$, admits a *-Ricci soliton whose potential vector field coincides with ξ, then its *-Ricci tensor is anti-commuting (i.e. $\phi \circ S^* + S^* \circ \phi = 0$), m is even and the hypersurface is an open part of a tube around a totally geodesic $\mathbb{C}P^{m/2}$.*

2.2. *Classes of almost contact metric manifolds*

The notion of *-Ricci soliton is also studied in the case of manifolds equipped with special structure.

In [16], A. Ghosh and D. S. Patra proved that if a Sasakian metric is a *-Ricci soliton, then it is either positive Sasakian, or null-Sasakian. Additionally, they showed that if a complete Sasakian metric is an almost gradient *-Ricci soliton, then it is positive-Sasakian and is isometric to a unit sphere $S^{(2n+1)}$. Finally, they classified nontrivial *-Ricci solitons on a non-Sasakian (κ, μ)-contact manifold.

In [21], P. Majhi et al. studied *-Ricci solitons on Sasakian manifolds and proved that if a Sasakian 3-manifold M admits *-Ricci soliton, then it has constant scalar curvature, and the potential vector field V is Killing. Furthermore, the vector field V is an infinitesimal automorphism of the contact metric structure on M.

In [23], K. Mandal and S. Makhal studied *-Ricci solitons and *-gradient Ricci solitons on three-dimensional normal almost contact metric manifolds (cf. [1]). In particular, they proved that certain non-cosymplectic normal almost contact metric manifolds which admit a *-Ricci soliton are *-Ricci flat.

In [10], S. Day and S. Roy studied *-η-Ricci soliton on Sasakian manifolds and proved some curvature properties on Sasakian manifold admitting *-η-Ricci soliton. They obtained some significant results on *-η-Ricci solitons in Sasakian manifolds satisfying certain conditions.

In [15], S. Dwivedi and D. S. Patra introduced the notion of almost *-Ricci-Bourguignon soliton (or almost *-RB soliton) and found its geometric characterizations on Sasakian manifolds. They presented several interesting sufficient conditions under which a gradient almost *-RB soliton or an almost *-RB soliton on a Sasakian manifold is isometric to an Euclidean sphere or *-Ricci flat (in particular, trivial or *-Einstein). They proved the following theorems.

Theorem 2.8. *A complete Sasakian manifold admitting a gradient almost *-RB soliton structure is *-Ricci flat (trivial or *-Einstein), compact positive-Sasakian and isometric to the unit sphere.*

Theorem 2.9. *A complete Sasakian manifold of dimension greater than 3 which admits an almost *-Ricci-Bourguignon soliton structure (with non-constant λ) is isometric to the unit sphere.*

In [26], V. Rovenski and D. S. Patra, presented a geometric classification of Sasakian manifolds that admit an almost *-Ricci soliton structure (g, V, λ). They proved the following:

Theorem 2.10. *If a complete Sasakian manifold $M^{2n+1}(\phi, \xi, \eta, g)$ of dimension greater than three has an almost *-Ricci soliton structure (g, V, λ) with non-constant λ, then it is the unit sphere S^{2n+1}.*

In [31], V. Venkatesha et al. considered *-Ricci soliton in the framework of Kenmotsu manifolds. They proved that if (N, g) is a Kenmotsu manifold and g is a *-Ricci soliton, then the constant λ is zero. In the three-dimensional case, if N admits a *-Ricci soliton, they showed that N is of constant sectional curvature -1. Then they proved that if N admits a *-Ricci soliton whose potential vector field is collinear with the characteristic vector field ξ, then N is Einstein and the soliton vector field is equal to ξ. Finally, they concluded that if g is a gradient almost *-Ricci soliton, then either N is Einstein or the potential vector field is collinear with the characteristic vector field on an open set of N.

In [32], Y. Wang studied *-Ricci solitons on Kenmotsu 3-manifolds. He stated the following theorems.

Theorem 2.11. *If the triple (g, V, λ) on a Kenmotsu 3-manifold represents a *-Ricci soliton, then the manifold is of constant sectional curvature -1, hence is a 3-dimensional hyperbolic space $H^3(-1)$, and $\lambda = 0$.*

Theorem 2.12. *If the triple (g, V, λ) on a Kenmotsu 3-manifold represents a gradient *-Ricci soliton, then V coincides with the Reeb vector field.*

In 2019, X. Dai et al. [7] studied a non-Kenmotsu $(\kappa, \mu)'$ -almost Kenmotsu manifold of dimension $2n+1$ whose metric g is a *-Ricci soliton and proved the following:

Theorem 2.13. *If a triple (g, V, λ) on a non-Kenmotsu $(\kappa, \mu)'$-almost Kenmotsu manifold is a *-Ricci soliton, then the soliton constant $\lambda = 0$, and either the manifold is locally isometric to the product $H^{n+1} \times \mathbb{R}^n$ and the potential vector field is Killing, or the potential vector field is strict infinitesimal contact transformation.*

In [11], S. Dey et al. initiated the study of *-η-Ricci solitons within the framework of Kenmotsu manifolds as a characterization of Einstein metrics. They proved that a Kenmotsu metric as a *-η-Ricci soliton is an Einstein metric if the soliton vector field is contact. Furthermore, the authors gave a characterization of the Kenmotsu manifold or the nature of the potential vector field when the manifold satisfies gradient almost *-η-Ricci soliton.

In 2020, D. Dey et al. [8] characterized three-dimensional trans-Sasakian manifolds of type (α, β) admitting *-Ricci solitons or *-gradient Ricci solitons. They proved that under certain restrictions on the smooth functions α and β, a trans-Sasakian 3-manifold of type (α, β) admitting a *-Ricci soliton reduces to (α, β)-Kenmotsu manifold. They also show that a trans-Sasakian 3-manifold of type (α, β) admitting a *-gradient Ricci soliton which satisfies $\phi \operatorname{grad} \alpha = \operatorname{grad} \beta$ is either flat or *-Einstein or it becomes (α, β)-Kenmotsu manifold.

In [13], S. Day and Y.J. Suh gave some characterizations by considering almost *-η-Ricci-Bourguignon soliton as a Kenmotsu metric. They proved that an almost Kenmotsu manifold such that ξ belongs to $(\kappa, 2)$-nullity distribution, where $\kappa < -1$, acknowledges a *-η-Ricci-Bourguignon soliton satisfying certain relation, then the manifold is Ricci-flat and is locally isometric to $H^{n+1}(-4) \times \mathbb{R}^n$. Moreover, they showed that if the metric admits a gradient almost *-η-Ricci-Bourguignon soliton and ξ leaves the scalar curvature r invariant on a Kenmotsu manifold, then the manifold is an η-Einstein.

In [27], S. Roy and A. Bhattacharyya studied *-conformal Yamabe soliton, whose potential vector field is torse-forming on Kenmotsu manifold. They showed the nature of the soliton and found expression of the scalar curvature when the manifold admitting *-conformal Yamabe soliton on Kenmotsu manifold.

In 2023, Yoldaş et al. [33] investigated Kenmotsu manifold satisfying certain curvature conditions endowed with *-η-Ricci solitons. First they proved that for the projectively flat, the concircularly flat and the M-projectively flat Kenmotsu manifolds the *-Ricci tensor is symmetric and these flat manifolds are ϕ-Einstein, whereas the *-Ricci tensor of the concircularly flat Kenmotsu manifold is symmetric and the manifold is weakly ϕ-Einstein.

In [9], D. Dey and P. Majhi introduced the notion of *-gradient ρ-Einstein soliton on a class of almost Kenmotsu manifolds. If a $(2n+1)$-dimensional (κ,μ)-almost Kenmotsu manifold M admits *-gradient ρ-Einstein soliton with Einstein potential f, then the following holds:

- the manifold M is locally isometric to $H^{n+1}(-4)\times\mathbb{R}^n$;
- the manifold M is *-Ricci flat;
- the Einstein potential f is harmonic or satisfies a physical Poisson's equation.

In [29], S. Sarkar et al. studied conformal Ricci solitons and *-conformal Ricci solitons within the framework of paracontact geometry. They proved that if an η-Einstein para-Kenmotsu manifold admits conformal Ricci soliton and *-conformal Ricci soliton, then it is Einstein. Furthermore, they showed that a 3-dimensional para-cosymplectic manifold is Ricci flat if the manifold satisfies conformal Ricci soliton where the soliton vector field is conformal. They presented the following theorems.

Theorem 2.14. *Let M be a $(2n+1)$-dimensional η-Einstein para-Kenmotsu manifold where $n>1$. If the metric of the manifold represents a conformal Ricci soliton, then the manifold is Einstein.*

Theorem 2.15. *Let $M^{2n+1}(\phi,\xi,\eta,g)$ be an η-Einstein para-Kenmotsu manifold. If g represents a *-conformal Ricci soliton, then the manifold is Einstein with constant scalar curvature $-2n(2n+1)$.*

In 2022, S. Dey and N. B. Turki [12] studied the *-η-Ricci soliton and gradient almost *-η-Ricci soliton within the framework of para-Kenmotsu manifolds as a characterization of Einstein metrics. They proved the following.

Theorem 2.16. *Let $M^{2n+1}(\phi, \xi, \eta, g)$ be a para-Kenmotsu manifold. If the metric g represents a *-η-Ricci soliton and if the soliton vector field V is contact, then V is a strictly infinitesimal contact transformation and the manifold is Einstein.*

Theorem 2.17. *Let $M^{2n+1}(\phi, \xi, \eta, g)$ be a para-Kenmotsu manifold. If the metric g represents a gradient almost *-η-Ricci soliton, then either M is Einstein or there exists an open set where the potential vector field V is pointwise collinear with the characteristic vector field ξ.*

In [6], X. Dai proved a non-existence result for *-Ricci solitons on non-cosymplectic (κ, μ)-almost cosymplectic manifolds

In 2021, R. Ma et al. [22] investigated the *-Ricci operators on trans-Sasakian three-manifolds. They presented and proved conditions at which *-Ricci tensor on trans-Sasakian three-manifolds is symmetric and under which the *-Ricci operators are Reeb flow invariant.

In [17], Haseeb et al. studied the properties of ϵ-Kenmotsu manifolds if its metrics are *-η-Ricci-Yamabe solitons. They proved that an ϵ-Kenmotsu manifold endowed with a *-η-Ricci-Yamabe soliton is η-Einstein. They also gave necessary conditions for an ϵ-Kenmotsu manifold, whose metric is a *-η-Ricci-Yamabe soliton, to be an Einstein manifold.

In [25], Naik et al. considered a coKähler 3-manifold (M, g) admitting a *-Ricci soliton (g, V) and proved that either M is locally flat, or V is an infinitesimal contact transformation. Then they studied non-coKähler (κ, μ)-almost coKähler metrics as critical point equation metrics and proved that such a g cannot be a solution of critical point equation with non-trivial function f. Finally, they proved that a (κ, μ)-almost coKähler manifold (M, g) is coKähler if either M admits a divergence free Cotton tensor or the metric g is Bach flat.

In [14], S. Dey et al. studied certain types of metrics. They focused on *-conformal Ricci-Yamabe solitons, whose potential vector field is torse-forming on Kenmotsu manifold. Then they established certain conditions for solitons to be expanding, shrinking or steady and computed the scalar curvature when the manifold admits a *-conformal Ricci–Yamabe soliton on a Kenmotsu manifold. Finally, they presented an example of *-conformal RYS on three-dimensional Kenmotsu manifolds.

In [28], the authors studied the geometric composition of an n-dimensional Sasakian manifold with *-Ricci-Bourguignon soliton under Zamkovoy connection. Also, they presented an expression of the Laplace equation from the soliton equation when the potential vector field V of the soliton is of gradient type, in terms of Zamkovoy connection.

3. Conclusion

The study of manifolds, hypersurfaces etc. is of high interest to geometers and physicists due to their applications in geometry, physics, and relativity. At the same time, the study of Ricci solitons on Riemannian manifolds is an issue, which is of great importance in differential geometry and in physics as well. In these studies, researchers used fundamental tools such as the Riemannian curvature tensors, the shape operators, and the Ricci tensors, to understand the differential geometric properties of manifolds, hypersurfaces etc. Over the last few years, several new concepts have been introduced in the literature with the help of those tools, in order to describe other structures (for example different kind of manifolds). One of these concepts is the *-Ricci tensor and was extensively used in various papers to describe several kinds of manifolds.

References

[1] D.E. Blair, *Contact Manifolds in Riemannian Geometry*, Lecture Note in Math., Vol. 509, Springer-Verlag, Berlin-New York, 1976.

[2] T. E. Cecil and P. J. Ryan, Focal sets and real hypersurfaces in complex projective space, *Trans. Amer. Math. Soc.* **269** (1982), 481–499.

[3] X. Chen, Real hypersurfaces with *-Ricci solitons of non-flat complex space forms, *Tokyo J. Math.* **41**(2) (2018), 433–451.

[4] X. Chen, Real hypersurfaces with *-Ricci tensors in complex two-plane Grassmannians, *Bull. Korean Math. Soc.* **54**(3) (2017), 975–992.

[5] X. Chen, Real hypersurfaces of complex quadric in terms of *-Ricci tensor, *Tokyo J. Math.* **41**(2) (2018), 587–601.

[6] X. Dai, Non-existence of *-Ricci solitons on (κ,μ)-almost cosymplectic manifolds, *J. Geom.* **110**(2) (2019), paper No. 30, 7pp.

[7] X. Dai, Y. Zhao and U.C. De, *-Ricci soliton on $(\kappa,\mu)'$-almost Kenmotsu manifolds, *Open Math.* **17**(1) (2019) 874–882.

[8] D. Dey and P. Majhi, *-Ricci solitons and *-gradient Ricci solitons on 3-dimensional trans-Sasakian manifolds, *Commun. Korean Math. Soc.* **35**(2) (2020), 625–637.

[9] D. Dey and P. Majhi, On a class of almost Kenmotsu manifolds admitting an Einstein like structure, *Sao Paulo J. Math. Sci.* **15**(1) (2021), 335–343.

[10] S. Dey and S. Roy, *-η-Ricci soliton within the framework of Sasakian manifold, *J. Dyn. Syst. Geom. Theor.* **18**(2) (2020), 163–181.

[11] S. Dey, S. Sarkar and A. Bhattacharyya, *-η-Ricci soliton and contact geometry, *Ricerche Mat.* (2021), https://doi.org/10.1007/s11587-021-00667-0.

[12] S. Dey and N.B. Turki, *-η-Ricci soliton and gradient almost *-η-Ricci soliton within the framework of para-Kenmotsu manifolds, *Front. Phys.* **10** (2022), https://doi.org/10.3389/fphy.2022.809405.

[13] S. Dey and Y. J. Suh, Geometry of almost contact metrics as an almost *-η-Ricci-Bourguignon solitons, *Rev. Math. Phys.* **35**(7) (2023), Paper No. 2350012, 20pp.

[14] S. Dey, S. Roy and F. Karaca, Geometry of almost contact metrics as a *-conformal Ricci? Yamabe solitons and related results, *Int. J. Geom. Meth. Mod. Phys.* **20** (9) (2023), Paper No. 2350146, 21pp.

[15] S. Dwivedi and D. S. Patra, Some results on almost *-Ricci-Bourguignon solitons, *J. Geom. Phys.* **178** (2022), Paper No. 104519, 11pp.

[16] A. Ghosh and D. S. Patra, *-Ricci soliton within the framework of Sasakian and (κ, μ)-contact manifold, *Int. J. Geom. Methods Mod. Phys.* **15**(7) (2018), Paper No. 1850120, 21pp.

[17] A. Haseeb, M. Bilal, S. Chaubey and M. Khan, Geometry of Indefinite Kenmotsu Manifolds as *-η-Ricci-Yamabe Solitons, *Axioms* **11** (9) (2022), 461, 13pp, https://doi.org/10.3390/axioms11090461.

[18] T. Hamada, Real hypersurfaces of complex space forms in terms of Ricci *-tensor, *Tokyo J. Math.* **25** (2002), 473–483.

[19] T. A. Ivey and P. J. Ryan, The *-Ricci tensor for hypersurfaces in $\mathbb{C}P^n$ and $\mathbb{C}H^n$, *Tokyo J. Math.* **34** (2) (2011), 445–471.

[20] G. Kaimakamis and K. Panagiotidou, *-Ricci solitons of real hypersurfaces in non-flat complex space forms, *J. Geom. Phys.* **86** (3) (2014), 408–413.

[21] P. Majhi, U. C. De and Y. J. Suh, *-Ricci solitons and Sasakian 3-manifolds, *Publ. Math. Debrecen* **93**(1-2) (2018), 241–252.

[22] R. Ma and D. Pei, Reeb flow invariant *-Ricci operators on trans-Sasakian three-manifolds, *Math. Slovaca* **71**(3) (2021), 749–756.

[23] K. Mandal and S. Makhal, *-Ricci solitons on three-dimensional normal almost contact metric manifolds, *Lobachevskii J. Math.* **40** (2) (2019), 189–194.

[24] S. Montiel, Real hypersurfaces of a complex hyperbolic space, *J. Math. Soc. Japan* **37** (3) (1985), 515–535.

[25] D. M. Naik, V. Venkatesha and H. A. Kumara, Certain types of metrics on almost coKahler manifolds, *Ann. Math. Qué.* **47** (2)(2023), 331–347.

[26] V. Rovenski and D. S. Patra, Characteristics of Sasakian manifolds admitting almost *-Ricci solitons, *Fractal and Fractional* **7** (2)(2023), 156, 11pp, https://doi.org/10.3390/fractalfract7020156.

[27] S. Roy and A. Bhattacharyya, A Kenmotsu metric as a *-conformal Yamabe soliton with torse forming potential vector field, *Acta. Math. Sin.-English Ser.* **37** (2021), 1896–1908.

[28] S. Roy and S. Dey, Study of Sasakian manifolds admitting *-Ricci-Bourguignon solitons with Zamkovoy connection, *Ann. Univ. Ferrara* (2023), https://doi.org/10.1007/s11565-023-00467-4.

[29] S. Sarkar, S. Dey and X. Chen, Certain results of conformal and *-conformal Ricci soliton on para-cosymplectic and para-Kenmotsu manifolds, *Filomat* **35** (15) (2021), 5001–5015.

[30] S. Tachibana, On almost-analytic vectors in almost Kahlerian manifolds, *Tohoku Math. J.* **11**(1959), 247–265.

[31] N. Venkatesha, D. Mallesha and K.H. Aruna, *-Ricci solitons and gradient almost *-Ricci solitons on Kenmotsu manifolds, *Math. Slovaca* **69**(6) (2019), 1447–1458.

[32] Y. Wang, Contact 3-manifolds and *-Ricci soliton, *Kodai Math. J.* **43** (2) (2020), 256–267.

[33] H. Yoldaş, A. Haseeb and F. Mofarreh, Certain curvature conditions on Kenmotsu manifolds and *-η-Ricci solitons, *Axioms* **12** (2)(2023), 140, 14pp, https://doi.org/10.3390/axioms12020140.

Received February 19, 2024
Revised April 29, 2024

Modern Approaches to Differential Geometry
and its Related Fields 195 – 210

INVARIANT DUALLY FLAT STRUCTURES ON q-EXPONENTIAL FAMILIES

Hiroshi MATSUZOE

Department of Mathematics and Mathematical Science, Graduate School of Engineering, Nagoya Institute of Technology, Nagoya, Aichi 466-8555, Japan
E-mail: matsuzoe@nitech.ac.jp

In information geometry, the geometry of statistical models is primarily constructed based on the criteria of invariance under Markov embeddings and dual flatness for a pair of affine connections. For exponential families, invariance and dual flatness hold simultaneously. On the other hand, for non-exponential families, these criteria do not hold simultaneously. In this paper, after reviewing the invariant geometry of statistical models including Markov embeddings, an invariant and dually flat structure is constructed for several q-exponential families.

Keywords: Information geometry; statistical manifold; dually flat space; Markov embedding.

1. Introduction

Invariance under Markov embeddings and dual flatness for a pair of affine connections are important criteria for constructing geometric structures for statistical models. The mathematical structure of a statistical model should not depend on the choice of reference measures or coordinate systems of the sample space. These requirements for a statistical model are related to invariance. On the other hand, the convexity of the likelihood function and the duality of parameters are important in statistical inference. Dual flatness is required to discuss these properties.

Exponential families exhibit a dually flat structure which is invariant. In contrast, non-exponential families generally do not possess this property. For this reason, dual flatness has mainly been discussed in the context of information geometry (cf. [10]).

In this paper, we briefly review Chentsov's characterization of the invariant geometry on finite sample spaces under Markov embeddings [4]. Then, we consider the relationship between Chentsov's theorems [4] and Amari-Nagaoka's dual information geometry [1].

At the later part of this paper, we construct a statistical structure that is both invariant and dually flat for several q-exponential families.

2. Statistical manifolds

We assume that all the objects are smooth throughout this paper. In this section, we briefly review the geometry of statistical manifolds. For further details, see [1, Chapter 3] and [9].

Let (M, h) be a semi-Riemannian manifold. In this paper, we may assume that M is simply connected since we only discuss the local geometry of the manifold. Roman capital letters X, Y and Z indicate vector fields on M, unless otherwise stated.

Let ∇ be an affine connection on M. We define the *dual connection* (which is also called the *conjugate connection*) of ∇ with respect to h by

$$Xh(Y,Z) = h(\nabla_X Y, Z) + h(Y, \nabla^*_X Z).$$

From a direct calculation, we have that

$$(\nabla_X h)(Y,Z) - (\nabla_Y h)(X,Z) = h(T^*(X,Y) - T(Y,X), Z), \tag{1}$$

where T and T^* are the torsion tensor fields of ∇ and ∇^*, respectively. We say that ∇ is *torsion-free*, or ∇ has *zero torsion* if the torsion tensor field T vanishes everywhere on M. The following proposition can be shown using Equation (1).

Proposition 2.1 (cf. [8]). *If two of the following conditions are assumed, then the remaining holds:*

(1) ∇ *is torsion-free;*
(2) ∇^* *is torsion-free;*
(3) $C := \nabla h$ *is totally symmetric;*
(4) $\nabla^{(0)} := (\nabla + \nabla^*)/2$ *is the Levi-Civita connection with respect to* h.

We say that the pair (h, ∇) is a *statistical structure* on M if ∇ is torsion-free and if ∇h is totally symmetric $(0,3)$-tensor field on M. In this case, the triplet (M, h, ∇) is called a *statistical manifold* and $C := \nabla h$ is called the *cubic form* of (M, h, ∇). For the dual connection ∇^* of ∇ with respect to h, if (h, ∇) is a statistical structure, then (h, ∇^*) is also a statistical structure on M, and it is called the *dual statistical structure* of (h, ∇).

A statistical manifold can be constructed by giving a totally symmetric $(0,3)$-tensor field to a semi-Riemannian manifold.

Proposition 2.2 (cf. [7]). *Let (M, h) be a semi-Riemannian manifold, and C a totally symmetric $(0,3)$-tensor field on M. Denote by $\nabla^{(0)}$ the Levi-Civita connection with respect to h. For $\alpha \in \mathbb{R}$, we define $\nabla^{(\alpha)}$ by*

$$h(\nabla^{(\alpha)}_X Y, Z) := h(\nabla^{(0)}_X Y, Z) - \frac{\alpha}{2} C(X, Y, Z). \tag{2}$$

Then $\nabla^{(\alpha)}$ is a torsion-free affine connection on M, and $\nabla^{(\alpha)} h$ coincides with αC. In particular, the pair $(h, \nabla^{(\alpha)})$ is a statistical structure on M.

In information geometry, the equivalence of metrics and cubic forms under constant multiples is often important.

Proposition 2.3. *Let h be a semi-Riemannian metric on M and C a totally symmetric $(0,3)$-tensor field on M. For non-zero constants γ and χ, set $\tilde{h} = \gamma h$ and $\tilde{C} = \chi C$. For $\alpha' \in \mathbb{R}$, denote by $\widetilde{\nabla}^{(\alpha')}$ the α'-connection derived from $(\tilde{h}, \tilde{C})$. Then we have that*

$$h(\widetilde{\nabla}^{(\alpha')}_X Y, Z) := h(\nabla^{(0)}_X Y, Z) - \frac{\alpha' \chi}{2\gamma} C(X, Y, Z).$$

Therefore, if $\alpha' = \alpha\gamma/\chi$, then $\widetilde{\nabla}^{(\alpha')}$ coincides with $\nabla^{(\alpha)}$.

Let us now review the flatness of affine connections. Fix a local coordinate system $(x^1, \ldots, x^n)$ on $U \subset M$. For an affine connection ∇ on M, the *curvature tensor field* R of ∇ is defined by

$$R(X,Y)Z := \nabla_X \nabla_Y Z - \nabla_Y \nabla_X Z - \nabla_{[X,Y]} Z.$$

Its coordinate expression is

$$R\left(\frac{\partial}{\partial x^k}, \frac{\partial}{\partial x^l}\right) \frac{\partial}{\partial x^j} = \sum_{i=1}^{n} R^i_{jkl} \frac{\partial}{\partial x^i},$$

$$R^i_{jk\ell} = \left(\frac{\partial \Gamma^i_{\ell j}}{\partial x^k} - \frac{\partial \Gamma^i_{kj}}{\partial x^\ell}\right) + \sum_{m=1}^{n} \left(\Gamma^m_{\ell j} \Gamma^i_{km} - \Gamma^m_{kj} \Gamma^i_{\ell m}\right), \tag{3}$$

where Γ^k_{ij} $(i, j, k = 1. \ldots, n)$ are the *Christoffel symbols (of the second kind)* or the *connection coefficients* for the affine connection ∇ relative to the local coordinate system $x = (x^1, \ldots, x^n) \in U \subset M$.

We say that ∇ is *curvature-free*, or ∇ has *zero curvature* if the curvature tensor field R vanishes everywhere on M. An affine connection ∇ is said to be *flat* if ∇ is both curvature-free and torsion-free.

Let M be a manifold equipped with an affine connection ∇ on M. We say that a local coordinate system $(x^1, \ldots, x^n)$ on $U \subset M$ is *affine* if all the Christoffel symbols Γ^k_{ij} $(i, j, k = 1, \ldots, n)$ of ∇ vanish everywhere on

U. The following proposition shows the equivalence of the existence of a flat affine connection and an affine coordinate system.

Proposition 2.4. *Let M be a manifold equipped with an affine connection ∇ on M. The connection ∇ is flat if and only if there exists an affine coordinate system on a neighborhood for each point in M.*

Proof. This is a well-known result, but it is important in information geometry. We check an elementary proof.

For a given local coordinate system $(x^1, \ldots, x^n)$ on U, we show that there exists a local coordinate system $(y^1, \ldots, y^n)$ on V such that

$$\overline{\Gamma}_{ab}^{c} = \sum_{i,j,k=1}^{n} \frac{\partial x^i}{\partial y^a}\frac{\partial x^j}{\partial y^b}\frac{\partial y^c}{\partial x^k}\Gamma_{ij}^{k} + \sum_{i=1}^{n} \frac{\partial^2 x^i}{\partial y^a \partial y^b}\frac{\partial y^c}{\partial x^i} = 0 \tag{4}$$

on $U \cap V$. From the identity for the Jacobian matrix

$$\sum_{a=1}^{n} \frac{\partial x^i}{\partial y^a}\frac{\partial y^a}{\partial x^j} = \delta_j^i,$$

we have that

$$\sum_{a,b=1}^{n} \frac{\partial^2 x^i}{\partial y^a \partial y^b}\frac{\partial y^a}{\partial x^j}\frac{\partial y^b}{\partial x^k} + \sum_{a=1}^{n} \frac{\partial x^i}{\partial y^a}\frac{\partial^2 y^a}{\partial x^k \partial x^j} = 0.$$

From above equations, for $i, j, c = 1, \ldots, n$, we have that

$$\begin{aligned}
&\sum_{a,b=1}^{n} \frac{\partial y^a}{\partial x^i}\frac{\partial y^b}{\partial x^j}\left(\sum_{k,\ell,m=1}^{n} \frac{\partial x^\ell}{\partial y^a}\frac{\partial x^m}{\partial y^b}\frac{\partial y^c}{\partial x^k}\Gamma_{\ell m}^{k} + \sum_{k=1}^{n} \frac{\partial^2 x^k}{\partial y^a \partial y^b}\frac{\partial y^c}{\partial x^k}\right) \\
&= \sum_{k=1}^{n} \frac{\partial y^c}{\partial x^k}\Gamma_{ij}^{k} + \sum_{a,b=1}^{n}\sum_{k=1}^{n} \frac{\partial y^a}{\partial x^i}\frac{\partial y^b}{\partial x^j}\frac{\partial^2 x^k}{\partial y^a \partial y^b}\frac{\partial y^c}{\partial x^k} \\
&= \sum_{k=1}^{n} \frac{\partial y^c}{\partial x^k}\Gamma_{ij}^{k} - \sum_{a=1}^{n}\sum_{k=1}^{n} \frac{\partial^2 y^a}{\partial x^i \partial x^j}\frac{\partial x^k}{\partial y^a}\frac{\partial y^c}{\partial x^k} \\
&= \sum_{k=1}^{n} \frac{\partial y^c}{\partial x^k}\Gamma_{ij}^{k} - \frac{\partial^2 y^c}{\partial x^i \partial x^j}.
\end{aligned}$$

Therefore, $\overline{\Gamma}_{ab}^{c} = 0$ if and only if

$$\sum_{k=1}^{n} \frac{\partial y^c}{\partial x^k}\Gamma_{ij}^{k} = \frac{\partial^2 y^c}{\partial x^i \partial x^j}.$$

The system of partial differential equations is written as

$$\frac{\partial y^c}{\partial x^i} = u_i^c, \tag{5}$$

$$\frac{\partial u_j^c}{\partial x^i} = \sum_{k=1}^n \Gamma_{ij}^k u_k^c. \tag{6}$$

The integrability condition for the simultaneous partial differential equations (5) is given by

$$\frac{\partial^2 y^c}{\partial x^i \partial x^j} = \frac{\partial^2 y^c}{\partial x^j \partial x^i}.$$

This condition holds if $T = 0$. In fact, we have that

$$\frac{\partial^2 y^c}{\partial x^i \partial x^j} = \sum_{k=1}^n \Gamma_{ij}^k \frac{\partial y^c}{\partial x^k} = \sum_{k=1}^n \Gamma_{ji}^k \frac{\partial y^c}{\partial x^k} = \frac{\partial^2 y^c}{\partial x^j \partial x^i}.$$

Similarly, the integrability condition for (6) is given by

$$\frac{\partial^2 u_k^c}{\partial x^i \partial x^j} = \frac{\partial^2 u_k^c}{\partial x^j \partial x^i}.$$

This conditions holds if $R = 0$. In fact, from Equation (6), using Equation (3), we have that

$$\begin{aligned}\frac{\partial^2 u_k^c}{\partial x^i \partial x^j} &= \frac{\partial}{\partial x^i}\left(\sum_{\ell=1}^n \Gamma_{jk}^\ell u_\ell^c\right) = \sum_{\ell=1}^n \left(\frac{\partial}{\partial x^i}\Gamma_{jk}^\ell + \sum_{m=1}^n \Gamma_{jk}^m \Gamma_{im}^\ell\right) u_\ell^c \\ &= \sum_{\ell=1}^n \left(\frac{\partial}{\partial x^j}\Gamma_{ik}^\ell + \sum_{m=1}^n \Gamma_{ik}^m \Gamma_{jm}^\ell\right) u_\ell^c = \frac{\partial}{\partial x^j}\left(\sum_{\ell=1}^n \Gamma_{ik}^\ell u_\ell^c\right) \\ &= \frac{\partial^2 u_k^c}{\partial x^j \partial x^i}.\end{aligned}$$

Conversely, if all the connection coefficients of ∇ vanish everywhere, it immediately follows that the connection ∇ is flat. □

For the pair of dual affine connections ∇ and ∇^*, we have that

$$h(R(X,Y)Z,V) + h(Z, R^*(X,Y)V) = 0.$$

This implies that, on a statistical manifold (M, h, ∇), the connection ∇ is flat if and only if ∇^* is flat.

Definition 2.1. Let (M, h, ∇) be a statistical manifold. Denote by ∇^* the dual connection of ∇ with respect to h. We say that the triplet (h, ∇, ∇^*) is a *dually flat structure* on M if ∇ is flat. In this case, the quadruplet (M, h, ∇, ∇^*) is called a *dually flat space*.

For a dually flat space (M, h, ∇, ∇^*), we suppose that $\theta = (\theta^1, \dots, \theta^n)$ is a ∇-affine coordinate system on $\Theta \subset M$. Then there exists a ∇^*-affine coordinate system $\eta = (\eta_1, \dots, \eta_n)$ on $H \subset M$, $(\Theta \cap H \neq \phi)$ such that

$$h\left(\frac{\partial}{\partial\theta^i}, \frac{\partial}{\partial\eta_j}\right) = \delta_i^j$$

on $\Theta \cap H$. We call η the *dual coordinate system* of θ with respect to h.

3. Statistical models

In this section, we review geometry of statistical models.

Let $(\mathcal{X}, \mathcal{F}, P)$ be a probability space, and Ξ an open subset of $\mathbb{R}^n$. We say that S is a *statistical model* if S is a set of probability density functions on $\mathcal{X}$ with parameter $\xi = (\xi^1, \dots, \xi^n) \in \Xi$ such that

$$S := \left\{ p(\cdot;\xi) \,\middle|\, p(x;\xi) > 0 \text{ (for all } x \in \mathcal{X} \text{ and } \xi \in \Xi), \int_{\mathcal{X}} p(x;\xi)dx = 1 \right\}.$$

In this paper, we regard S as a manifold with a coordinate system $\xi = (\xi^1, \dots, \xi^n)$. For further details, see [1, Chapter 2]. In particular, we assume that differentials and integrals are interchangeable:

$$\int_{\mathcal{X}} \left(\frac{\partial}{\partial\xi^i} p(x;\xi)\right) dx = \frac{\partial}{\partial\xi^i} \int_{\mathcal{X}} p(x;\xi)dx = \frac{\partial}{\partial\xi^i} 1 = 0.$$

From the assumption that Ξ is an open subset of $\mathbb{R}^n$, we find that $\xi = (\xi^1, \dots, \xi^n)$ is a global coordinate system on S. Therefore, we often identify a point $p(\cdot;\xi) \in S$ by its coordinates ξ.

Let us define a statistical structure on S. Denote by $p_\xi = p(x;\xi)$ and $\ell_\xi = \log p(x;\xi)$ for simplicity. We define the *Fisher metric* g^F on S by

$$\begin{aligned} g_\xi^F(X,Y) = g_{p(\cdot;\xi)}^F(X,Y) &:= \int_{\mathcal{X}} (X \log p(x;\xi))\,(Y \log p(x;\xi))\, p(x;\xi)\, dx \\ &= \mathrm{E}_\xi[(X\ell_\xi)\,(Y\ell_\xi)], \end{aligned}$$

where $\mathrm{E}_\xi[*]$ is the expectation with respect to $p(\cdot;\xi)$. By assuming that g^F is positive definite and all components are finite, we regard g^F as a Riemannian metric on S.

We define a totally symmetric $(0,3)$-tensor field C^{AC} on S by

$$C_\xi^{AC}(X,Y,Z) = C_{p(\cdot;\xi)}^{AC}(X,Y,Z) := \mathrm{E}_\xi[(X\ell_\xi)\,(Y\ell_\xi)\,(Z\ell_\xi)].$$

The tensor field C^{AC} is called the *cubic form* or the *Amari-Chentsov tensor field* on S. For a fixed $\alpha \in \mathbb{R}$, we define the *α-connection* $\nabla^{(\alpha)}$ on S by

$$g^F(\nabla_X^{(\alpha)}Y, Z) := g^F(\nabla_X^{(0)}Y, Z) - \frac{\alpha}{2} C^{AC}(X,Y,Z).$$

From Proposition 2.2, the connection $\nabla^{(\alpha)}$ is torsion-free and $\nabla^{(\alpha)} g^F$ is totally symmetric.

Definition 3.1. Let S be a statistical model, g^F the Fisher metric on S, and $\nabla^{(\alpha)}$ an α-connection on S. We call the pair $(g^F, \nabla^{(\alpha)})$ an *invariant statistical structure* on S.

We will consider why this pair is called invariant later (Theorem 4.1).

Let S_e be a statistical model, and Θ an open subset of $\mathbb{R}^n$. We say that $S_e = \{p(\cdot;\theta) \mid \theta \in \Theta\}$ is an *exponential family* if each probability density function $p(\cdot;\theta) \in S_e$ is given by

$$p(x;\theta) = \exp\left[z(x) + \sum_{i=1}^{n} \theta^i F_i(x) - \psi(\theta)\right], \tag{7}$$

where $z, F_1, \ldots, F_n$ are functions on the sample space $\mathcal{X}$, ψ is a function on the parameter space Θ, and $\theta = (\theta^1, \ldots, \theta^n)$ is a local coordinate system of S_e. We call θ a *natural coordinate system*.

Proposition 3.1 ([1], Chapter 3). *Suppose that $S_e = \{p(\cdot,\theta) \mid \theta \in \Theta\}$ is an exponential family with the form (7). Then the following holds.*

(1) *± 1-connections $\nabla^{(1)}$ and $\nabla^{(-1)}$ are flat.*
(2) *$\theta = (\theta^1, \ldots, \theta^n)$ is a $\nabla^{(1)}$-affine coordinate system on S_e.*
(3) *Set $\eta_i := \mathrm{E}_\theta[F_i]$. Then $\eta = (\eta_1, \ldots, \eta_n)$ is a $\nabla^{(-1)}$-affine coordinate system on S_e and the dual coordinate system of θ with respect to g^F.*
(4) *ψ is the Hessian potential of g^F and C^{AC} with respect to θ, that is,*

$$g^F_{ij}(\theta) = \frac{\partial^2}{\partial\theta^i \partial\theta^j}\psi(\theta), \qquad C^{AC}_{ijk}(\theta) = \frac{\partial^3}{\partial\theta^i \partial\theta^j \partial\theta^k}\psi(\theta).$$

The connection $\nabla^{(e)} := \nabla^{(1)}$ is called the *exponential connection* and $\nabla^{(m)} := \nabla^{(-1)}$ is called the *mixture connection*. The connections $\nabla^{(e)}$ and $\nabla^{(m)}$ are mutually dual with respect to g^F. The $\nabla^{(-1)}$-affine coordinate system η is called the *expectation coordinate system* on S.

Example 3.1. (Gaussian densities) Suppose that the sample space $\mathcal{X} = \mathbb{R}$ and the parameter space $\Xi = \{(\mu,\sigma)\} | -\infty < \mu < \infty,\ 0 < \sigma < \infty\}$. The set of all the Gaussian density functions on $\mathbb{R}$ is an exponential family:

$$S = \{p(\cdot;\mu,\sigma) \mid (\mu,\sigma) \in \Xi\},$$

where

$$p(x;\mu,\sigma) = \frac{1}{\sqrt{2\pi}\sigma} \exp\left[-\frac{(x-\mu)^2}{2\sigma^2}\right].$$

In fact, by setting

$$\theta^1 = \frac{\mu}{\sigma^2}, \quad \theta^2 = -\frac{1}{2\sigma^2}, \tag{8}$$

$$z(x) = 0, \quad F_1(x) = x, \quad F_2(x) = x^2,$$

$$\psi(\theta) = \frac{\mu^2}{2\sigma^2} + \log(\sqrt{2\pi}\sigma) \; = \; -\frac{(\theta^1)^2}{4\theta^2} + \frac{1}{2}\log\left(-\frac{\pi}{\theta^2}\right), \tag{9}$$

we have that

$$p(x; \mu, \sigma) = \exp\left[\theta^1 x + \theta^2 x^2 - \psi(\theta)\right].$$

Hence the set of all the Gaussian density functions is an exponential family.

The parameter $\theta = (\theta^1, \theta^2)$ defined by (8) is a $\nabla^{(1)}$-affine coordinate system, and the Fisher metric of S with respect to θ coordinate has the following components:

$$g^F(\theta) = -\frac{1}{\theta^2}\begin{pmatrix} 1 & -\theta^1/\theta^2 \\ -\theta^1/\theta^2 & ((\theta^1)^2 - \theta^2)/(\theta^2)^2 \end{pmatrix}. \quad \square$$

Example 3.2. (probability simplex) For $n \in \mathbb{N}$, set $\Omega_n := \{0, 1, \ldots, n\}$ and

$$H_n := \left\{ \eta = (\eta_1, \ldots, \eta_n) \;\middle|\; \eta_i > 0 \; (i = 1, \ldots, n), \; \sum_{j=1}^{n} \eta_j < 1 \right\}.$$

Obviously, H_n is an open subset of $\mathbb{R}^n$. The statistical model for the set of all probability mass functions on Ω_n is given by

$$S_n := \{p(\cdot; \eta) \mid \eta \in H_n\}, \tag{10}$$

where

$$p(0; \eta) = 1 - \sum_{j=1}^{n} \eta_j, \quad p(i; \eta) = \eta_i \quad (i = 1, \ldots, n).$$

We call S_n the *probability simplex* on Ω_n. By setting $\eta_0 = p(0; \eta)$ and

$$\theta^i = \log p(i; \eta) - \log p(0; \eta) = \log\frac{\eta_i}{\eta_0} \quad (i = 1, \ldots, n), \tag{11}$$

$$z(\omega) = 0, \; F_i(\omega) = \delta_i(\omega) \quad (\omega \in \Omega_n),$$

$$\psi(\theta) = \log p(0; \eta) = \log\left(1 + \sum_{j=1}^{n} e^{\theta^j}\right),$$

we have that

$$\log p(\omega;\theta) = \sum_{j=1}^{n} \theta^j F_j(\omega) - \log\left(1 + \sum_{j=1}^{n} e^{\theta^j}\right).$$

Hence the probability simplex is an exponential family.

The parameter $\theta = (\theta^1, \ldots, \theta^n)$ defined by (11) is a $\nabla^{(1)}$-affine coordinate system. The Fisher metric g^F with respect to the θ coordinate has the following components:

$$g^F_{ij}(\theta) = \begin{cases} \dfrac{e^{\theta^i}\left(1 + \sum_{j=1}^{n} e^{\theta^j} - e^{\theta^i}\right)}{\left(1 + \sum_{j=1}^{n} e^{\theta^j}\right)^2} = \eta_i(1-\eta_i) & (i = j) \\[2ex] -\dfrac{e^{\theta^i+\theta^j}}{\left(1 + \sum_{j=1}^{n} e^{\theta^j}\right)^2} = -\eta_i\eta_j & (i \neq j) \end{cases}.$$

The cubic form C^{AC} has the following components:

$$C^{AC}_{iii}(\theta) = \frac{e^{\theta^i}\left(1+\sum_{\ell=1}^{n} e^{\theta^\ell} - e^{\theta^i}\right)\left(1+\sum_{\ell=1}^{n} e^{\theta^\ell} - 2e^{\theta^j}\right)}{\left(1 + \sum_{\ell=1}^{n} e^{\theta^\ell}\right)^3} = \eta_i(1-\eta_i)(1-2\eta_i),$$

$$C^{AC}_{iij}(\theta) = C^{AC}_{iji}(\theta) = C^{AC}_{jii}(\theta) = -\frac{e^{\theta^i+\theta^j}\left(1 + \sum_{\ell=1}^{n} e^{\theta^\ell} - 2e^{\theta^1}\right)}{\left(1 + \sum_{\ell=1}^{n} e^{\theta^\ell}\right)^3}$$
$$= -\eta_i(1-\eta_i)\eta_j \qquad (i \neq j),$$

$$C^{AC}_{ijk}(\theta) = \frac{2e^{\theta^i+\theta^j+\theta^k}}{\left(1 + \sum_{\ell=1}^{n} e^{\theta^\ell}\right)^3} = 2\eta_i\eta_j\eta_k \qquad (i \neq j \neq k \neq i).$$

Since θ is a $\nabla^{(1)}$-affine coordinate system, from (2), the connection coefficients of $\nabla^{(\alpha)}$ have the following components:

$$\Gamma^{(\alpha)i}_{ii}(\theta) = \frac{1-\alpha}{2} \cdot \frac{\left(1 + \sum_{\ell=1}^{n} e^{\theta^\ell} - 2e^{\theta^j}\right)}{1 + \sum_{\ell=1}^{n} e^{\theta^\ell}} = \frac{1-\alpha}{2}(1-2\eta_i), \tag{12}$$

$$\Gamma^{(\alpha)i}_{ij}(\theta) = \Gamma^{(\alpha)i}_{ji}(\theta) = -\frac{1-\alpha}{2} \cdot \frac{e^{\theta^i}}{1 + \sum_{\ell=1}^{n} e^{\theta^\ell}} = -\frac{1-\alpha}{2}\eta_j \quad (i \neq j), \tag{13}$$

$$\Gamma^{(\alpha)i}_{jk}(\theta) = 0 \qquad (i \neq j,\ i \neq k). \tag{14}$$

□

4. Markov embedding and invariant statistical structure

In this section, we consider the invariant statistical structure on finite sample spaces. For the extension of Chentsov's theorem to continuous sample spaces, see [2], for example.

From the viewpoint of statistics, a statistical structure on S_n should be invariant under the change of reference measures on the sample space Ω_n. Let us formulate such an invariant property for S_n.

Definition 4.1. (Markov embedding) Let $n, \ell \in \mathbb{N}$ $(n \leq \ell)$, and let $Q_{(i)} = (Q^0_{(i)}, Q^1_{(i)}, \ldots, Q^\ell_{(i)})$ $(i = 0, 1, \ldots, n)$ be a probability mass function on Ω_ℓ.

(1) A family $\{Q_{(0)}, Q_{(1)}, \ldots, Q_{(n)}\}$ of probability mess functions on Ω_ℓ is a *Markov partition* if

$$\Omega_\ell = \bigsqcup_{i=0}^{n} \mathrm{supp}(Q_{(i)})$$

holds, where $\mathrm{supp}(Q_{(i)})$ is the support of $Q_{(i)}$.

(2) A map $f_{n,\ell} : S_n \to S_\ell$ is a *Markov embedding* if, for arbitrary point p in S_n, there exists a Markov partition $\{Q_{(0)}, Q_{(1)}, \ldots, Q_{(n)}\}$ such that

$$f_{n,\ell}(p) = \sum_{i=0}^{n} p(i) Q_{(i)}.$$

We define invariance of tensor fields and affine connections as follows.

Definition 4.2. (Markov invariance) For $n \in \mathbb{N}$, let $g^{[n]}$ be a symmetric $(0,2)$-tensor field on S_n, $C^{[n]}$ a symmetric $(0,3)$-tensor field on S_n, and $\nabla^{[n]}$ an affine connection on S_n. Suppose that $f_{n,\ell}$ is a Markov embedding of S_n to S_ℓ. Denote by $f_* = (f_{n,\ell})_*$ the differential of $f_{n,\ell}$, for simplicity. We say that series $\{g^{[n]}\}_{n\in\mathbb{N}}$, $\{C^{[n]}\}_{n\in\mathbb{N}}$ and $\{\nabla^{[n]}\}_{n\in\mathbb{N}}$ are *invariant* under Markov embeddings, respectively, if

$$g^{[n]}_p(X, Y) = g^{[\ell]}_{f_{n,\ell}(p)}(f_*X, f_*Y),$$

$$C^{[n]}_p(X, Y, Z) = C^{[\ell]}_{f_{n,\ell}(p)}(f_*X, f_*Y, f_*Z),$$

$$f_*\left(\nabla^{[n]}_X Y\right)_p = \left(\nabla^{[\ell]}_{f_*X} f_*Y\right)_{f_{n,\ell}(p)}$$

hold for all $n, \ell \in \mathbb{N}$ with $n \leq \ell$, and $p \in S_n$.

The uniqueness of invariant tensor fields under Markov embeddings was showed by Chentsov [4].

Theorem 4.1 ([4] Theorem 11.1). *For $n \in \mathbb{N}$, let $g^{[n]}$ be a symmetric $(0,2)$-tensor field on S_n, and $C^{[n]}$ a symmetric $(0,3)$-tensor field on S_n. Suppose that $\{g^{[n]}\}_{n\in\mathbb{N}}$ and $\{C^{[n]}\}_{n\in\mathbb{N}}$ are invariant under Markov embeddings. Then $g^{[n]}$ and $C^{[n]}$ coincide with the Fisher metric g^F and the cubic form C^{AC} on S_n, respectively, up to constant multiplications.*

For the proof of this theorem, see [3]. See also [6], which contains a survey of Chentsov's theorem.

In information geometry, invariance of tensor fields is often discussed. In fact, a family of invariant affine connections is obtained from g^F and C^{AC}. However, Chentsov also showed the invariance of affine connections on S_n independently of the invariance of tensor fields.

Theorem 4.2 ([4] Theorem 12.2). *For $n \in \mathbb{N}$, let $\nabla^{[n]}$ be an affine connection on S_n. Suppose that the series $\{\nabla^{[n]}\}_{n\in\mathbb{N}}$ is invariant under Markov embeddings. Then $\nabla^{[n]}$ is described by formulas:*

$$\nabla^{[n]}_{\partial_i}\partial_i = \gamma(1-2\eta_i)\partial_i, \tag{15}$$

$$\nabla^{[n]}_{\partial_i}\partial_j = -\gamma(\eta_i\partial_j + \eta_j\partial_i) \quad (i \neq j), \tag{16}$$

where $\partial_i := \partial/\partial\theta^i, \eta_i := p(i)$, and $\gamma \in \mathbb{R}$.

For the proof of this theorem, see [6, Theorem 11].

We remark that the Chentsov's invariant connection $\nabla^{[n]}$ on S_n coincides with the α-connection $\nabla^{(\alpha)}$ with $\gamma = (1-\alpha)/2$. In fact, from formulas (12), (13) and (14), for distinct i, j, we have that

$$\nabla^{(\alpha)}_{\partial_i}\partial_i = \sum_{k=1}^{n}\Gamma^{(\alpha)k}_{ii}\partial_k = \Gamma^{(\alpha)i}_{ii}\partial_i = \frac{1-\alpha}{2}(1-2\eta_i)\partial_i = \nabla^{[n]}_{\partial_i}\partial_i,$$

$$\nabla^{(\alpha)}_{\partial_i}\partial_j = \sum_{k=1}^{n}\Gamma^{(\alpha)k}_{ij}\partial_k = \Gamma^{(\alpha)i}_{ij}\partial_i + \Gamma^{(\alpha)j}_{ij}\partial_j = -\frac{1-\alpha}{2}(\eta_j\partial_i + \eta_i\partial_j) = \nabla^{[n]}_{\partial_i}\partial_j.$$

5. Geometry for q-exponential families

In this section, we discuss the geometry of q-exponential families. A q-exponential family is a generalization of an ordinary exponential family. There have been many studies on the dual flatness of q-exponential families. Therefore, we focus our attention their invariant statistical structures.

Let us define the q-exponential and the q-logarithm. Suppose that q is a fixed positive real number. We define the *q-exponential function* by

$$\exp_q x := \begin{cases} [1+(1-q)x]^{\frac{1}{1-q}}, & q \neq 1 \quad (1+(1-q)x > 0), \\ \exp x, & q = 1 \end{cases}. \tag{17}$$

Taking the limit $q \to 1$, the q-exponential function recovers the ordinary exponential function. The inverse of the q-exponential function is the *q-logarithm function*, and it is given by

$$\log_q x := \begin{cases} \dfrac{x^{1-q}-1}{1-q}, & q \neq 1 \quad (x > 0), \\ \log x, & q = 1 \end{cases}.$$

In this section, we assume that the condition $1+(1-q)x > 0$ in (17) is always satisfied. Hence q-exponential and q-logarithm function are always inverse of each other.

Let Ξ an open subset in $\mathbb{R}^n$. For $q > 0$, we say that a statistical model $S_q = \{p(\cdot;\xi) \mid \xi \in \Xi\}$ on the sample space $\mathcal{X}$ is a *q-exponential family* if each probability density function $p(\cdot;\xi) \in S_q$ is given by

$$p(x;\theta) = \exp_q \left[z(x) + \sum_{i=1}^{n} \xi^i F_i(x) - \psi(\xi) \right] \tag{18}$$

where $z, F_1, \ldots, F_n$ are functions on the sample space $\mathcal{X}$, ψ is a function on the parameter space Ξ, and $\xi = (\xi^1, \ldots, \xi^n)$ is a local coordinate system of S_q. It is known that S_q admits a dually flat structure. (See [10] and the references therein.) In this case, ξ is an affine coordinate system of some flat affine connection.

Example 5.1. (probability simplex) The probability simplex S_n on Ω_n is a q-exponential family. In fact, from the definition of probability simplex (10), set

$$\begin{aligned} \xi^i &= \log_q p(i;\eta) - \log_q p(0;\eta) = \log_q \eta_i - \log_q \eta_0, \qquad (19) \\ \psi(\xi) &= -\log_q \eta_0. \end{aligned}$$

Then, we have that

$$\log_q p_q(\omega;\xi) = \frac{1}{1-q}\left\{p^{1-q}(\omega;\xi) - 1\right\} = \sum_{i=1}^{n} \xi^i \delta_i(x) - \psi(\xi).$$

This implies that the probability simplex S_n is a q-exponential family. The natural coordinate system ξ in Equation (19) is an affine coordinate system with respect to some affine connection, but the corresponding dually flat structure is not invariant [10, Theorems 2 and 3].

Since S_n is both an exponential family and a q-exponential family, the dually flat structure is given in Example 3.2. □

Example 5.2. (q-Gaussian densities) For $1 < q < 3$, the q-Gaussian density function on $\mathbb{R}$ is defined by

$$p(x;\mu,\sigma) = \frac{1}{Z_{q,\sigma}}\left[1 - \frac{1-q}{3-q}\frac{(x-\mu)^2}{\sigma^2}\right]^{\frac{1}{1-q}} \qquad ((\mu,\sigma) \in N_q),$$

where $N_q = \{(\mu,\sigma)\} | -\infty < \mu < \infty,\ 0 < \sigma < \infty\}$ is the parameter space, and $Z_{q,\sigma}$ is the normalization defined by

$$Z_{q,\sigma} = \frac{\sqrt{3-q}}{\sqrt{q-1}}\ \mathrm{Beta}\left(\frac{3-q}{2(q-1)}, \frac{1}{2}\right)\sigma.$$

The set of all the q-Gaussian density functions $S_q = \{p(\cdot;\mu,\sigma) \,|\, (\mu,\sigma) \in N_q\}$ is a q-exponential family. In fact, by setting

$$\begin{aligned}
&\xi^1 = \frac{2}{3-q}\, Z_{q,\sigma}^{q-1} \cdot \frac{\mu}{\sigma^2}, \qquad \xi^2 = -\frac{1}{3-q}\, Z_{q,\sigma}^{q-1} \cdot \frac{1}{\sigma^2},\\
&z(x) = 0, \qquad F_1(x) = x, \qquad F_2(x) = x^2,\\
&\psi(\xi) = -\frac{(\xi^1)^2}{4\xi^2} - \frac{Z_{q,\sigma}^{q-1} - 1}{1-q},
\end{aligned}$$

we have that

$$\log_q p(x;\mu,\sigma) = \xi^1 x + \xi^2 x^2 - \psi(\xi).$$

A q-Gaussian distribution coincides with a Student's t-distribution with degree of freedom $\nu = (3-q)/(q-1)$ when $1 < q < 2$. □

Let us consider the invariant statistical structure on S_q. The Fisher metric and the α-connection were given in [5].

Proposition 5.1 ([5]). *For $1 < q < 2$ and $\alpha \in \mathbb{R}$, let $S_q = \{p(\cdot;\mu,\sigma)\}$ be the set of all the q-Gaussian density functions. Then the Fisher metric g^F and the α-connection $\nabla^{(\alpha)}$ on S_q are given as follows:*

$$g^F_{ij}((\mu,\sigma)) = \frac{1}{q\sigma^2}\begin{pmatrix} 1 & 0 \\ 0 & 3-q \end{pmatrix}, \tag{20}$$

$$\Gamma^{(\alpha)\,2}_{11} = \frac{(2q-1)-(2-q)\alpha}{(3-q)(2q-1)\sigma},$$

$$\Gamma_{12}^{(\alpha)\,1} = \Gamma_{21}^{(\alpha)\,1} = -\frac{(2q-1)+(2-q)\alpha}{(2q-1)\sigma}, \tag{21}$$

$$\Gamma_{22}^{(\alpha)\,2} = -\frac{(2q-1)+2(2-q)\alpha}{(2q-1)\sigma}, \tag{22}$$

$$\Gamma_{11}^{(\alpha)\,1} = \Gamma_{12}^{(\alpha)\,2} = \Gamma_{21}^{(\alpha)\,2} = \Gamma_{22}^{(\alpha)\,1} = 0.$$

The α-scalar curvature of $(S_q, g^F, \nabla^{(\alpha)})$ is given by

$$K^{(\alpha)} = -\frac{q\{(2q-1)^2-(2-q)^2\alpha^2\}}{(3-q)(2q-1)^2}. \tag{23}$$

From Equation (23), we find that $(g^F, \nabla^{(\alpha)}, \nabla^{(-\alpha)})$ on S_q is a dually flat structure when $\alpha = (2q-1)/(2-q)$. In addition, $(g^F, \nabla^{(\alpha)})$ and $(g^F, \nabla^{(-\alpha)})$ are invariant statistical structures on S_q. We elucidate the invariant dually flat structure on S_q in more detail.

Theorem 5.1. *For $1 < q < 2$ and $\alpha \in \mathbb{R}$, let $S_q = \{p(\cdot;\mu,\sigma)\}$ be the set of all the q-Gaussian density functions, and $(g^F, \nabla^{(\alpha)})$ an invariant statistical structure on S_q. Set $A = (2q-1)/(2-q)$ and*

$$\theta^1 = \frac{2}{3-q}\cdot\frac{\mu}{\sigma^2}, \qquad \theta^2 = -\frac{1}{3-q}\cdot\frac{1}{\sigma^2}. \tag{24}$$

Then the following holds:

(1) *The Fisher metric g^F with respect to θ has the following components*

$$g_{ij}^F(\theta) = -\frac{3-q}{2}\cdot\frac{1}{2\theta^2}\begin{pmatrix} 1 & -\dfrac{\theta^1}{\theta^2} \\ -\dfrac{\theta^1}{\theta^2} & \dfrac{(\theta^1)^2-\theta^2}{(\theta^2)^2}\end{pmatrix}; \tag{25}$$

(2) *The triplet $(g, \nabla^{(A)}, \nabla^{(-A)})$ is an invariant dually flat structure on S_q and $\theta = (\theta^1, \theta^2)$ is a $\nabla^{(A)}$-affine coordinate system;*

(3) *The Hessian potential of g^F with respect to θ is given by*

$$\Psi(\theta) := -\frac{3-q}{2q}\left\{\frac{(\theta^1)^2}{4\theta^2} + \frac{1}{2}\log(-\theta^2)\right\}; \tag{26}$$

(4) *The cubic form C^{AC} of S_q is given by*

$$C_{ijk}^{AC}(\theta) = \frac{2-q}{2q-1}\cdot\frac{\partial^3}{\partial\theta^i\partial\theta^j\partial\theta^k}\Psi(\theta). \tag{27}$$

Proof. The representation of the Fisher metric (25) is obtained by the coordinate transformation $(\mu,\sigma) \mapsto (\theta^1,\theta^2)$ and (20). If $\theta = (\theta^1,\theta^2)$ is an affine coordinate system, we can obtain (25) by differentiating (26). Hence it is clear that Ψ is the potential of the Fisher metric g^F.

The fact that θ is an affine coordinate system is straightforward, but requires some calculation. When $\alpha = (2q-1)/(2-q)$ $(=: A)$, from (21) and (22), the non-zero connection coefficients are only $\Gamma^{(A)1}_{12}$, $\Gamma^{(A)1}_{21}$ and $\Gamma^{(A)2}_{22}$. Denote by $\overline{\Gamma}^{(A)k}_{ij}$ $(i,j,k = 1,2)$ the connection coefficients of $\nabla^{(A)}$ on S_q with respect to θ-coordinates. From (24), we have that

$$\frac{\partial\sigma}{\partial\theta^1} = \frac{\partial^2\sigma}{\partial\theta^1\partial\theta^1} = 0, \qquad \frac{\partial^2\mu}{\partial\theta^1\partial\theta^1} = 0.$$

Therefore, using Equation (4), we obtain

$$\begin{aligned}\overline{\Gamma}^{(A)1}_{11} &= \frac{\partial\mu}{\partial\theta^1}\frac{\partial\sigma}{\partial\theta^1}\frac{\partial\theta^1}{\partial\mu}\Gamma^{(A)1}_{12} + \frac{\partial\sigma}{\partial\theta^1}\frac{\partial\mu}{\partial\theta^1}\frac{\partial\theta^1}{\partial\mu}\Gamma^{(A)1}_{21} + \frac{\partial\sigma}{\partial\theta^1}\frac{\partial\mu}{\partial\theta^1}\frac{\partial\theta^1}{\partial\mu}\Gamma^{(A)1}_{21}\\ &\qquad + \frac{\partial^2\mu}{\partial\theta^1\partial\theta^1}\frac{\partial\theta^1}{\partial\mu} + \frac{\partial^2\sigma}{\partial\theta^1\partial\theta^1}\frac{\partial\theta^1}{\partial\sigma}\\ &= 0.\end{aligned}$$

Similarly, we can show that each of the other coefficients $\{\overline{\Gamma}^{(A)k}_{ij}\}$ always vanishes.

Since the parameter θ is a $\nabla^{(A)}$-affine coordinate system, Ψ is the potential of C^{AC} multiplied by A. Therefore, we have the formula (27). □

6. Concluding remarks

In this paper, we constructed an affine coordinate system different from that used in previous studies. As a consequence, from formulas (9) and (26), we find that the two invariant statistical structures for Gaussian and q-Gaussian densities are essentially the same due to the freedom of constant multiplication in tensor fields, as was stated in Proposition 2.3.

Using the affine coordinate system defined in (24), the q-Gaussian density can be written as follows:

$$p(x;\theta) = \frac{1}{Z_{q,\sigma}}\exp_q\left[\theta^1 x + \theta^2 x^2 - \frac{(\theta^1)^2}{4\theta^2}\right].$$

This representation is different from the original definition of q-exponential family (18). In particular, the normalization of probability densities is different. In non-exponential families, normalization of probability measures is a non-trivial problem, which affects properties of statistical models.

Acknowledgments

The author was partially supported by JSPS Grant-in-Aid for Scientific Research (KAKENHI) Grant Number JP19K03489 and JP23K03088.

References

[1] S. Amari and H. Nagaoka, *Methods of information geometry*, Transl. Math. Monogr., **191**, Amer. Math. Soc., Providence, RI, 2000. Translated from the 1993 Japanese original by Daishi Harada.

[2] N. Ay, J. Jost, H. V. Lê and L. Schwachhöfer, *Information geometry*, Ergeb. Math. Grenzgeb. (3), **64**. Results in Mathematics and Related Areas. 3rd Series. A Series of Modern Surveys in Mathematics, Springer, Cham, 2017.

[3] L. L. Campbell, An extended Čencov characterization of the information metric, *Proc. Amer. Math. Soc.* **98** (1986), 135–141.

[4] N. N. Čencov, *Statistical decision rules and optimal inference*, Transl. Math. Monogr. **53**, Amer. Math. Soc., Providence, RI, 1982. Originally published in Russian, Izdat. "Nauka", Moscow 1972.

[5] B. S. Cho and S. Y. Jung, A note on the geometric structure of the t-distributios, *Journal of the Korean Data & Information Science Society* **21** (2010), 575–580.

[6] A. Fujiwara, Hommage to Chentsov's theorem, *Inf. Geom.* **7** (2024), S79–S98.

[7] S. L. Lauritzen, Statistical manifolds, *in Differential geometry in statistical inference, IMS Lecture Notes Monogr. Ser.*, **10**, Institute of Mathematical Statistics, Hayward, CA, 1987, 163–216.

[8] H. Matsuzoe, Geometry of statistical manifolds and its generalization, in *Topics in contemporary differential geometry, complex analysis and mathematical physics*, World Sci. Publ., Hackensack, NJ, 2007, 244–251.

[9] H. Matsuzoe, Statistical manifolds and affine differential geometry, in *Probabilistic approach to geometry*, *Adv. Stud. Pure Math.* **57**, Math. Soc. Japan, Tokyo, 2010, 303–321.

[10] H. Matsuzoe and M. Henmi, Hessian structures and divergence functions on deformed exponential families, in *Geometric theory of information*, Signals Commun. Technol. Springer, Cham, 2014, 57–80.

Received June 5, 2024
Revised June 13, 2024

AFTERWORD

On the last day of ICDG2023, we had a chance to visit the city of Ancient Olympia, 120 km west of Patras.

The area of Ancient Olympia's municipality, birthplace of the Olympic Games, constitutes an International Centre of Athleticism, Culture and Peace. Its unique archaeological site with the stadium (declared by UNESCO as a World Heritage Site), the Archaeological Museum, The International Olympic Academy, the Archaeological Museum of the Olympic Games as well as the Museum of the Modern Olympic Games make Olympia an area of international interest. The lighting of the Olympic flame is held several months before the beginning of the Olympics in the actual site of Olympia, near the temple of Hera.

The Mayor of Ancient Olympia city, Mr. Georgios Georgiopoulos kindly offered us a professional tour guide at the archaeological site and the museums of the Municipality. We also had the chance to meet him in his office, where an extensive reference was made to the long-standing relationship between Ancient Olympia and the sister town, Inazawa in Japan. In Inazawa, on every 13th of January of the lunar calendar, a festival called *Hadaka matsuri* is held. It is a festival for exorcizing evil spirits which started about 1,250 years ago. On that day, many practically naked men rush to the holy man of the year in front of *Konomiya* shrine to get their luck by touching him. Since participants of old Olympic games play their games nakedly to show that they are fair, not only the long histories of these cities for more than 2000 years but also this point that binds them. In the meeting, various possibilities for cultural and educational exchanges between these two cities were discussed.

Andreas Arvanitoyeorgos
Department of Mathematics, University of Patras

www.ingramcontent.com/pod-product-compliance
Lightning Source LLC
Jackson TN
JSHW011605180225
79191JS00002B/19
* 9 7 8 9 8 1 1 2 9 6 7 0 3 *